Diane Lonewell Pike

SOCIOLOGICAL THEORY
CONSTRUCTION

SOCIOLOGICAL THEORY CONSTRUCTION

JACK GIBBS
University of Texas, Austin

The Dryden Press Inc.
901, North Elm Street
Hinsdale, Illinois 60521

To Sylvia, Laura, Douglas,
Catherine, Margaret, and Mayme

PREFACE

I wrote this book under no illusion that it would be well received. My pessimism stems from a conviction that sociologists are not inclined to adopt any mode of formal theory construction, let alone the scheme proposed here.

Throughout the history of sociology, no major theorist has stated a theory formally, and I fear that the tradition will not be abandoned. Certainly the tradition makes it difficult to propose a mode of formal theory construction, as it cannot represent a synthesis or codification of conventional practices. Formal theory construction is unconventional, and that condition created an insoluble problem in writing this book. Any mode of formal theory construction will strike the typical sociologist as exotic because it is alien to extant sociological theories, and that reaction is especially likely in this case. I often found myself introducing notions that are unrelated to actual practice, even to the point of describing types of theories that cannot be illustrated by reference to the literature. So unless sociologists are willing to contemplate a radical departure from convention, the proposed scheme doubtless will be ignored.

At the risk of being greedy, I plead for indulgences. Some parts of the book are difficult and complicated. I truly tried to simplify and clarify those parts but with little success. So, presumptuous though it may be, at various points in the book I tacitly ask the reader to struggle with an idea. Since the scheme itself is set forth in Chapters 5-7, all of Part I may strike the reader as inessential. In any case, I beseech the reader to be patient in reading Chapters 1-4. Finally, several ideas and problems enter into different aspects of theory construction; as a consequence, time and again in my writing I return to certain subjects. If the repetition is inordinate, forgive my poor judgment.

Some three years ago Dick Hill encouraged me to write on theory construction, and I fear that he got more than he bargained for. Since Hill did not have a clear idea as to what the product would be, he is exonerated; nonetheless, I am grateful for his encouragement.

I also thank Miss Alice Laine, who did a great deal of difficult typing for me. Several of my colleagues (Ivan Belknap, Walter Firey, Richard Henshel, Dick Hill, Dale McLemore, Sheldon Olson, and Gideon Sjoberg) were kind enough to read the last draft. I truly appreciate their many helpful comments.

My wife gets all of the credit for our marriage surviving this book. She was tolerant even though recognizing that I am eccentric but not a genius.

Jack P. Gibbs
University of Texas at Austin

CONTENTS

SOCIOLOGICAL THEORY CONSTRUCTION

PART ONE: BACKGROUND

CHAPTER 1 PRELIMINARY CONSIDERATIONS

This book reflects the author's conviction that formal theory construction is essential for the progress of sociology, but it is refreshing to say that the subject has not been neglected. More books on theory construction were published between 1965 and 1970 than throughout the field's history before that period.[1] The intent, then, is not to stimulate interest in the subject, as that would be gratuitous.

Recent works on theory construction have been deficient in one respect—they ignore basic issues in sociology.[2] A mode of theory construction can be promulgated without reference to controversies, but it is unrealistic to presume that some particular mode will be accepted by all sociologists. On the contrary, there is no effective consensus in the field as to the appropriate mode of theory construction and no trend in that direction. The basic issues which divide sociologists preclude widespread acceptance of formal theory construction even in principle, let alone any particular scheme. Yet that is all the more reason to confront the issues. However, it is not done here in the naive presumption that the issues are resolvable; there is no mechanism, no final arbitrator, for resolution. Nonetheless, one should "choose sides" in advancing a mode of theory construction, and the reader can take it or leave it.

It will suffice momentarily to characterize the proposed mode of theory construction as positivistic. However, since there are many divergent conceptions of positivism (see Ayer, 1959; Bergmann, 1954; and Mises, 1951), the characterization should be clarified. Briefly, three arguments underlie the proposed scheme: (1) theories should be stated formally; (2) theories should be testable; and (3) predictive power should be the primary criterion for assessing theories. If the reader rejects any of those arguments, he need read no further.

The Notion of a Theory

Sociologists probably agree that testable theories are desirable, but it is an innocuous consensus. The proliferation of untestable theories surely signifies that testability is not taken seriously.

While testability is not a criterion of a theory, the rarity with which sociological theories are tested surely tells something about conceptions of theory in the field.[3] As observed by Sjoberg and Nett:

> One group of social scientists would identify, at least implicitly, any kind of conceptualization with theory. Such concepts as "status," "role," "culture," or "public opinion," when defined and utilized in the interpretation of research materials, are equated with social theory. In this sense, theory comes to be defined as almost any thought process as opposed to the observations of facts per se. Although all

theory involves conceptualization, and although concepts are the building blocks for theory, to equate conceptualization in general, or isolated concepts in particular. with scientific theory seems inutile (1968:29).

Sjoberg and Nett do not indicate why such a conception is inutile. Certainly one implication should be recognized: it is virtually inevitable that some theories will be untestable. If conceptualizations are taken as theories, how can they possibly be subjected to tests? A conceptualization may be useful (for instance, investigators can use it to identify events, things, or properties); but that is true of any definition, and definitions are not testable. The point is especially relevant in assessing Parsonian theory; if it is nothing more than conceptualization, then demands for tests of it are pointless.[4]

Having analyzed alternative conceptions of theory, Sjoberg and Nett formulate their own: "it refers to a set of logically interrelated 'propositions' or 'statements' that are 'empirically meaningful,' as well as to the assumptions the researcher makes about his method and his data" (1968:30). The definition ignores the crucial consideration that not all propositions or statements are empirical assertions (see Ayer, 1946). Certainly the words "empirically meaningful" do not clarify the import of the definition. All conceptualizations (for example, classificatory schemes) are empirically meaningful in one sense or another, but in itself a conceptualization is not an empirical assertion.[5] So Sjoberg and Nett's definition would extend to statements that *in principle* are not testable, as witness their commentary: "our definition is broad enough to encompass such classificatory schemes as those of Parsons . . . as well as the more rigorous logico-deductive schemes" (1968:30). One may applaud such a broad conception, but it perpetuates the tradition of formulating untestable theories, as it admits "theories" that are untestable in principle.

An alternative conception.   A theory is a set of logically interrelated statements in the form of empirical assertions about properties of infinite classes of events or things. Several notions that enter into the definition require clarification.

Note especially that the component assertions of a theory are distinct from definitions or analytic statements.[6] As such, the assertions do not simply stipulate a label to identify instances of a relation between properties. To illustrate, if one defines *homo sapiens* as a class of featherless, bipedal animals, then the definition does not assert an empirical relation between properties (that is, featherless and bipedal); it is not even an assertion that such an animal does, has, or will exist. Now consider this statement: All featherless animals are bipedal. The statement does not claim that featherless animals exist, but it clearly asserts that if one does exist it is bipedal. Hence the assertion could be refuted by pointing to an animal that is featherless but not bipedal, and for that reason the statement is an empirical assertion or, more conventionally, a synthetic proposition. By contrast, the definition only signified that any featherless, bipedal animal is to be identified as an instance of *homo sapiens*; as such, neither in principle nor in practice can it be refuted.

The emphasis on *empirical* assertions may suggest that definitions are not components of a theory. Ordinarily a theorist does more than articulate assertions; he also defines some of their constituent terms. However, a theory could comprise only assertions, in which case the theorist presumes that the constituent terms need not be defined to identify events, things, and properties.

When a theorist does offer definitions, they are components of the theory even though untestable. Moreover, a theory may be such that none of its component statements, not even the empirical assertions, are actually testable—directly or indirectly. A statement may be testable only in principle, but it is an empirical assertion nonetheless. We can take an absurdity as an illustration: If a living unicorn is ever found, it will weigh over 400 pounds. The statement is not likely ever to be confirmed or refuted, but it is no less an empirical assertion, as it says something about unicorns that is not true by definiton.[7] Consider an unconditional statement: Some human beings are more than ten feet tall. The statement could never be falsified, but it is still an empirical assertion because it is neither true nor false by definition.

The general point is that even though theories comprise something more than definitions, it does not follow that all of them are actually testable or falsifiable even in principle. One may demand theories that are falsifiable in principle (see Popper, 1965) or, more stringently, theories that are actually testable;[8] but neither demand is incorporated in the present definition. Yet the definition is less inclusive than it appears; in particular, a classificatory scheme, comprising as it does a series of labels and definitions, is not a theory (for elaboration on this point, see Rudner, 1966:28). At the same time it is recognized that, given a classificatory scheme, assertions may be made about members of particular classes (for example, no member of a particular class exists), but the assertions can be distinguised from the classificatory scheme itself. One can also distinguish between classificatory schemes in general and taxonomies, a distinction that is blurred in the literature. A taxonomy is a special kind of classificatory scheme; it comprises empirical assertions about members of specified classes.[9] As such, it is a theory insofar as the classes are interrelated logically.

The notion of a formally constructed theory. This book does not offer a survey of sociological theories; rather, it proposes a mode of formal theory construction.[10] Although the author's conception of a formally constructed theory and a mode of formalization is made evident in subsequent chapters, a brief summary is in order.

A formally constructed theory is divided into two parts: the intrinsic and the extrinsic. The intrinsic part comprises statements in the form of assertions about properties of infinite classes of events or things. Each intrinsic statement is identified uniquely (for example, by a number or letter), and the statements are distinguised as to type (for example, axiom, postulate). The extrinsic part of a theory comprises definitions of intrinsic terms (those in the intrinsic part) along with formulas, procedural instructions, and specifications of kinds of data. To summarize, a formally constructed theory is one in which the components are differentiated and identified systematically, with argumentation excluded.

A formal theory is articulated in accordance with some mode of construction, which itself is methodological rather than substantive. Reduced to its essentials, a mode of formal theory construction stipulates: (1) major divisions or parts of a theory, (2) basic units of a theory (for example, statements in the form of empirical assertions), (3) criteria by which basic units are distinguished as to type and identified, (4) rules by which statements are derived from other statements, (5) the procedure for tests of statements derived from the theory, (6) rules for the interpretation of tests, and (7) criteria for assessing theories.

Modes of theory construction differ in various respects, and there may be several contenders in a field. No mode is right or wrong, and the superiority of one over another is a matter of opinion. One mode may come to be thought of as conventional, but that condition is alien to sociology, where there are no truly distinct modes of formal theory construction.[11] Hence the exposition of a theory should be preceded by a reference to the mode of construction; but should a particular mode of formal theory construction come to be conventional, its use need not be explained or defended.

Historical Considerations

Sooner or later all fields face the question: What should be done with the intellectual ancestors? Whitehead argued that a field is doomed if the ancestors are not forgotten, and his dictum is surely an indictment of sociology. Given the attention their works receive, Marx, Durkheim, Pareto, and Weber are still very much alive. What have those great names to do with formal theory construction? By any reasonable definition, their works are the antitheses of formal theory construction. On the whole, the grand theories are conglomerations of ideas so vague that no evidence would lead to their rejection. The foremost defect is in terminology. Evidence can be brought to bear on a theory systematically only to the extent that *some* of the component terms are empirically applicable, meaning the extent to which independent investigators can agree in applying a term to identify particular events, things, or properties. But the major terms of the "grand" sociological theories, if they are defined at all, are not defined so as to promise empirical applicability. Consequently, insofar as there have been any tests of the grand theories, the results are debatable. Because the theories are vague, there is no effective consensus as to what would be an appropriate test or what would constitute negative evidence.

The foregoing does not mean that all component terms of a theory must be empirically applicable or that all component statements must be testable. But in the typical grand theory, testable and untestable statements are not distinguished; rather, it is left to the audience not only to decide what component statements are testable but also to formulate a test procedure. In any case, if only some of the component statements are testable, systematic evidence cannot be brought to bear *on the theory as a whole* without formally deriving testable assertions; and that mode of formalization is alien to the works of the grand theorists. They preferred the _discursive_ exposition of a theory, in which empirical assertions are interspersed with conceptual analyses, arguments, and digressions over hundreds of pages.[12] The discursive mode makes it difficult to identify the central assertions of a theory and no less difficult to determine which statements are derived and how they are derived.[13] Consequently, apart from the problem of empirical applicability, the amorphous character of the typical grand theory alone precludes systematic tests.[14]

Without doubt, sociology is indebted to Marx, Durkheim, Pareto, and Weber; they posed what proved to be significant questions and structured the work of subsequent generations. Nonetheless, it is questionable to label their ideas as theories and ludicrous to regard them as formally constructed theories. So the advocate of formal theory construction indulges in self-deception if he thinks of his work as a continuity. On the contrary, it is a radical departure from tradition, and, like any other advocated change, it will have opposition.

Many sociologists never will grant the desirability of formal theory construction.[15] They see sociology as one of the humanities, and hence their goal is not theory construction (formal or otherwise) but "scholarly works." For them, the method and aspirations of Toqueville are forever sufficient.

Some sociologists see a special wisdom in grand theories, and that manifestation of ancestor worship is an even more formidable barrier to formal theory construction. Since none of the masters employed a mode of formal theory construction, praise of them implies (especially to students) that it is not needed. Indeed, once a sociologist identifies Marx, Durkheim, Pareto, or Weber as the model theorist, his attitude toward formal theory construction is likely to be not merely indifference but antipathy.[16]

Contemporary advocates of formal theory construction appear blissfully unaware of opposition. They are concerned with alternative strategies; but that concern overlooks the larger issue—the desirability of *any* mode of formal theory construction. Although an affirmative argument will not win the day, it is nonetheless preferable to ignoring the issue or failing to recognize that formal theory construction is alien to sociological tradition.[17]

On the Desirability of Formal Theory Construction

Four objections to formal theory construction are identified. The first reflects glorification of grand theory, the objection being that formalization does not enhance the significance of a theory. Valid though it may be, the objection does not recognize that consensus in assessing the significance of a theory, formal or otherwise, is problematical. Even more pertinent, it is difficult to see how formalization could possibly make a theory insignificant. For that matter, the significance of a theory may not be fully realized without formalization. Even if formalization has no effect on significance one way or the other, the parsimony achieved is sufficient justification in itself.

Another objection to formal theory construction is that it may be used to manufacture pretentious theories. The tradition of grand theory again is manifested, the idea being that theories should reflect brilliance or profundity and be judged by criteria alien to the physical sciences.[18] To be sure, a mode of formal theory construction may lead sociologists to make assertions uncritically, but it does not compel the audience to suffer fools gladly. After all, a theory will not be accepted merely because it is stated formally, and there is no better way than formalization to reveal pretentiousness. A discursive theory which may appear to have great scope and significance might be seen as merely pretentious after formalization has removed the verbiage. In that connection, one must surely wonder what the outcome would be if Parsonian "theory" were restated formally; in any event, it cannot be assessed properly without a restatement.

The discursive exposition of a theory does serve a purpose: it enables the theorist to make his assertions appear plausible, that is, to make a case for his theory. Hence, sociologists may object to formal theory construction because it excludes argumentation. Although sociologists generally are fond of argumentation, it is difficult to see what rhetoric adds to a theory. If empirical validity should be central in assessing theories, then the outcome of tests— not rhetoric—is decisive. Indeed, should a theorist convince his audience by forceful argumentation, it is nothing more than a personal triumph. Whatever one may mean by truth or empirical validity, surely neither can be equated with conviction (see Popper, 1965).

A mode of formal theory construction does not preclude argumentation altogether. Two versions of a theory can be presented, the first in the conventional style, which virtually demands skill in rhetoric, and the second as a formal restatement. Alternatively, each component empirical assertion can be preceded or followed by argumentation, with the assertions subsequently brought together as the formal version of the theory. In either case, attempts to test a theory should focus only on the formal version, whether the audience regards the arguments given as feckless or compelling.

A further objection to the formalization of a theory is that it in no way insures empirical validity. This objection does not consider the role of tests in assessing empirical validity. Specifically, systematic tests will not be realized until sociological theories are stated formally. Contrary to radical operationalism, sociological theorists will continue to employ terms that do not designate measurable phenomena. Any doubts on that score should be eliminated by contemplating five terms: cohesion, differentiation, stratification, normative consensus, and anomie. To take just one term, how is normative consensus to be defined and measured at the national level? Anyone who thinks the question can be answered satisfactorily is either not a sociologist or one who is a stranger to research. The argument is that major sociological terms cannot be defined with precision, nor do they designate measurable phenomena. Consequently, when such a term is used in a theoretic statement, the statement is not testable. Untestable statements can be used to derive testable predictions, but those derivations are not defensible unless made in accordance with a mode of formal theory construction.

The Philosophy of Science

Imbued with the notion of a unity in science, sociologists commonly look to the philosophy of science in seeking or assessing a mode of theory construction. The present work rejects that orientation, because the philosophy of science literature offers very little in the way of specific modes of theory construction, and there is no assurance that any of the schemes are suitable for sociology. A mode of theory construction articulated by a philosopher of science is likely to be so abstract that it provides only very general guidelines. Additionally, such a scheme is not likely to reflect a concern with the "conditions of work" and practical problems in sociology (see Kyburg, 1968; and Woodger, 1939).

One major theme of this book is that the conditions of work and practical problems vary so much among disciplines and from time to time that a *general* mode of theory construction is not feasible. More specifically, each mode should be formulated for a particular field at a particular stage of its development; otherwise, it will not provide solutions for practical problems. To illustrate, some sociological specialties are concerned primarily with phenomena at the macroscopic level, and limited research resources necessitate extensive use of published data (such as census statistics) in those specialties. Consequently, unless sociologists are content with untestable theories, they should formulate theories with a view to the availability of data. However, when practical considerations dictate the use of published data, various problems are introduced that can be solved only by a special mode of theory construction, one not provided by the philosophy of science.

Consistent with the foregoing, the applicability of the proposed scheme is extremely limited. It is restricted to sociology and slanted in the direction of specialties that analyze macroscopic social phenomena. The conditions of work and practical problems in those specialties are different from those of other sciences, even from other sociological specialties, and it is an illusion to suppose that one mode of theory construction can transcend the difference.

The argument should not be construed as a rejection or a belittlement of the philosophy of science. Focusing as they do on science in general, philosophers cannot be expected to treat the practical problems of each sociological specialty, let alone provide solutions (see Lazarsfeld, 1962:470). It is unrealistic even to expect methodologists in sociology to provide the solutions. They will come, if at all, only from individuals working in the specialties, who know the practical problems and, as Einstein put it (1954:290), feel "where the shoe pinches." Finally, the argument does not deny a unity of science; but the unity does not lie in procedures, techniques, methods, instrumentation, or a common mode of theory construction. Rather, what scientists share is a concern with order in the universe and a particular criterion— predictive power—for assessing statements about order. In any case, no appeal to the philosophy of science will solve a field's practical problems, and it is a mistake to suppose that those problems can be solved without regard to theory construction.

The Content of Sociology and Formal Theory Construction

Although the proposed scheme is slanted toward certain sociological specialties, that limitation or emphasis does not imply invidious comparisons. Briefly, the scheme does not dictate the content of sociology; that is, it does not limit the field to particular units of investigation. What sociologists choose to analyze— countries, organizations, particular types of social relations—is a substantive consideration and, as such, divorced from formal theory construction.

Whatever the units, sociologists analyze properties of them, and only in that connection does the proposed mode of theory construction touch on the content of sociology. Specifically, the scheme is limited to *quantitative properties*, such as the degree of urbanization. But that limitation excludes very little in the substantive content of sociology, which is to say that most properties of social units are quantifiable. It is true that many sociological terms appear to designate a qualitative property, but often that is only because the term is used uncritically. When an attempt is made to formulate an empirically applicable definition of a term that designates a qualitative property, typically it becomes obvious that the property is quantitative. Consider, for example, the term "norm." All definitions suggest that a given norm (for instance, polyandry) is either present or absent in any social unit, meaning that it is a qualitative property. However, on what basis is one justified in declaring polyandry to be the norm in a social unit? If a norm is defined in reference to actual behavior, then in no social unit are all marriages polyandrous, but the proportion of marriages that are polyandrous is clearly a quantitative attribute of a social unit. Now it could be argued that a norm is not a matter of actual behavior but rather the evaluation of acts or conditions as acceptable or unacceptable, morally right or wrong, and so on. That conception of norm leads to a difficult question: How does one know that polyandry is acceptable in a given social unit? If it is a matter of the opinions expressed by members of a social unit, then it is extremely unlikely that all members approve of polyandry. It can, of course, be stipulated that a type of act or condition is accepted if more than a certain proportion of the members voice approval of it, but any such stipulation would be arbitrary. However, arbitrariness is eliminated by simply regarding the proportion approving a type of act or condition as a quantitative property. Finally, consider the idea that a norm can be identified by reference to law. Whatever merit that notion may have, it does not speak to the question of "extralegal" norms. Moreover, it ignores the sustained controversy over the meaning of law or a law (see Gibbs, 1968). If a law is defined by reference to statutes, the definition is alien to nonliterate peoples and also to Anglo-American jurisdictions. Further, unless one is prepared to accept just any statute as a law, then the question of regularity of enforcement is raised, and any stipulation of a numerical criterion to eliminate so-called dead laws is bound to be arbitrary. Indeed, any attempt to identify a norm by reference to the character of actual reactions to a particular type of act, whether the reactions be considered legal or extralegal, cannot escape arbitrary distinctions. To summarize, the term "norm," like so many others in sociology, appears to designate a qualitative phenomenon, but only insofar as the notion is accepted uncritically. Upon close scrutiny, the normative properties of social units are predominantly quantitative.

The foregoing is intended to emphasize one point—that most major sociological terms designate quantitative properties, although that may not be recognized readily. This is not to say that there are no qualitative properties of social units or that such properties should be ignored. However, it must be emphasized that the proposed mode of theory construction is applicable only to quantitative properties.

Notes for Chapter 1

[1] See, for example, Zetterberg, 1965; Willer, 1967; Stinchcombe, 1968; Dubin, 1969; and Blalock, 1969. Some of those authors might deny that their work pertains to formal theory construction, but in each case the obvious intent is to make theory construction more systematic, which is the *raison d'etre* of formal theory construction. Prior to 1965 virtually all works on sociological theory focused primarily on particular theories and substantive questions rather than on modes of theory construction (see, for instance, Borgatta and Meyer, 1956; Coser and Rosenberg, 1964; Gross, 1959; Martindale, 1960; Parsons, *et al.*, 1961; Timasheff, 1961; Becker and Boskoff, 1957; and Loomis and Loomis, 1961). This is not to suggest a recent radical change in the sociological literature. On the contrary even since 1965, the focus remains on particular theories and substantive questions, with scant attention to formal theory construction (see McKinney and Tiryakian, 1970; Wallace, 1969; Gross, 1967; DiRenzo, 1966; Berger, *et al.*, 1966; Cohen, 1968; Sorokin, 1966; Borgatta, 1969; and Merton, 1967).

[2] The stance is exemplified by one of Stinchcombe's statements: "Thus the point of view of this book is deliberately eclectic. The reader will find no cases of one 'approach' contradicting or conflicting with another 'approach' " (1968:4).

[3] In sociology the diverse conceptions of theory virtually defy description (see Zetterberg, 1965:6-8). The typical definition of "a theory" is so vague that it is difficult to imagine what would not qualify as instances. For example, Dubin says: "I will use the terms *theory, theoretical model, model,* and *system* interchangeably. All these terms will stand for a closed system from which are generated predictions about the nature of man's world— predictions that, when made, the theorist agrees must be open to some kind of empirical test" (1969:8-9). But subsequently he says that a set of logically related propositions is "not a theory or theoretical model" (1969:172). If a set "generates predictions," there is no reason why it would not qualify as a

theory, especially since logically related propositions could be formulated as a "closed system." Like many sociologists, Dubin offers a definition that evidently does not completely reflect his conception of a theory. The conception may include notions (a theory must "explain," it must be profound, it must tell us something about human nature) that are not made explicit in the definition. Hence what one regards as instances of theory may be inconsistent with his explicit definition.

[4] "The theory of action is a conceptual scheme for the analysis of the behavior of living organisms" (Parsons and Shils, 1951:53). Rudner (1966:34) has pointed out that Parsons makes inconsistent statements about his "theory" of action.

[5] The point is especially overlooked in characterizations of Weber's work as "theory." With the possible exception of the *Protestant Ethic and the Spirit of Capitalism* (1930), Weber's "theories" (1968) are actually conceptual schemes and descriptions of "historical types."

[6] The assertions also are distinct from normative or prescriptive propositions (that is, statements as to what "ought to be"); as such, the present definition refers to only one kind of theory—empirical or "scientific."

[7] Needless to say, the distinction between empirical assertions and definitions (or between analytic and synthetic statements) is difficult to draw, and the problem is not solved by the conventional idea that an empirical assertion or synthetic statement is "testable." For one thing, the notion of a test is vague and entails several debatable questions. Does a "test" refer to experience or to assertions about experience? Is an assertion "empirical" even though testable only in principle? Is an assertion testable only in the sense of being falsifiable? The last question is crucial in considering Popper's dictum (1965) that statements or theories are not empirical unless falsifiable. Thus, he would not regard the statement "Some crows are white" as empirical because it cannot be falsified. However, it might be verified, and the verification would not be purely linguistic, that is, the statement is not true by definition. Popper's argument confounds two questions. First, *should* an empirical assertion or theory be falsifiable? Second, is falsifiability a criterion of an empirical assertion or theory? Contrary to Popper's perspective, one may respond affirmatively to the first question but negatively to the second. Certainly there is an alternative criterion: a statement is an empirical assertion if it is neither true nor false by definition. Of course, the words "by definition" are subject to varying interpretations, but Popper's conception of an empirical assertion does not escape that problem, for he would not argue that "falsifiability" is in no sense a question of semantics or syntax.

[8] The extreme position is stated by Willer: "A theory is an integrated set of relationships with a certain level of validity" (1967:9). According to that

conception, a set of statements is not a theory unless actually tested and validated to a "certain level." Willer stipulates that "perfect validation" is not a criterion, but he fails to specify any particular level, and any such specification would be arbitrary.

[9] Both taxonomies and typologies are classificatory schemes with more than one *fundamentum divisionis*. One can formulate a typology without making empirical assertions, but sociologists commonly do not recognize such distinctions. Rather, they use one of three words—taxonomy, typology, classification—generically, and hence it appears that the three are interchangeable.

[10] Until quite recently virtually all treatments of formal theory construction in the sociological literature were limited largely to symbolic logic and, as such, divorced from issues and problems in the field. See, for example, Hochberg, 1959.

[11] It cannot be said that sociologists even agree in their conceptions of formal theory construction or "formal theory." Whereas the proposed scheme emphasizes divisions of a theory, types of terms, types of statements, and rules of derivation, in a recent paper on formalization Blalock (1970) pays scant attention to any of those subjects. He evidently sees formal theory construction largely as a matter of statistical or mathematical technique. Glaser and Strauss (1967:32) contrast "formal" and "substantive" theory, but the distinction is by no means clear. Evidently, level of abstraction is the only consideration. Using their examples, a theory on deviant behavior would be "formal," whereas a theory on delinquency would be "substantive." If the only contrast is level of abstraction, as the example clearly suggests, then the typification of a particular theory as substantive or formal is arbitrary. In any case, the conception of a "formal" theory by Glaser and Strauss is quite different from that advocated here. Specifically, whether a theory is "formal" depends not on its content or level of abstraction but on its mode of construction.

[12] Zetterberg (1968) has created the impression that Pareto employed a formal mode of theory construction. To the contrary, Pareto's major sociological work (1963) exemplifies the discursively formulated theory. In addition to the usual intermingling of conceptual analyses, empirical assertions, and rhetoric, Pareto's promulgation of a theory includes anecdotes, proverbs, historical accounts, and seemingly interminable other digressions. The distinction between analytic and synthetic statements is blurred persistently in Pareto's writings. Even when one of Pareto's statements appears to be clearly synthetic, it is typically difficult to determine whether or not he intended it to be a testable assertion. Finally, given one of Pareto's statements that might be construed as testable, it is difficult to answer three questions. First, was the statement derived? Second, if so, from what statements was it derived? Third, by what rules was the statement derived?

[13] Contemplate this question: By what rules and from what statements did Marx derive his so-called laws of capitalism? Far from being derived systematically, some of those laws (1915) are more nearly extrapolation of trends in nineteenth-century Europe, Britain in particular. In any case, Marx's "theory" is actually a vast polemic scattered over thousands of pages and dozens of publications, with no consistent effort to distinguish synthetic and analytic statements or to derive assertions systematically.

[14] See, for instance, Davis' analysis (1955) of Malthus' theory and Johnson's critique (1965) of Durkheim's theory on suicide. Without exception, each such examination of a classical work has raised grave doubts about the logical structure of the theory. However, as evidence that the logical structure of contemporary theories are no less debatable, see Huaco's commentary (1963) on the Davis-Moore theory.

[15] Kinch (1963) is one of the few sociologists who have systematically examined the merits of formal theory construction.

[16] The glorification of "grand theory" is unabashed in Nisbet, 1966; Parsons, et al., 1961; Aron, 1965; 1970; Parsons, 1949; and Bierstedt, 1960. It is by no means surprising that some of these writers are critical of formal theory construction, logico-deductive systems in particular (see Nisbet, 1966).

[17] Actually, the advocates of formal theory construction do not have a sufficient following in sociology to make it a major issue. Even among sociologists with an abiding interest in theory, formal theory construction receives scant attention (see Merton, 1967).

[18] In commenting on Weber's *Protestant Ethic and the Spirit of Capitalism*, Bierstedt expressed the sentiment clearly: "We admire this work not because it is 'true'—indeed its truth escapes all of the ordinary canons of scientific verification—but because of the excellence of its conception, the erudition of its argument, and the general sociological sophistication that informs it" (1960:7).

CHAPTER 2 SPECIAL PROBLEMS

A set of assertions becomes a theory only when it is made public, and public reaction determines its fate. Some critics demand that a theory identify the cause or causes of a phenomenon, while others require that a theory "explain." The difference is significant, for it suggests disagreements as to the purpose of a theory, and that is certainly true in sociology.

This chapter assesses causation and explanation as notions, with emphasis on controversies. The subject is crucial in contemplating theory construction, for the way one constructs a theory reflects his conception of its purpose. Moreover, the subject entails issues that are irresolvable, and for that reason the notions of causation and explanation are designated as special problems.

Whatever its function or purpose, a theory comprises assertions about the empirical world, that is, about classes of things or events perceived by the theorist as external to himself. In any theory some of the symbols (or terms) refer to events or things, but they designate infinite classes and not particular instances. However, a theory has no connection with the empirical world unless individuals other than the theorist agree in applying some of its constituent symbols to identify particular events or things. Stated more generally, a theory is connected with the empirical world to the extent that its constituent symbols are empirically applicable. Since sociological terminology promises very little empirical applicability, it is treated as another special problem.

Causation

The notion of causation is perhaps the most controversial in the history of science. Sociologists have been drawn to the controversy, but they commonly look to philosophers of science as arbiters. That perspective is rejected here, for the debate in the philosophy of science remains inconclusive and the issue should be limited to sociology.[1] The problem is not merely that of accepting or rejecting the notion of causation, since the question suggests (erroneously) that there can be only one conception. Which conception of causation sociologists adopt (if any) should depend on the practical problems and conditions of work in the field.

No conception of causation is complete unless it speaks to this question: What kind of research findings are evidence that one phonomenon is the cause or a cause of another? Without an answer, the use of the word "cause" is not justified in formulating theories or even in isolated assertions. To be sure, one may think in terms of causation without using the word in stating a theory, but then the notion is private and not part of a theory. Suppose that a philosopher of science does answer the question pertaining to causation and research findings, but stipulates an experimental procedure that is alien to the conditions of work in a particular sociological specialty. Sociologists working in that specialty would gain nothing by adopting that conception of causation.

Blalock and Costner's conception of causation.   Two methodologists, Blalock and Costner, have considered causation at length. As sociologists with interests in theory construction, their perspective is especially relevant.

In an early publication Blalock recognized some of the issues and problems entailed in the notion of causation, but he explicitly refused to define causality formally. Instead, he stated: "it indeed may turn out wise to treat the notion of causality as primitive or undefined" (1964:9). That statement does not circumvent a crucial question: If the word "causes" appears in an assertion, what kind of evidence would refute it? If the word is left undefined, then the question cannot be answered. It is pointless to argue that a definition is unnecessary; it would be unnecessary only if everyone agrees as to what constitutes evidence of causation, but the history of the controversy (see Bunge, 1959, and Theodorson, 1967) reveals a lack of consensus. The argument is not that one must define cause or causes so that everyone will agree with the definition; but it is pointless to use the word and leave it undefined.

Having avoided a formal definition of causation, Blalock makes two revealing statements. First, "conjunction is not sufficient to distinguish a causal relationship from other types of association" (1964:9). Second, "our conception of causality should not depend on temporal sequences, except for the impossibility of an effect preceding its cause" (1964:10). As these statements indicate, Blalock offers only a negative or incomplete definition of causation, and in subsequent publications (1968b, 1969, 1970), he abandons all pretense of a definition.

At various points Blalock equates the words "causing," "producing," and "forcing"; but that usage only substitutes one nomic expression for another. Consider Blalock's statement: "X is a direct cause of Y if and only if a change in X produces a change in the mean value of Y" (1964:19). Is the word "produces" any less debatable than "causes"? Blalock himself subsequently admits: "All we can observe is a change in X followed by a change in Y" (1964:19). But substituting "followed by" for "produces" introduces an ambiguity. An instance of a change in X is certain to be followed *sooner or later* by an instance of change in Y, so without specification of a definite time lag the term "followed by" is vacuous. For that matter, the assertion would be incomplete without specifying a spatial relation between X and Y; otherwise the assertion means that Y will change *somewhere* in the universe, which is virtually a truism. To illustrate, if one asserts that an increase in the unemployment rate is followed by an increase in the crime rate, then reports of events in Los Angeles relative to events in Bombay could be offered to refute or support the assertion. Since sociologists usually imply that a social unit is the spatial context of a relation between variables there is no particular problem. Nonetheless, a conception of causation is incomplete unless it incorporates the notion of a spatial and temporal relation.

Ostensibly, Blalock avoids a formal definition on the ground that causation cannot be demonstrated or empirically verified. Invoking that ancient argument, however, does not avoid a difficult question. Granted that causation cannot be observed, on what basis does one infer a causal relation? Given Blalock's refusal to define causation, one must surely wonder how any inference can be other than intuitive.

Of course, Blalock's failure to offer a formal definition may reflect his awareness of issues, but those issues will not go away if causation is left undefined and Hume's argument is ignored. In particular, Blalock's statements suggest that causation can be inferred from a space-time relation between variables, but he does not specify the type of relation, and his observations on constant conjunction and temporality (as quoted above) make his perspective even more ambiguous. The ambiguity is understandable; but any conception of causation in terms of strictly space-time relations will be rejected by numerous sociologists. Phenomenologists and followers of Weber, to mention only two schools, are unlikely to accept research findings as evidence of social causation unless those findings pertain to the motives, values, and/or perception of the actors, a consideration which negates the idea that causation can be inferred from space-time relations alone. After all, as causalists never tire of saying, a space-time relation is nothing more than a correlation.

Although some sociologists reject causal inferences from space-time relations alone, they fail to identify a specific kind of evidence as an alternative. By negative definitions, MacIver (1942) creates the impression that one infers causation by intuition, and the criterion is conviction.[2] Finally, even if everyone should agree that causation can be inferred from space-time relations alone, there is no reason to anticipate additional agreement in designating a particular type of relation as decisive.

Whereas Blalock recognizes issues and problems but offers no definition of causality, Costner offers a definition (in a 1964 publication with Leik) but ignores the issues and problems. "An asymmetric causal relationship is a relationship that would be verified by an experiment in which the first-named variable is manipulated while controlling for or randomizing all other variables, whereas the relationship would not be verified by an experiment in which the second-named variable is manipulated while controlling for or randomizing all other variables" (Costner and Leik, 1964:824). The definition is difficult to follow, since it is silent as to what would constitute verification. Evidently, if the second-named variable changes after manipulation of the first-named variable but not the reverse, then the relation is causal.

By implication, Costner and Leik's definition designates a spatial context, the experimental situation; but nothing is said or implied about a time lag and its duration. Equally important, they refer to an experiment and thereby ignore a crucial point— sociology is primarily observational rather than experimental; so the idea of manipulating one variable while controlling for or randomizing all others is largely alien to the conditions of work in the field. Hence the question: What is gained by a definition of a causal relation that is not applicable to much of sociology?

Costner and Leik's definition can be modified so that it is applicable in an observational science. Given two variables, X and Y, then X is the cause of Y if a change in X is always followed by a change in Y, some changes in Y are not followed by a change in X, and all changes in Y are preceded by a change in X. The definition is hardly unconventional, but it is acceptable only if adopted uncritically. Again, unless space-time limitations are stipulated, an instance of a change in X is certain to be both preceded by and followed by an instance of a change in Y. Consequently, an assertion of a relation between two variables is ambiguous unless it stipulates definite space-time limitations. But at this stage in sociology theorists are not willing or able to stipulate a particular time lag in making assertions about empirical relations. Even if definite space-time limitations should be stipulated, it is most improbable that the theory could be tested systematically. One painful condition of work in sociology is limited opportunities for longitudinal research, that is, the ability to make diachronic comparisons. Further, in those rare instances when a relation can be analyzed over time, typically it is difficult to decide which variable is changing first. When all variables are constant during a period and then changes occur, lead and lag may be conspicuous; but that condition seldom is realized.

In a 1969 publication Costner devotes scant attention to causation, but he does describe a one-way causal relation as: "a change in X leads to a change in Y, but not the reverse" (Costner, 1969:246). The statement may appear to apply to observational as well as experimental sciences, but all it does is substitute a nomic term, "leads to," for "causes." Further, even if Costner had employed the words "followed by" rather than "leads to," the statement would not escape the foregoing problems.

Opposing arguments.   As already suggested, a causal language is excluded from the proposed mode of theory construction.[3] The reasons are summarized as follows: (1) it is doubtful that a definition of causation can be formulated that will truly satisfy anyone, let alone everyone; (2) numerous conceptions of causation and related rules or procedures of inference are alien to some sociological specialties; and (3) theories can be constructed without employing a causal language.

Needless to say, there are several counterarguments, one being that a rejection of a causal language denies reality. Critics may joust with Hume by maintaining that we can observe the cause of a particular sound, death, act, and so forth. That may or may not be, but for sociological purposes, it is largely irrelevant. Can we "observe" the cause of the division of labor in the United States, the constitutional monarchy in the United Kingdom, or the high armed robbery rate in California? If a sociologist thinks we can observe such causes and agree in our observations, then his colleagues should be grateful for instructions.

Numerous sociologists (such as Blalock, 1964, 1969; Sjoberg and Nett, 1968) emphasize that individuals think in causal terms, but the argument that everyone thinks of all empirical questions in causal terms is denied. Moreover, even if such thinking is universal, it does not follow that scientists, especially social scientists, agree in their conceptions of causation.[4] For that matter, Blalock and Costner notwithstanding, one may think in causal terms and yet omit the word "causes" in stating a theory. Should the word be used, tests of the theory are precluded unless the theorist stipulates rules by which causation is to be inferred from space-time relations, and that stipulation entails all of the issues and problems previously considered.

Another counterargument pertains to Blalock's "causal models" (1964, 1969). Described briefly, the idea is that correlations can be analyzed so as to identify some of the variables as causes and others as effects. The description oversimplifies, but if it distorts the purpose of Blalock's work, surely one must wonder what that purpose is.

The correlation coefficients that enter into Blalock's causal model are empirical in the sense that they symbolize research findings, and therefore it might appear that Blalock has equated causation with a particular type of space-time relation between variables. However, since he does not offer an explicit definition of causation, it is difficult to see how any set of correlation coefficients can be construed as indicative of causation. In any case, a glaring discrepancy between Blalock's causal models and his statements about causation is created by his focus on synchronic correlations. Since synchronic correlations do not refer to changes in variables, or to a time lag, it is difficult to see how they reflect the temporal dimension of causation. Blalock himself declares (1964:10) that the effect cannot precede the cause in time,[5] but that quality of a relation is not reflected by a synchronic correlation.

Blalock implicitly denies that causation is reflected only by invariant empirical relations or, stated another way, by unity coefficients of correlation. That denial is realistic, as invariant empirical relations and unity coefficients of correlation are virtually unknown in sociological research. However, the subject introduces a difficult question: How close must the empirical relations be to justify a causal inference? To restate with reference to coefficients of correlation: What absolute or relative magnitude of a coefficient must obtain before an inference of a causal relation is justified? An answer to either question would be arbitrary, and the problem is more complex than might appear.[6] The magnitude of a correlation between two variables may vary considerably from one set of data to the next, especially in sociological research. Thus it is not sufficient to specify evidence of a causal relation in an isolated instance; additionally, one must stipulate what constitutes sufficiently *consistent* evidence, and any such stipulation is bound to be arbitrary or alien to sociological findings.

As a counterargument against rejection of the word "causes" in theory construction, critics may appeal to a defense of the notion by someone outside of sociology, such as Bunge (1959). The appeal is irrelevant for two reasons. First, the use of the word "causes" in theory construction is rejected but thinking in causal terms is another matter. Second, no conception of causation is relevant if it ignores the conditions of work and practical problems in sociology.[7] As for Bunge, one question survives his work: What kind of research evidence is to be construed as indicative of causation? Committed to the idea that causality cannot be equated with any particular type of space-time association, Bunge does not even speak to the question. He does formulate a conception of causation (1959:52) but uses nomic terms (such as, producing) that are no less debatable and no more empirically applicable than the word "causes" itself.

A qualification. The present argument is likely to be construed as a denial of "causation" or rejection of causal thinking. Either interpretation would be a distortion, as only one thing is rejected—the use of "causes" or a similar nomic term in theory construction. Far from enhancing a theory, such terminology creates problems that, at least in sociology, defy solution.

Above all, the argument is confined to sociology. As for other fields, the acceptance or rejection of a causal language is to be decided not by the philosophers of science but by those who construct theories; so the rejection of a causal language is relative. A causal language can perhaps be used in sociology at some distant future, but a mode of theory construction should be spplicable now.

Explanation

Any theory makes assertions, and critics will question more than their "truth." It is not sufficient that the assertions be "true" in one sense or another; additionally, so some writers maintain, the theory must explain phenomena. As put pointedly by Homans: "A theory is nothing—it is not a theory—unless it is an explanation" (1964a:812).

The idea that a theory must offer an explanation differs from the demand that it identify causes. Causal assertions are commonly thought of as explanations, but some explanations do not assert causation (see Nagel, 1961:78; and Kaplan, 1965).

The logical form of an explanation.   One persistent theme in commentaries on explanation is manifest in the following statements.

> hence, the scientific explanation of a fact consists, from a *logical* point of view, in showing that it is an *instance* of a general law. In turn, the scientific explanation of a uniformity or regularity will, when available, consist in its deduction from a higher-level law, that is, in its subsumption under a statement of greater generality. (Bunge, 1959:288)

> Any incorporation of a fact—be it a particular instance of a law or the law itself—into a deductive system in which it appears as a conclusion from other known laws is, by virtue of that incorporation, an explanation of that fact or law. (Braithwaite, 1953:349)

> the sentences of the *explanans* are supposed to imply logically the *explanandum*. (Rudner, 1966:63)

> To give a *causal explanation* of an event means to deduce a statement which describes it, using as premises of the deduction one or more *universal laws*, together with certain singular statements, the *initial conditions*. (Popper, 1965:59)

less general propositions, which investigators have found to hold good empirically, may be deduced from the general ones under specified given conditions. To deduce them is to explain them. (Homans, 1961:12)

Although the universality of the deductive pattern may be open to question, even when the pattern is projected as an ideal, it is hardly disputable that many explanations in the sciences—and indeed the most comprehensive and impressive systems of explanation—are of this form. Moreover, many explanations that ostensibly fail to realize this form can be shown to exemplify it, when the assumptions that are taken for granted in the explanations are made explicit; and such cases must count not as exceptions to the deductive model, but as illustrations of the frequent use of enthymematic arguments. (Nagel, 1961:29)

All of the foregoing statements emphasize one point—explanation requires deduction. That terse characterization may be misunderstood, for there is a tendency to think of explanations as assertions, such as: "John committed suicide because he was unemployed." The assertion is not an explanation, for it is not deduced. Given certain rules, the assertion could be deduced from two other statements: (1) All unemployed men commit suicide. (2) At the time of his death, John was an unemployed man. The explanation of John's suicide is thus not the initial assertion but rather in the deduction of that assertion, meaning that explanation is the deduction of a statement and not the statement itself. True, a statement in the form of an assertion (John killed himself because he was unemployed) may be considered as an explanation, but it actually only suggests an explanation.

Needless to say, it would simplify the matter to describe an explanation as a statement about a particular thing or event that is deduced in part from an assertion about a uniform relation between classes. But uniformities themselves supposedly are subject to explanation, which may consist of nothing more than the derivation of the uniformity. Reconsider a previous statement: All unemployed men commit suicide. Now that uniformity, whether taken as "true" or as an assertion, can be deduced (again supposing certain rules) from two other statements: (1) All men commit suicide after a change in status. (2) Becoming unemployed is a change in status. So the alleged uniformity is explained without making any statements about it, meaning that the *deduction* of a putative uniformity is an explanation of it.

The foregoing commentary on explanation is not offered in the expectation that it will be accepted. On the contrary, there is no consensus on the logical form of an explanation, or on what constitutes an "adequate" explanation. As to logical form, the agreement suggested by the foregoing quotations is superficial, since simply equating explanation with deduction is an oversimplification that conceals issues. It could be argued that any answer to a "why" question is an explanation, whether formally deduced or not (see Nagel, 1961:15; Scriven, 1962; and Nettler, 1970). Even if it is agreed that explanation necessarily entails deduction, there are still two debatable questions. First, must the rules of deduction and the premises be made explicit? Second, is an explanation "logically" valid regardless of the rules of deduction? If the first question is answered affirmatively, the reply is inconsistent with common practice among laymen, scholars, and scientists. In answering the second question, a reply that "conventional" rules of deduction must be employed is not adequate. Even philosophers write as though there is and can be only one kind of logic, but that is clearly not the case.[8] Common conceptions notwithstanding, logic is simply any set of rules for relating symbols, and such rules are limited only by human imagination. True, most philosophers and scientists appear to hold that a deduction is "logically" valid only if governed by certain rules (to the exclusion of others), but that argument is questionable. It suggests that "the" rules of deduction are known and agreed upon by everyone; but that is not the case. There are even divergent conceptions of deduction (see Sadovskij, 1970), and there is no way to resolve those differences. Debates take the form of an infinite regression on the logic of logic, and certainly no rules of deduction can be defended on the ground that they assure "empirical" truth.

It may appear that the foregoing commentary raises false issues concerning the logical form of explanation; false in that the arguments posed are not actually debated with spirit in the socio-logical literature. That is so, but the consensus reflected in the literature is only indicative of a superficial treatment of the sub-ject and indifference to issues and problems.[9] As suggested earlier, sociologists are content to equate explanation with deduction, but that characterization is debatable and does not speak to several questions.[10] As just one illustration from the sociological litera-ture, examine a statement by Sjoberg and Nett: "Although theories have explanatory power, the notion of a theory, as we have defined it, encompasses more than explanation: It includes in addition the underlying premises" (1968:32). If explanation is deduction, how can the "underlying premises" be excluded?

The question of substance. The foregoing issues are not spurious but they are largely irrelevant. Even if agreement on the form and *logical adequacy* of explanation should be realized, there is a greater and more difficult problem—one concerning substance or content. Whereas the rules of deduction create the logical form, the substance of an explanation has to do with the extralogical terms in the constituent statements.

Brown (1963:41) considers a "logical derivation" as only one of nine methods or types of explanations that are distinguished by "the different kinds of information used in the course of the explanation." As such, the emphasis in his classification is on sub-stance, not logical form. But the emphasis is realistic in that it touches on the central issue. As put by Brown: "It is the content of the premises and conclusions that occupies their attention . . . and not the machinery of derivation" (1963:42).

The emphasis on substance is so marked that critics com-monly ignore the logical form of explanation altogether. In reject-ing a theory a critic may say that it "does not offer an explana-tion." If, however, an asserted uniformity is derived as a component of a theory, or if predictions about particular events or things can be derived from it, it is difficult to see how the theory fails to offer an explanation. More often than not, a critic actually rejects a theory because he regards the explanation as inadequate. So the central question becomes: What constitutes an adequate or satisfactory explanation? A more controversial question cannot be raised in sociology.

Of sociological theorists, none has been more explicit than Homans in equating explanation with deduction. However, Homans' pronouncements on explanation are misleading, for he demands (1964a) that sociological theories "bring men back in," meaning that explanations of social phenomena must be couched in terms of needs, motives, values, and perceptions. Hence, even if a theory does offer an explanation, Homans will reject it unless the explanation is "adequate"; but his criterion of adequacy has to do with substance, not logical form. There is no way that one can prove Homans' criterion as incorrect; it is beyond constructive argumentation. If he is prepared to reject all but reductionistic theories of social phenomena, then nothing can alter his perspective. That many sociologists disagree with Homans is of no significance, since many others evidently agree with him (see Inkeles, 1959).

Disagreements over explanatory adequacy transcend the issue of reductionism. Sociologists debate the merits of various kinds of explanation, for example, mechanistic, functional, teleological, historical, genetic, and the notion of *Verstehen*.[11] The contenders are not analyzed here because they have been treated at length elsewhere (see Brown, 1963; Nettler, 1970) and, more importantly, because further debate is pointless. The argument has gone on for generations without a sign of resolution, which is to say again that sociologists do not agree as to what constitutes an adequate explanation of social phenomena.

Rather than consider each of the various kinds of explanation, one illustration of the debate will suffice. Consider the following conception of an adequate explanation: A particular thing or event is explained by identifying it as an instance of a class and by stipulating the conditions in which instances of that class will appear or be found; the adequacy of the explanation depends on the accuracy of predictions about particular instances that are deduced from it. Observe that the conception does encompass the notion of deduction, that is, predictions are to be deduced in part from asserted relations between classes of phenomena.

Even if sociologists should agree that the foregoing conception does exemplify the logical form of an explanation, many of them would not accept the criterion of adequacy. For some, especially the followers of Weber and phenomenologists, not even consistently correct predictions would compel them to accept such an explanation as adequate. The criterion would be dismissed as "positivistic," and all the more so since no reference is made to the motives, values, or perceptions of the actors. Further, since the condition stipulated in the *explanans* need not be a process, the conception will not placate sociologists who demand historical or genetic types of explanation. Finally, since the *explicandum* need not be described as meeting some "need" or as "contributing" to the maintenance of a system, the criterion of adequacy is alien to functionalism.

Even sociologists who eschew functionalism (for instance, Catton, 1967) may reject the criterion of adequacy, as it does not refer to causation. In other words, as suggested previously, some sociologists demand "causal" explanations; and that demand introduces all of the previously examined controversies pertaining to a causal language in theory construction.

The issue cannot be resolved by holding with Bunge (1959), that a "scientific" explanation entails deduction from a scientific "law." His terminology combines a definition of an explanation with a criterion of adequacy; in other words, Bunge is not prepared to regard just any deduction as explanation and hence his reference to "scientific".[12] But his terminology only gives rise to a controversial question: What is a scientific law? Answers to that question are notoriously divergent,[13] and equating a law with an invariant relation makes the notion alien to sociology. Moreover, if some scientific laws are not causal, then deductions from them will not be accepted by sociologists who demand causal explanations.

Ultimate, complete, and unique explanation. All of the foregoing issues are exacerbated by the tendency for debates over a particular explanation to regress infinitely. When correct predictions are deduced from the assertion of a relation between classes, the uniformity itself becomes subject to question. In turn, should the uniformity be deduced, then the critics of the theory will question the premises, and the regression continues indefinitely unless an improbable agreement is realized on some "ultimate explanation." Few contemporary scientists accept the notion of an ultimate explanation, or an "ultimate cause." Nonetheless, debates tend to move in that direction, especially in sociology, where critics of a theory commonly employ only one criterion— conviction.[14] In a field where the promulgation of untestable theories is a tradition, there is scarcely any basis for accepting or rejecting explanations other than personal opinion.

Explanations commonly are given in response to "why" questions, and the notion of explanation is most likely to be defended for that reason. However, that defense overlooks the variety of "why" questions that can be asked of virtually any phenomenon. Consider some questions about cancer: Why does the reported incidence of cancer vary with age? Why does reported incidence vary internationally? Why does it vary by marital status? But the list is only illustrative; questions about cancer are seemingly infinite, and the same can be said for any social phenomenon.

If an explanation is inadequate unless it provides acceptable answers to all possible questions, then no theory can ever be an adequate explanation. Of course, it could be argued that each "why" question calls for an independent explanation, but the argument is contrary to common practice. Traditionally, a theory is challenged by raising various questions and demanding answers to each. To illustrate, suppose a theory is constructed so as to derive the following asserted uniformity: Among countries, the degree of urbanization varies directly with the suicide rate. Now suppose further that consistently correct predictions are derived from the asserted uniformity. If so, the theorist may claim that he has "explained" international variation in the suicide rate. Some critics would reject the explanation as implausible (Homans' men are missing); others would reject it as "not causal"; and still others might observe that the asserted uniformity says nothing about suicide and marital status. The theorist would gain little by pointing out that the theory is not concerned with variation in the suicide rate by marital status. Rightly or wrongly, critics reject a theory by posing questions that it does not even purport to answer. They are unrealistic in demanding answers to all possible questions, but the practice will not be abandoned as long as the notion of explanation is central in assessing theories.

Suppose that critics come to realize that all theories will be rejected if "complete" explanations are demanded. The alternative is that a theory need not answer all questions, only some. However, given a variety of questions, which must be answered? Surely any designation of a certain proportion would be arbitrary, and it would be ludicrous to suppose that sociologists would agree as to the "important" questions, let alone acceptable answers.

Debates over contending explanations cannot be resolved by appealing to logic, as the logical form of explanation does not preclude contenders. While it is attractively simple to say that a phenomenon is explained when deduced (Homans, 1961, 1964a), the dictum ignores a question: What if the phenomenon can be deduced from different premises? Should one regard an explanation as adequate only if it is unique, how can it be known that a given phenomenon can be deduced from only one set of premises? The question introduces the *dilemma of deduction*, which is no more resolvable than the dilemma of induction. One can, of course, ignore the deductive dilemma by simply declaring that explanatory adequacy does not require assurance of uniqueness. But what if an explanation is not unique? It will not do to reply that the adequacy of the contending premises becomes crucial, for that reply does not answer the question. Again, how is adequacy to be judged? Nor will it do to reply that an explanation is adequate if the premises are "true." Scientists often reason from the truth of the consequences to the probable truth of the premises, and they do so for two most understandable reasons. First, unlike formal logic, the premises of a synthetic theory are not true by definition; and, second, they may refer to phenomena that cannot be observed. But suppose that they do refer to observable phenomena. If the premises are generalizations, as they must be in a scientific theory, the dilemma of induction precludes establishing their truth. Nonetheless, for the sake of argument, suppose that the truth of premises can be established and then contemplate the deduction of some phenomenon from a set of "true" premises. There is nothing that would preclude the deduction of the *same* phenomenon from another set of "true" premises. Given such a situation, which of the explanations would be adequate and which inadequate? The question cannot be answered without introducing substantive criteria (that is, considerations divorced from the logical form of explanation), and there is no effective consensus in sociology on those criteria.

Rejection of the notion.   Sooner or later in any field a crucial question arises: On what basis are contending theories to be judged? The question is not to be dismissed lightly, for it is difficult to imagine progress in a field lacking effective consensus as to the appropriate criteria for assessing theories.

The notion of explanation should be rejected because it does not further consensus in the assessment of sociological theories; if anything it precludes consensus that is, since sociologists disagree in their conceptions of explanation, especially of "adequate" explanation.[15] The conclusion will not be well received. There will be cries that the quest for explanation reflects something basic about human nature and hence should not be rejected. To be sure, laymen probably would find the rejection incomprehensible; but, from the viewpoint of science, the typical layman has several undesirable habits, and science is not simply an extension of the layman's mentality. On the contrary, with each generation science becomes more and more divorced from "common sense"; so it is not unreasonable to suggest abandoning the notion of explanation.

The mention of science may lead critics to observe that scientists do view their theories as "explanations." Although this is true in most instances, it is irrelevant. How one chooses to view his theory is a private matter, but how one proposes to assess theories in general is a public issue. Unfortunately, criteria for assessment of sociological theories remain privatized and hence the practitioners persistently argue about the unarguable.

Terminological Problems

As stated previously, unless investigators achieve agreement in applying *some* component terms of a theory to identify things, events, and properties, then the theory has no connection with the empirical world and is not testable. The proliferation of untestable sociological theories indicates that very few sociological terms are empirically applicable or that theorists use such terms reluctantly.

Since sociologists give at least lip service to the desirability of testable theories, it is most unlikely that theorists deliberately eschew the use of empirically applicable terms. So terminology is taken as *the* problem in formulating a sociological theory.

Historical considerations. The paucity of empirically applicable terms in the sociological vocabulary reflects the absence of a "natural history" stage in the field. In a natural history stage the emphasis is on observation and classification, and those activities give rise to a set of conventional term as labels for events, things, and properties. The natural history stage ordinarily is not an exciting period; "naturalists" only pose questions (sometimes the wrong ones), and the answers come later in theory construction. But the natural history stage paves the way by providing subsequent theorists with extralogical "primitive" terms, meaning terms that are empirically applicable without a definition.

The absence of a natural history stage is manifested in two terminological problems. First, definitions of a major sociological term (for example, social stratification) are so divergent that none can be regarded as conventional. Second, given a typical definition, it is difficult to see how the terms in the *definiens* can be taken as primitive. Had sociology passed through a natural history stage, there might now be much greater agreement in the definitions, and those definitions would promise much more empirical applicability. In any case, definitions of major sociological terms ordinarily accomplish very little, as the words in the *definiens* are not empirically applicable without further definitions, and the regression is seemingly infinite because the field lacks conventional primitive terms.

However unfortunate the absence of a natural history stage may be, it is difficult to imagine the field having developed otherwise. From the outset the concern was with theory and theorists occupy center stage today, which is to say that sociologists who have worked in the naturalist tradition have gone largely unrecognized and unrewarded.

The initial terminology of sociology did not emerge from extensive and systematic observations; rather, the great men in the field's history simply borrowed terms from the humanities and commenced immediately to formulate theories. Consequently, given the veneration of grand theory, sociology's terminology has remained a hodgepodge. New terms are introduced continually, and the great names in the field are known largely for those innovations; but seldom is a new term introduced along with a definition that promises empirical applicability. Indeed, the introduction of terms without a concern for empirical applicability is a tradition, and with monotonous regularity the sociologists of one generation attempt to define the terms employed by theorists of the preceding generation.

The argument is not that sociology's development should be reversed, so that it would now undergo a natural history stage and emerge with a revised terminology. In a sense the field has reached a linguistic point of no return. Numerous major terms (such as social integration, status) are entrenched in the field's vocabulary and, however little empirical applicability they offer, sociologists will not abandon them. While one may contemplate redefining the major terms so as to achieve empirical applicability, the idea is questionable. Those terms already have a general meaning for sociologists, who are not likely to accept alien definitions. For that matter, given the paucity of primitive terms in the field's vocabulary, it is most unlikely that a sociologist can provide empirically applicable definitions of the major terms. That point will be lost on those sociologists who demand that each term either be defined "operationally" or rejected. It is idle to suppose either that the major sociological terms can be defined so as to make them empirically applicable or that sociologists will abandon them. Hence the only solution is to use the terms without regard to empirical applicability, but that usage will be ineffective without formal theory construction.

The complexity of social phenomena.   Some commentators (see Blumer, 1954, 1956) imply that sociology's terminological problems virtually defy solution, the idea being that social phenomena are so complex that any symbolization of them distorts reality. Complexity cannot be denied, but it is hardly informative to invoke that notion in analyzing terminological problems. Even if social phenomena are fundamentally different from the subject matter of the physical sciences, it does not follow that sociology's terminological problems cannot be solved. What it does suggest is that the terminology must also be fundamentally different.

At the outset note that *some* terms in the physical sciences supposedly designate phenomena that can be experienced directly. To use Northrop's terminology (1947), the terms are "concepts by intuition," meaning they designate that which is experienced directly or immediately apprehended, for instance, a particular color. Such terms are primitive in that they are undefined and enter into the definition of other terms; nonetheless, they are empirically applicable in that observers may achieve agreement in allegations about experience. Above all, primitive terms are the foundation of the physical science language, as all notions supposedly can be made intelligible by relating them to primitive terms.

The despair of some sociologists stems from recognition that the field has no concepts by intuition and from the conviction that the nature of social phenomena precludes them. The writer agrees, but it does not follow that the terminological problems of sociology defy solution; what it means, instead, is that the problem has been misconstrued.

The first step toward solution is to reject the notion that some sociological terms must designate phenomena that can be experienced directly or immediately apprehended. The argument is not a radical Berkeleianism, that is, events, things, or properties have no existence independently of experience. Nor does the rejection stem from recognition that the notion of experience is ambiguous, especially as regards the difference between experience and perception, sensation, or observation. Rather, the rejection reflects one consideration—the primitive terms of sociology designate phenomena that can be known only inferentially. It should be obvious that no one experiences a revolution, a division of labor, social stratification, a social class, normative consensus, social integration, a community, or social cohesion, to mention only a few "phenomena." The counterargument is, of course, that such notions are defined by "empirical" terms, meaning those that actually are employed in making observations, such as occupation, resident, wife, parent, store, police officer, execution, marriage, household, and so forth. But taking just one of those "empirical" terms, has anyone ever experienced or observed an occupation?

Certainly one may apply an occupational label to an individual, as in the following statement: John Jones is a taxi driver. But it does not follow that one "sees" a taxi driver. Even if one declares that any individual driving a taxi is a taxi driver but only at the time that the individual is actually driving, the terms in the implied definition scarcely designate phenomena that can be experienced directly. Assuming that one sees an individual behind the wheel of a car, it does not follow that he is driving the automobile, not even if it is in motion. Further, no one "experiences" a taxi in the same sense that one experiences a color.

The gist of the foregoing is that no conventional sociological term designates an isolated act (which is about as close to direct experience as a sociologist ever comes); rather, each term designates a *series or set* of acts, none of which can be experienced directly. So while it is conventional to say that the terms designate things, events, or properties, the terms do not designate that which is experienced directly but rather that which is "known" inferentially.[16] Thus, one may identify an automobile as a taxi by "observing" a symbol on its door, but the inference is that the automobile is being used or will be used to transport individuals in exchange for money or some other compensation.

There are two obvious problems when sociological terms are used to designate events, things, or properties. First, as indicated previously, the use of a term reflects an inference from more than one act or "event," but the different types of acts or events that could lead to the same inference approach infinity. Consider someone saying: John Jones resides in that house. Clearly he did not see John Jones "residing"; rather, the statement expresses a conclusion, but we would not know how it was reached. On several occasions the individual may have "observed" John Jones entering the building, accepting mail near the building, eating in the building, mowing the lawn around the building, unlocking the door of the building, and so forth. Given all of those possibilities, it would not be clear why the individual made the statement about John Jones; indeed, the individual himself may not be able to account for it.

The inferential character of sociological terms is not an insoluble problem, for rules of inference can be stipulated in definitions. But even if such rules could be phrased in words that designate the directly experienceable, which of the various types of acts or events are to be recognized in a definition? For example, in defining the term "resident," an incredible variety of acts or events could be specified—eating, sleeping, receiving mail, mowing the lawn, unlocking the door, and so on. Obviously, a definition cannot cover all such possibilities, but including some and excluding others may appear arbitrary, and arbitrariness is suggested even more if distinctions are quantitative (for example, an individual must sleep in a place for thirty nights to be considered as a resident of that place). Of course, in framing a definition one may ignore alternatives and stipulate only one act or event; for instance, friends are individuals who have shaken hands, a resident of a place is anyone who last swallowed food in that place. Such definitions are viewed by some sociologists as arbitrary distortions of reality, and that is a common reaction to so-called operational definitions. More importantly, however, there is a widespread belief that social reality is so complex that no terminology can describe it satisfactorily.

As we have seen, some sociological terms can be defined by reference to one particular type of act or event, but such a definition may appear to be an arbitrary distortion of reality. And there is still another problem. One could deny that no sociological term refers to the directly experienceable. The argument is that typification of even one isolated act entails empathy, the attribution of "meaning," and/or inferences as to intent. Extending the argument, no physical movement of a human being can be identified as an act, let alone as an instance of a particular type, without inferences. That argument is accepted, but it poses no insoluble terminological problem.

One major theme in the present work is that sociology's terminological problems are insoluble only if one demands primitive terms that designate the directly experienceable. That demand is alien to the notion of empirical applicability. More specifically, the notion entails the assertion that independent investigators can achieve substantial agreement in applying *some* terms to identify what they take to be events, things, or properties, even though the terms cannot be applied without inferences. Whether a term designates the directly experienceable is not relevant; what matters, rather, is the agreement achieved by individuals who use the term.

Moreover, the empirical applicability of a term or of its definition is a matter of discovery, not something known *a priori*. If substantial agreement is realized in using a term to identify what the investigators take to be particular things, events, or properties, what would a definition of it accomplish? As for nonprimitive terms, if substantial agreement is realized among independent investigators in applying a definition, of a nonprimitive term, there is no terminological problem.

Of course, the notion of empirical applicability will not be accepted by everyone, and several objections can be anticipated. A critic may demand a list of empirically applicable terms and definitions that can be used by sociologists. There is no such list, for the identification of empirically applicable terms is a matter of discovery, but that does not negate the notion of empirical applicability as a criterion. Then critics may argue that no sociological term can be defined so as to realize empirical applicability, but that is opinion, nothing more. Finally, some critics may object that a definition may distort reality even though it is empirically applicable. The presumption is that we know "reality"; on the contrary, whether a definition distorts reality *cannot be known outside the context of a theory*. It is true that a theory may have minimal predictive power even though all of its terms have maximum empirical applicability, which is to say that a theorist should be concerned with more than empirical applicability in framing a theory. But if the adequacy of the theorist's terms and definitions are judged independently of his theory, the judgment is merely opinion.

Other criticisms will represent a demand for certainty. Critics may point out that substantial agreement between two independent investigators does not insure agreement among others. What the criticism calls for is a resolution of the dilemma of induction. That dilemma haunts all fields (see Greer, 1969:117), not just sociology, and demanding certainty when uncertainty is an inevitable feature of science cannot be a constructive demand. For that matter, the demand suggests that empirical applicability is not a matter of theory. On the contrary, when a theorist formulates what purports to be a testable theory, he tacitly asserts that some of the component terms are and will continue to be empirically applicable, but no research could ever verify that assertion. Unless that perspective is adopted, sociological theorists will continue to regard the terminological problem as a technical matter divorced from theory construction. Indeed, precisely that attitude has been manifested for generations in sociology, and the outcome is painfully obvious—a proliferation of untestable theories.

The present argument does not amount to a declaration that a term is empirically inapplicable unless maximum agreement is realized among independent observers in all instances. Empirical applicability is a matter of degree, but it does not follow that the predictive power of a theory is minimal if some of its constituent terms are not empirically applicable to the maximum degree. On the contrary, considerable predictive accuracy may be realized without maximum empirical applicability, so an acceptable level of empirical applicability is a relative consideration. It depends on what the theorist will settle for in the way of predictive accuracy, and that in turn should depend on contending theories. Further, there is always the possibility that the predictive accuracy can be enhanced by altering the definitions of a theory so as to realize greater empirical applicability. The maximum degree may never be realized, but who is prepared to argue that a theory can ever be regarded as complete and final? Certainly the proposed strategy is preferable to the idea that terminological problems must be solved before undertaking theory construction, an idea that would have a crippling effect in any field.

Finally, it may be alleged that empirical applicability as a terminological criterion is contrary to the nature of science. Appeals to the nature of science, including any made here, are innocuous, as opinions in sociology on that subject approach infinity. Further, in proposing a mode of theory construction, the conditions of work and practical problems of a particular field are far more relevant than abstract conceptions of science. Nonetheless, critics may allege that science deals with experience, and hence a field's terminology must be reducible to experience. If by experience one means that which is given directly or immediately apprehended, then the present argument may be construed as alien to science. The reply to such criticism is brief. Science does not deal with experience; rather, it deals with public assertions. One may construe those assertions as reports of experience, but experience is forever private and communicated only through assertions or allegations.[17] In anticipating the charge of solipsism, the present characterization of science is not a denial of a transcendent reality; but as a public enterprise science deals with reality only through assertions.

Quantitative properties and the question of validity.   If a constituent term of a theory designates a quantitative property, no definition of that property will make the term empirically applicable. Investigators cannot describe a particular instance of that property, let alone agree in their descriptions, unless they apply a formula to data. Accordingly, tradition in sociology notwithstanding, a theory is incomplete unless the theorist stipulates not only formulas but also a procedure for obtaining the requisite data. Such stipulations are the most difficult steps in theory construction, and sociologists have compounded the difficulty by demanding that the "validity" of formulas (or "measures") be demonstrated.

The demand is rejected because it makes the notion of validity independent of theory. On the contrary, insofar as the notion is entertained at all, the validity of any component of a theory cannot be demonstrated outside the context of that theory. To illustrate, suppose that someone constructs what purports to be a test of intelligence and then proposes to "validate" the instrument by examining the relation between scores of students on the test and their school grades. However conventional the strategy may be, it evades a question: What is the rationale for expecting a correlation between intelligence test scores and grades? The proposal implies a theory concerning the correlates of intelligence or, viewed the other way, the correlates of grade assignments; and it is pointless to deny that implication. Indeed, there is every argument for making the theory explicit; otherwise, the impression is created that the validity of a formula, procedure, or operation can be ascertained by some technique independently of theory. On the contrary, an attempt at validation is nothing more than a test of an implicit theory; and once that is realized the utility of validity as a notion is questioned.[18] Briefly, there are no valid formulas, procedures, or operations; rather, there are only theories, expressed or implied, with varying degrees of predictive accuracy.

It could be argued that positive tests of a theory validate the component formulas and procedures; but do negative tests invalidate them? The terms of a theory commonly designate more than one formula or procedure and, as we shall see, it may be questionable to attribute negative tests to a particular formula or procedure. For that matter, negative tests may not be due to any of the formulas or procedures. So the very notion of validity is dubious, and nothing is lost by abandoning it.

Reliability. Given a formula and directions for application, the values computed are interpreted as symbolizing a quantitative property. Two questions commonly are posed in assessing such values. First, do the values actually represent what they purport to represent? Second, how well or to what extent do they designate or describe what they purport to represent? The first question traditionally is associated with the notion of validity, and the related notion in the second question is "reliability."

Although the notion of reliability is accepted, it is not free of ambiguities.[19] In tests of a theory that deals with quantitative phenomena there are: (1) formulas, (2) numerical data to which the formulas are applied, and (3) a series of values computed by application of the formula to the data. Customarily, one speaks or thinks of the values as being reliable or unreliable to some extent; however, a particular value may be judged as unreliable because the formula has not been applied as stipulated and/or because the data to which it has been applied are not reliable. To illustrate, suppose that a formula and the procedure for its application are specified in the statement of a theory, with the specification including instructions as to how the requisite data are to be obtained. Now suppose that census data on occupations are designated, and tests of the theory entail international comparisons. Suppose further that two investigators independently apply the formula to the same census data and it is discovered that the two values are markedly divergent. Certainly one would question both sets of values. Now assume the converse: that there is no divergence. Would it follow that the census data are reliable? Not at all! But observe again that marked divergence between the two sets of values would cast doubts on their reliability without even considering the data. So two considerations enter into the notion of reliability: (1) *agreement* in values computed by independent investigators in applying a particular formula to the same body of data and (2) *credibility*, that is, the extent to which the data are believed to represent what they purport to represent. Correlatively, negative tests of a theory could be produced by (1) errors in the application of the formula and/or (2) errors in gathering data.

Since both agreement and credibility enter into the notion of empirical applicability, the conventional concern with reliability can be subsumed under that notion. However, whereas reliability commonly is regarded as a "technical" rather than a theoretical problem, that is not so for empirical applicability.

Given agreement between values computed by two investigators, there is no assurance that either value is "correct," meaning that the same value would have been reported by a third and fourth investigator. So empirical applicability is a "theoretical" matter if only because the notion entails an inevitable uncertainty; that is, the reliability of values reported by investigators is always questionable. Nonetheless, in formulating a theory that comprises assertions about quantitative properties, a theorist must specify formulas and procedures for their application. Each such specification is a tacit *assertion* of empirical applicability, and that assertion is a part of the theory, not a technical consideration as sociological theorists in the grand tradition would have it.

The idea of empirical applicability entails a consideration not ordinarily associated with the notion of reliability. In specifying formulas and procedures for their application, the theorist asserts not only that sufficient agreement in the values computed by independent investigators will be realized, but also that the requisite data can be secured. If requisite data are not available in published form or cannot be gathered because of limited resources, the formulas are not empirically applicable.

It is not sufficient for a theorist to specify formulas; he also should spell out the procedure for their application. If the theorist does not stipulate the use of published data in tests, then he must stipulate the procedure by which data are to be gathered. In either case, however, he asserts that the data are credible. The word "credible" is appropriate, as no one can know with certainty the extent to which data represent what they purport to represent.

Consider a stipulated procedure for gathering data on the marital composition of a population. Given that stipulation, an investigator (or team of investigators) may report that in a particular population at a specified time there were the following marital statuses and numbers of individuals in each: single, 3,490; married, 8,920; widowed, 1,723; divorced, 42. The meaning of the words and numbers depends on, *inter alia*, the procedure ostensibly followed in gathering the data. But no critical examination of that procedure can provide assurance that the data represent what they purport to represent. To be sure, if independent investigators had applied the same procedure to the same population at the same time, then a comparison of their reports would suggest some conclusion as to the reliability of both sets of data. Indeed, it is difficult to imagine reaching a truly defensible conclusion without comparing independent sets of data; but still no comparison can confirm that data actually represent what they purport to represent.

Suppose that two investigators report identical data on marital composition after allegedly applying the same procedure to the same population. Even so, one cannot be certain that a third investigator would not have reported markedly divergent data. Hence the reliability of data is a theoretical matter, as it entails uncertainty. But it is a theoretical matter in another sense. Contrary to the tradition in sociology, a theorist should not assume that someone else (the ever faithful "technician") will stipulate the procedure for gathering data to test a theory.[20] If someone else must do it, then the theorists can disavow the negative tests by alleging that the "wrong" data were employed, and hence the theory cannot be falsified. So should the theorist fail to stipulate a procedure for gathering data, then the theory is simply incomplete. In other words, procedure is part of a theory, and it is inconceivable that test results do not depend on it. Imagine a theorist saying: Correct predictions can be derived from this theory regardless of the procedure employed to gather data. This is surely impossible, but in effect it is precisely what sociological theorists have tacitly presumed throughout the field's history. Now if it is recognized that the predictive accuracy of a theory depends on the procedures for obtaining data, who else but the theorist should specify that procedure? After all, it is his theory. True, in specifying a procedure, the theorist only asserts that the data will be reliable, but that assertion is a part of the theory.

What has been said of procedures for gathering data applies in much the same way to the use of published data in tests of a theory. The credibility of published data is particularly difficult to judge, especially in sociology. Nonetheless, in specifying that particular *kinds* of published data are to be used in tests of a theory, the theorist asserts that the data are and will continue to be reliable (at least to a sufficient extent), and that assertion is part of the theory.

One may question the treatment of reliability as a terminological problem, but the two cannot be divorced. It is inconceivable that the empirical applicability of a formula in no way depends on the terms that specify kinds of data. Similarly, in specifying a procedure for gathering data the goal is to maximize the likelihood that different investigators will gather the same data, but the realization of that goal depends in part on terminology.

As constantly emphasized, reliability is a theoretical matter. However, the emphasis does not mean that related problems can be solved by formal theory construction. On the contrary, no mode of formalization should give instructions for devising formulas, nor specify procedures for gathering data, nor pass judgment on the credibility of particular kinds of published data. Such matters are substantive, and decisions can be made only in formulating particular theories. However, a theorist can draw on the works of others; so specialists in research methods should continue to devise formulas, invent procedures for gathering data, and concern themselves with the credibility of published data. The division of labor is defensible, but in the final analysis it is the theorist's responsibility to make judgments about the empirical applicability of formulas and the credibility of *kinds* of data.

The problem of research resources.   Even if sociological theorists should abandon their traditional indifference to terminological problems, there is no assurance that testable theories would be realized. Appearances to the contrary, it is not sufficient for a theorist to stipulate formulas and procedures for gathering data. The application of a procedure requires resources—time, money, and personnel—and the amounts required to gather some kinds of sociological data would be astronomical. Contemplate the cost of gathering international data by field studies or surveys on occupational composition, the volume of interaction, participation in voluntary associations, number of births and deaths, marriages, strikes, occupational aspirations, or educational achievement. Cost would not be limited to gathering one body of data, as repeated tests of a theory is an unquestionable goal. So it should be abundantly clear that tests of a theory may be precluded for one simple reason: the field's resources are limited.

Some sociologists would grant that the field's resources drastically limit opportunities for testing theories, but they may not consider those limits as having any bearing on theory construction or as a terminological problem. The divorcement of those subjects reflects a firmly fixed notion of a division of labor—the theorist need not test his creations; others will do it. That notion is accepted readily, but it does not follow that a theorist should not be concerned with research resources. He should, indeed, adopt a practical course and formulate theories that are testable despite limited research resources, and in so doing limited research resources are treated as a terminological problem. If competent, the theorist will anticipate what researchers can and cannot do with terms, especially those that designate formulas and procedures for their application.

As suggested previously, in some sociological specialties it may be feasible to use published data for tests of a theory, and competent theorists will be aware of such data. But again the problem is terminological. Specifically, it will not do for the theorist to say, in effect: If data cannot be gathered as stipulated, use published data in testing the theory. The *kind* of published data must be stipulated in such a way that agreement will be realized when independent investigators apply formulas to published data on the same population. The point may appear trivial, but a theorist should not presume that investigators will somehow know which published data are to be utilized, and one goal in theory construction is to minimize discretion in tests. The stipulation of kinds of published data is not a "technical" consideration for still another reason. As indicated previously, the reliability of published data is especially questionable, and only the theorist should decide whether to use a particular kind of published data in tests of a theory. Tests are distinct from the theory itself, but the decision to use a particular *kind* of data is part of the theory.

Linguistic differences.   When a theorist uses a term, the universe of its application is not limited to any particular country, culture, or point in time, unless one holds to the quaint idea that one can formulate a theory restricted to, say, the United States on March 12, 1971. One goal in theory construction is generalization, but sociology has no "universal" technical terms, that is, theories are stated in a historical or natural language (such as English or French). Accordingly, given a theory stated in English and taking one of the component assertions as a generalization, how is it to be applied to non-English speaking populations? True, the generalization need not apply to all populations; it may be restricted to certain types (for example, literate societies); but even a restriction to English-speaking populations would not solve the linguistic problem.

The conventional meaning of terms is not fixed, as witness change in the meaning of "political liberal" in English-speaking populations during less than two centuries. This, then, is the problem: Whereas the meaning of terms in an assertion should remain fixed, the conventional lay meaning may change. So it is questionable to assume that such English words as marriage, police, city, and hospital (terms employed by sociologists and laymen alike) will have the same meaning in the year 2400 as today, or even that in 2400 any population will speak the English language. That consideration alone precludes reference to the linguistic behavior of a population as a guide in testing what purports to be a nomothetic assertion. To illustrate, suppose that someone makes the following assertion: No married persons commit suicide. The assertion may appear to be patently false, but how could it be demonstrated? One could select an English-speaking population and pose two questions for members of the population concerning any event: Was the event a suicide? Was the individual involved in the event married? Sooner or later an affirmative response to both questions would be made, and that response could be construed as negative evidence for the assertion. But there is an assumption—that members of the population attribute the same meaning to the key terms, married and suicide, as did the individual making the assertion. That assumption is questionable even today, and it would be even more questionable in the year 2400. But illustrations of the problem need not be restricted to change in the meaning of terms. Consider posing the foregoing questions to Kapauku Papuans; obviously, few of them would even comprehend either question. So we can reach a general conclusion: The meaning of sociological terms cannot be determined by linguistic usage in any particular population.

The conclusion offers no solution to the terminological problem, nor does the idea of "translation." Even if a universal multilingual dictionary could be consulted, its use would be debatable. Given the equation of two terms in such a dictionary, for instance, suicide = *Selbstmord*, the basis for equating them would be either unknown or questionable if known, especially since the empirical applicability of either term in English-speaking or German-speaking populations is conjectural. Systematic evidence to justify the equation could be realized only by analyzing the agreement achieved by German-speaking and English-speaking observers in applying the terms to identify the same events. Such comparisons in all language are not feasible, and the results could not be conclusive if only because (1) some terms in one language have no equivalent in another; (2) new languages may appear; and (3) the equation of two terms may be justifiable at one point in time but not later, that is, the conventional meaning may change in one language but not in the other, or both meanings may change divergently.

The scope of the linguistic problem is such that the implications cannot be described fully, so two general observations must suffice. First, the problem is not hypothetical or "academic"; its manifestations are conspicuous and, though largely ignored today,[21] the problem will become more evident as truly comparative research is attempted. Second, no solution is in sight. As evidence in support of the first observation, recall debates in the literature as to the presence of law among nonliterate peoples, the universality of the nuclear family, and the prevalence of com-

munism among nonliterate peoples.[22] With monotonous regularity such debates degenerate into semantic squabbles, and the
reason is painfully obvious. The empirical applicability of key
terms—law, nuclear family, and communism—is questionable even
in reference to English-speaking populations. Further, as definitions of those terms become precise they become peculiar to
Anglo-America, and as they become less ethnocentric they become
so vague as to minimize empirical applicability in any population.

As for solutions, there are several alternatives but none prom-
ising. Certainly efforts by sociologists to develop universal taxonomies have not proven to be fruitful. Parsons' conceptualizations
are relevant in that connection; more than other sociological
theorists, he is concerned with universals. But his work illustrates
the most conspicuous defect of that genre. The empirical applicability of Parsons' terminology is dubious even in English-speaking
populations, so there is no basis whatever to expect that it can be
used effectually in cross-cultural studies. What has been said of
Parsonian terminology is also true of sociological terms in general.
The more abstract terms are not ethnocentric, but it is precisely
those terms that promise very little empirical applicability. One
could undertake to define all major sociological terms so as to
make them universally applicable, but it is not clear how that goal
can be realized. For that matter, we have very little systematic
knowledge about the empirical applicability of English terms in
English-speaking populations, let alone universally.

Certainly no significance should be attached to the use by
anthropologists of a European language to describe non-Western
social units, for that practice developed uncritically. In particular,
anthropologists do not conduct independent field studies and
compare the results to determine the empirical applicability of
their descriptive terminology, and for that reason alone their
reports on non-Western peoples are debatable. The commentary
does not question the honesty or the competence of anthropologists, but integrity and systematization of observations alone cannot solve the linguistic problem. Because of the paucity of work
on the problem, it is not known in what way and how much
linguistic considerations limit the value of anthropological reports,
but a skeptical attitude is justified.

One proposed solution to the linguistic problem is especially controversial. Rather than employ a natural language (such as English) for descriptive purposes, sociologists and anthropologists should develop a technical vocabulary in which the "basic" terms designate specific types of acts (see Harris, 1964). One immediate objection is obvious. Even if the basic terms are technical, definitions of them would entail the use of a natural language, which defeats the purpose. In any event, neither sociologists nor anthropologists are likely to abandon conventional terminology, and there are reasons other than the weight of tradition. The change would entail setting aside the vast body of data now employed in lectures, texts, and theory construction; and gathering "new" data would require astronomical resources. Further, assuming that some data are necessary to formulate theories, theory construction would have to be halted until the new terminology is employed in descriptive studies. Certainly it would be difficult to salvage existing theories.

Even the creation of a radically different terminology is improbable. The variety of specific types of acts appears infinite, and if a basic term is used to identify each, the number of terms would be unmanageable. Some types of acts could be ignored, of course, but there would be little basis other than intuition for recognizing some and not others. For that matter, however radically different it may appear, it is inconceivable that a terminology can be independent of all natural languages.

The foregoing does not mean that the linguistic problem defies solution, or that it must be solved before progress in sociology can be achieved. Indeed, a demand for an immediate solution would have a crippling effect on research and theory construction.

Operationalism in sociology.   Given the emphasis on terminological problems and the demand for testable theories, critics are likely to regard the present perspective as a plea for "operationalism." The characterization is innocuous, for conceptions of operationalism are obviously divergent, even among sociologists who are commonly identified as "operationalists." In any case, taking commonly held conceptions as criteria, the present perspective departs from the putative tenets of operationalism in several respects.

The idea that all component terms of a theory must be defined "operationally" is rejected categorically. A term may be used effectively in theory construction even though the theorist is unable or unwilling to define it, operationally or otherwise. The emphasis in operationalism on testable theories is accepted; but a theory may be testable even though some of its component terms designate vague, undefined notions.

Critics will be mistaken should they equate the notion of empirical applicability with operationalism. It would be a mistake if only because after several decades the meaning of an "operational definition" is not clear. Operationalists would understand the meaning of empirical applicability and perhaps approve the notion, but it does not follow that an empirically applicable definition entails an "operation." A definition may be empirically applicable even though it incorporates no reference to procedure or instruments, which is to say that the content or "kind" of definition is not relevant. A definition is only a means to an end— empirical applicability—and the end justifies the means. Had the operationalists in sociology grasped that point, they would have made less of a fetish of instrumentation or physical operations. That inordinate emphasis did not clarify the meaning of an "operational" definition, and it appears that operationalists were simply calling for "good" definitions, meaning those that promise empirical applicability.[23] Certainly the notion of empirical applicability has one advantage: it incorporates recognition that some terms in any field's vocabulary must be primitive, meaning that they are not defined, operationally or otherwise. Finally, however "operational" a definition may be, it is not considered as empirically applicable if limited resources preclude its use.

Notes for Chapter 2

[1] See Madden, 1960:201-240; Pap, 1962:Part 4; Nagel, 1961:chap. 10; Braithwaite, 1953:293-341; Bunge, 1959; and Feigl and Brodbeck, 1953:387-437.

[2] As expressed by Calhoun: "Though MacIver maintains that causation is correlation plus something more, he fails to tell what the something is, save implicitly by reference to the feeling people have when they think of one event's 'causing' another" (1942:715).

[3] Conceptions of "causation" in the sociological literature on theory construction are commonly gross oversimplifications that ignore issues and problems (for example, Stinchcombe, 1968:28-38). Of recent works on theory construction in sociology, only Dubin (1969:91) questions a causal language.

[4] Disagreements are not conspicuous for one simple reason—social scientists rarely make their conceptions of causation explicit, let alone confront the issues and problems. As an example, see Lerner, 1965.

[5] Blalock (1964:10) evidently subscribes to Bunge's idea (1959:62) that the cause need not precede the effect. But that argument only adds still another irresolvable issue to the debate, since numerous philosophers and scientists insist that cause and effect are separated by a finite time interval. In other words, contrary to Bunge (1959:63), they argue that the principles of antecedence and the notion of causation are not independent.

[6] It is hardly surprising that sociologists who invoke the notion of causation typically ignore issues and problems. Thus, Hirschi and Selvin (1970:225) characterize the idea of a perfect relation as a "false" criterion of causation. However, even if that characterization be accepted, they do not answer the crucial question: As regards the degree of association between variables, what is the "true" criterion of causation?

[7] It does not follow that a conception of causation is acceptable if consistent with the conditions of work and practical problems in sociology. The related philosophical or methodological issues transcend particular fields, and they survive Blalock and Costner's treatment of causation. What has been said of Blalock and Costner's treatment of causation also applies to that of Simon (1957, 1968).

[8] "For apart from 'philosophy' there is perhaps no name of a branch of knowledge that has been given so many meanings as 'logic'" (Bochenski, 1961:2). It is true that Bochenski speaks of "different varieties of one logic" and not "different logics" (1961:14), but the distinction is debatable. In any case, he does not present an explicit and complete definition of "logic" (as opposed to negative or residual definitions). As for his idea (1961:2) that logic is concerned with the problems introduced in Aristotle's *Organon*, the suggested definition is ethnocentric and inconsistent with Bochenski's subsequent statement that "every new variety of logic contains new logical problems" (1961:14).

[9] As an illustration of such indifference, see Meehand, 1968, and Smelser, 1968. With the exception of Homans 1969:81), Smelser's indifference to issues and problems entailed in the notion of explanation has been tolerated by his audience.

[10] One school of thought denies that explanation necessarily entails deduction (see Brodbeck, 1962, for references and a critique). In any case, there is still the question of the *logical* requirements for an adequate explanation (see

Feyerabend, 1961, and Hempel, 1965:247-248). Those requirements are debated (see Brodbeck, 1962) and the issues will not disappear by simply equating explanation with deduction.

[11] Contemplate Stinchcombe's statement: "I have a firm conviction that some things are to be explained one way, some another" (1968:4). Sociology should be much to Stinchcombe's liking, for that is precisely the way things are explained in the field. As to divergent kinds of explanation, Stinchcombe evidently presumes that sociologists agree as to which kind is adequate. The history of the field is quite to the contrary.

[12] If Bunge introduced the qualification to suggest that there are no particular issues or problems associated with the notion of "scientific explanation," it is misleading (see Rescher, 1970:97-162; Feigl and Maxwell, 1962:28-272).

[13] Dubin's comment is apt: "It is fascinating . . . to consult the index of current and classic works on the philosophy of science. The delightful ambiguity and downright confusion in the analysis of scientific law is enough to make a 'mere empiricist' out of any budding scientist" (1969:87). Yet, paradoxically, Dubin would have sociologists use the term "law" in theory construction. No less paradoxical, some philosophers of science attempt to make a most debatable notion the "fundamental concept" for science (Braithwaite, 1953:2).

[14] The criterion is consistent with the inclination of sociologists to indulge in and tolerate *ad hoc* explanations. For example, in a few passages Gouldner (1970:125-134) explains why sociology did not flourish in England, why functionalism became so predominant in British anthropology, and why Malinowski was prominent in British anthropology. He offers such explanations without deriving them from any generalization whatever. Gouldner evidently knows the "domain" and "background" assumptions of sociologists very well. In any case, he joins a legion of sociologists who write as though they can provide an adequate explanation of a particular event or thing without reference to any generalization about infinite classes (see Brodbeck, 1962).

[15] The disagreement is not peculiar to sociology, and there are issues beyond those considered here (see Nagel, 1961; Hempel, 1965; Brodbeck, 1962; and Rescher, 1970).

[16] Types of terms commonly distinguished by philosophers of science are not particularly relevant in contemplating the terminological problems of sociology. For example, the distinction between universal and individual concepts (see Popper, 1965:64) does not speak directly to the question of empirical applicability. Of all the typologies, that proposed by Northrop (1947) is probably the most relevant, but his distinction between two major classes—concepts by postulation and concepts by intuition—is dubious when applied to sociological terms. Since no sociological term designates that which can be experienced directly or apprehended immediately, the field has no concepts

by intuition. Northrop does distinguish types of "concepts by postulation," but those distinctions would be most debatable when applied to sociological terms. Thus, concepts by intellection designate factors that can be neither sensed nor imagined, concepts by imagination designate factors that can be imagined but not sensed; concepts by perception designate factors that are in part sensed and in part imagined; and logical concepts by intuition are abstractions from the totality of sense awareness (Northrop, 1947:94). Given those distinctions, what of population density, status, voluntary association, suicide rate, and social stratification as terms used by sociologists? Clearly Northrop's distinctions would be difficult to apply, and the difficulty questions the idea that a particular typology of terms can be useful in all fields.

[17] The distinction between experience and assertions or allegations about it often is ignored in methodological dictums, such as Popper's statement that "it must be possible for an empirical scientific system to be refuted by experience" (1965:41). If the refutation of an "empirical scientific system" is a matter of experience, then the criterion is personal and private.

[18] "From the logical or theoretical standpoint, a measure is said to be valid to the degree that it measures what it is supposed to measure" (Blalock, 1968a:13). When it comes to phenomena that cannot be experienced in a direct sense (as is the case for virtually all sociological phenomena), how can one purport to measure them without a theory or at least an empirical assertion? Indeed, how can we reach a conclusion about the degree to which such a measure does "measure what it is supposed to measure" without reference to a theory? For the most recent round of sterile debates over validity and references to the literature see Deutscher, 1969; Ajzen, *et al.*,1970; and Lastrucci, 1970.

[19] Some of the ambiguities stem from the idea that there is a distinction between reliability and validity. In the case of a particular value, it is difficult to see what the distinction would be. If the value does not represent what it purports to represent (if it is invalid), how could it possibly be reliable? The only alternative is to restrict the notion of validity to formulas or procedures and to speak of particular values as reliable or unreliable. But contemplate the paradox: a "valid" formula, measure, or procedure that generates unreliable values. So, even if the notion of validity should be retained, the distinction between it and reliability is dubious. Needless to say, the distinction has been questioned before (see Lastrucci, 1970).

[20] The tradition is not questioned by most sociologists who write on theory construction (see Blalock, 1969; Dubin, 1969; Stinchcombe, 1968; Willer, 1967; and Zetterberg, 1965).

[21] With very few exceptions (see Cicourel, 1964) methodologists in sociology have paid scant attention to linguistic phenomena as data, research tools, or as an aspect of theory construction.

[22] See Hoebel, 1954; Buchler and Selby, 1968; and Malinowski, 1959.

[23] Of course, operationalists may argue that only "instrumental" definitions are empirically applicable, but that is opinion, nothing more. In any event, had they stressed empirical applicability as the criterion of a "good" definition rather than harping on the distinction between "verbal" definitions and "physical operations," operationalism might have been spared many criticisms (see Sjoberg, 1959).

CHAPTER 3 CRITERIA FOR ASSESSING SOCIOLOGICAL THEORIES

Some sociologists would deny that the field lacks consensus as to the appropriate criteria for assessing theories, but consider the following men and contemplate their work: Talcott Parsons and George Homans, Erving Goffman and Otis Dudley Duncan, Harold Garfinkel and James Coleman, Robert MacIver and George Lundberg. It will not do to say that the only contrast is one of substantive interest. The works of those men clearly suggest divergent criteria for judging sociological theories and so the question: If consensus prevails in sociology, how did all of them become prominent? The answer lies in a not-so-secret scandal: sociologists wittingly or unwittingly write for cliques.

Pointing to the obvious, sociologists are divided into camps—positivists, functionalists, and phenomenologists, to mention only three. The major divisive issue is the appropriate criteria for assessing theories, but it is not conspicuous because many sociologists studiously refrain from stipulating criteria or debating the question. As an instance, Talcott Parsons has never explicitly stated the criteria by which his theories are to be judged. The advantages of reticence are obvious, for critics of the silent theoretician are certain to employ the wrong criteria in assessing his work, and it is better to be perennially misunderstood than found wanting by one's own standards. However, while a theorist profits from remaining silent, the outcome for the field is merely avoidance of a crucial issue.

Evidence of Dissensus

In opposition to the assertion of dissensus in sociology, one may ask: But what of Durkheim and Weber? The question suggests that we parade Emile and Max to demonstrate consensus in the recognition of greatness. Invoking those illustrious names only reveals that sociologists have some *esprit de corps*. Certainly it would be preposterous to declare that contemporary sociologists accept the criteria of both Weber (1947:88-120; 1949) and Durkeim (1950) for assessing theories. Numerous sociologists reject Weber's notion of an adequate explanation of social phenomena, but Homans (1961) and Inkeles (1959) say the same with regard to Durkeim, and so it goes. For ritualistic purposes the admiration of both Weber and Durkheim is justified, but when it comes to criteria for evaluating theories, only methodological schizophrenia would permit accepting the sociology of both men. Indeed, Weber and Durkheim's criteria are so divergent that they are cited as exhibit "A" for the present argument.

If the reader demands something more than general observations on dissensus in sociology, then he should consider the debate over *Verstehen*,[1] Sorokin's description of the logico-meaningful explanation (1943, 1947, 1957), Catton's enumeration (1966) of the principles of "naturalistic" sociology, Wilson's defense (1970) of "interpretive" sociology, and Blumer's criticism (1956) of what he takes to be conventional sociology. But one need not search for evidence of dissensus; several writers have compiled it, though perhaps unwittingly. For example, Brown (1963), Duke (1967), and Nettler (1970) have documented divergent types of explanations in sociology. Conceivably, sociologists could agree that each type of explanation (for example, functionalistic) is appropriate for certain classes of social phenomenon; but that is not the situation. The typical sociologist prefers one type of explanation to the exclusion of others, and contrasts in those preferences are beyond argumentation.

In rebuttal one may point out that there are alternative criteria for assessing theories in any science. Thus, in the philosophy of science literature five criteria are stipulated (Frank, 1956:3-36): predictive power or agreement with observations, parsimony or simplicity, logical consistency, plausibility, and fertility. But it does not follow that difference in opinion concerning those criteria are as marked in any field as in sociology. In particular, who is prepared to deny that in most scientific fields there is effective consensus as to the primacy of one particular criterion—predictive power or agreement with observation?

Dissensus in sociology transcends conventional criteria for assessing theories, which is to say that sociologists invoke criteria that are virtually peculiar to the field. Consider the notion of "bringing men back in" (Homans, 1964a) as a criterion. It is a substantive consideration in addition or in opposition to conventional criteria (predictive power, simplicity, and so on); but that consideration will not prevent Homans and his followers from invoking reductionism in appraising theories. Evidently, regardless of its predictive power or simplicity, any theory that does not "bring men back in" will be rejected by Homans.

To mention only one other exotic criterion, the followers of C. Wright Mills, predominantly conflict sociologists, demand that a theory deal with a "significant" social issue or historical trend.[2] A more subjective criterion cannot be imagined.

The criteria of "significance" and "bringing men back in" are substantive rather than formal. As such, when one employs them, he asks in effect: How does the theory square with my preconceptions of society, social life, and human nature? If the theory and the preconceptions are divergent, then so much the worse for the theory.[3] Such preconceptions are in keeping with "Newton-before-Ptolemy" sociology—the idea that we can have an adequate theory of society or "human nature" without first having valid theories about social mobility, voluntary associations, deviant behavior, and so forth. Unless a "grand" theory represents a synthesis of such special theories, it cannot be other than a preconception, and the ultimate absurdity is to employ such a preconception to evaluate special theories.

While some sociologists would grant that their field lacks consensus on criteria for assessing theories, they might disagree with the present interpretation of that condition. Opposing interpretations could take four forms.

One counterargument can be anticipated from positivists and Catton's "naturalists" (1966). In either quarter the identification with natural science is so intense that sociologists ignore anarchy in their own field and their ardent opponents (see Hayek, 1955). Stated another way, their commitment to the methodology of the physical sciences is so intense that deviation from that methodology is, to them, inconceivable. For that reason naturalists and positivists dismiss their opponents as isolated aberrations and fail to recognize them for what they are—members of established and viable schools of thought. Hence positivists or naturalists will be inclined to regard the present argument as a gross exaggeration. But contemplate two questions. First, if "naturalism" or "positivism" is dominant in the field, how do sociologists who reject both perspectives rise to prominence? Second, how is it that formal awards are given to sociological works that are patently alien to both positivism and naturalism?

Another interpretation does not minimize dissensus in sociology; rather, it is viewed as transitory. There are two versions of this interpretation, one of which takes dissensus as symptomatic of sociology's youth, that is, as a normal stage in any field's development. The appeal to sociology's youth is wearing thin, especially since it appears that there was greater consensus as to criteria for assessing theories in the 1930s than today. Certainly the issue was treated more overtly in that decade, for example in the debates between Lundberg and MacIver.[4]

Still another interpretation makes no reference to developmental states in the sciences. The argument is that a vast, impersonal competitive process is under way in sociology, and the outcome will be consensus through selective survival. That interpretation is questionable and for reasons other than trends since the 1930s. True, there are competing schools, but selective survival is only a remote possibility. Each sociologist can continue publishing for a clique, with deans and even some departmental chairmen none the wiser. Above all, there is no mechanism by which the competitive process could result in selective survival; if anything, competition is institutionalized to work against that outcome.

Consider the editorial policies of sociological journals. It is no secret that editors commonly opt for eclecticism in assessing papers. This policy both reflects recognition of dissensus in the field and perpetuates that condition.

Even the recruitment policies of departments unwittingly sustain dissensus in the field. Departmental chairmen are prone to seek a balance in recruitment. The balance sought is not in terms of substantive interests alone but also in the going "brands" of sociology, and the divergent brands entail the issue at hand. Accordingly, as long as a departmental balance in perspective is considered desirable, there will be room for all. The point is that if a sociologist longs for the day that the last functionalist, phenomenologist, or positivist is dead or fired, he should be prepared to live for centuries.

Although some sociologists will acknowledge dissensus as a condition of their field, they are concerned with what they believe are more pressing problems. One widespread belief is that the quality and/or quantity of sociological data preclude progress in the field; hence the concern with techniques for gathering data, perfection of measures, and computer mania. Certainly the data problem is not merely a technical consideration, but the adequacy of data is relative to criteria for assessing theories. Accordingly, if there is no consensus on criteria, how can sociologists agree as to what kind of data are needed, or even that the problem is crucial? That question is likely to be overlooked by positivists, who blithely assume that they could sell their perspective to the field "if they only had the data." They should attempt to convince a phenomenologist or ethnomethodologist with so-called hard data.

What has been said of the data problem applies also to theory construction. There seems to be a growing conviction that sociologists simply do not know how to formulate theories. However, anyone who aspires to lead sociology out of the wilderness by advocating some mode of theory construction is in for a rude shock; the point being that how one formulates a theory depends on his criteria for assessing theories. If a theorist is not primarily concerned with simplicity, logical consistency, or predictive power as criteria, he will not be impressed by any plea for formal theory construction or any mode of deriving testable assertions. It is not through inadvertence or ignorance that phenomenologists, Weberians, ethnomethodologists, Parsonian sociologists, and functionalists refrain from employing a mode of formal theory construction. They are not having any of it.

A final interpretation of dissensus in sociology views it as neither transitory nor undesirable. The argument can be extended to a declaration that divergent criteria for assessing sociological works are actually desirable. However charitable that point of view may be, it ignores the implications of dissensus. For one thing, continuity in research and theory is limited to cliques, with the further consequence that the resources of the field cannot be organized. Moreover, the ultimate goal of any field—a synthesis of all special theories—is precluded. Dissensus also promotes the habit of writing for a particular audience, and, consequently, sociological journals cannot develop a common theme or editorial consistency. Finally, while a concern with image can be overdone, dissensus in the field makes a spectacle of sociology in the eyes of the public and especially those involved with the other sciences.

Predictive Power as the Primary Criterion

Although sociologists agree on the desirability of testable theories, tests play a minimal role in the assessment of sociological theories. But it is not sufficient to demand testable theories, for tests alone will not insure consensus in assessing theories. Specifically, some sociologists argue that test results must somehow demonstrate causation or justify a causal inference, but they do not agree as to the kind of test results that are to be taken as evidence of a causal relation. As for the idea that test results indicate whether or not the theory offers an adequate explanation, it simply ignores dissensus among sociologists as to what constitutes an adequate explanation.

Now the argument is not that test results should be ignored in assessing sociological theories; far from it. But in examining test results one question should be paramount: What do the results suggest about the predictive power of the theory? As the question indicates, predictive power should be the primary criterion by which sociological theories are assessed.

The rationale. Observations on predictive power as the primary criterion seldom go beyond platitudes about science, or they merely suggest that the achievement of predictive power is an end in itself. Dubin (1969:9) and others notwithstanding, predictive power is not an end in itself, nor is it necessarily a step toward control. The question of control as the ultimate goal of science involves ideological issues that can and should be avoided.

The rationale for the predictive power criterion is that it provides the most effective basis for consensus in the assessment of theories. To be sure, disputes over the predictive power of particular theories can occur, perhaps in any field; nonetheless, the criterion is more effective than alternatives for resolving differences of opinion, particularly alternatives formulated by sociologists. As a case in point, Sorokin's logico-meaningful explanation (1943, 1947, 1957) is inherently subjective and cannot be made public.[5] If one sociologist asserts a logico-meaningful relation between phonomena but another denies that interpretation, how can the difference be resolved? The question alludes to what should be obvious—that which is logical and meaningful to one sociologist may not be so to others. Similarly, with reference to *Verstehen*, what is understandable to one sociologist may be a mystery to another, or their "interpretations" may conflict. How can such a difference be resolved? If the answer makes reference to empirical observations, it is essentially a designation of predictive power as the final arbiter. Moreover, were predictive power recognized at the outset as the primary criterion, a theory would not have to be restated to resolve differences in opinion. In any event, sociologists should face the question squarely: What criterion other than predictive power is capable of resolving differences in opinion? The assumption is, of course, that sociologists want their field to be something other than a debating club,[6] as those advocates of dialectics would have it (see Gross, 1961).

As suggested previously, predictive power is not an end in itself. Given a theory, scientists do not run about making predictions as though they were prophets. A theory should lead to predictions, but each prediction is made in tests of the theory, that is, only under certain conditions, and neither predictions nor tests are ends in themselves. Predictions are made only for the purpose of assessing the theory.

To all of the foregoing one may ask: What does predictive power have to do with the enterprise of science? Stated another way, what is the purpose or function of a theory? Regardless of the field, the purpose or function of theory need not be described in terms of causation or explanation; it can be taken as the identification or creation of order, and success can be judged by predictive power.

Some fears. Given dissensus over predictive power, there must be fears concerning that criterion, and an attempt to allay them should be made. For one thing, the criterion evidently raises the specter of operationalism and crude empiricism, in brief, atheoretical sociology. Yet the criterion is hardly indicative of an atheoretical perspective, nor does it demand that all concepts in a theory be "operationalized" (whatever that word may mean). The only demand is that a theory generate testable assertions. Is that demand unrealistic? The criterion will work against the premature publication of an idea as a theory, that is, before the theorist actually derives testable assertions. It also would force sociologists to distinguish conceptual analysis from substantive theory and cease passing the former off as the latter.

Some aspects of sociological work are essential even though not assessable in terms of predictive power, conceptual analysis being an instance. But acceptance of predictive power as the primary criterion for judging theories would not preclude publication of conceptual pieces, nor in any way belittle their importance.

Similary, it is recognized that exploratory research may be a step toward a theory. So neither conceptual analysis nor reports of exploratory research would be barred from the literature; the only requirement is that such works be labeled for what they are and not paraded as theory.

Then there is the fear that the predictive power criterion will lead to premature rejections of theories. The fear stems from recognition that the quality of sociological data makes consistently positive tests of a theory highly improbable. However, the criterion does not require rejection of theories, prematurely or otherwise, and it does not demand positive results in all tests. Theories can and should be evaluated in competitive terms;[7] as such, the question is not the absolute predictive power of a theory but, rather, its power relative to contenders.

The notion of predictive power does raise one irresolvable issue. Given a series of tests, the results are indicative of predictive accuracy, but there is no assurance as to what future tests will reveal. In brief, the criterion does not resolve the dilemma of induction, and critics may reject it for that reason. If so, however, they ignore two considerations. First, there is evidently no solution to the dilemma of induction, and a demand for that which cannot be realized—certainty—is pointless. Second, given the history of the physical sciences, it is unrealistic to argue that the dilemma precludes progress in a field.

Aspects of predictive power. As suggested previously, a theory may be retained even though consistently accurate predictions are not realized in tests. Nonetheless, some sociologists may reject the primacy of predictive power on the ground that more than accuracy should be considered in assessing a theory. The argument is correct as far as it goes, but accuracy is only one of seven aspects of predictive power.

The very notion of accuracy presupposes that a theory has been tested; hence "testability" is one dimension of predictive power. Inspection of a theory may suggest a conclusion as to its potential testability, but such conclusions are debatable. Accordingly, in assessing theories, "testability" refers to the number of actual tests. As for the rationale, suppose there are two contending theories, both of which have been tested at least once and the results suggest that the predictive accuracy of the two is approximately equal. However, suppose that one theory has been tested 200 times but the other has been tested only once. If invidious comparisons are to be made, who is prepared to ignore the number of tests? To be sure, the difference does not justify rejection of the theory that has been tested only once, and given unlimited research resources it would be tested another 199 times before comparing the two. But research resources are never unlimited, and for that reason alone invidious comparisons of theories are a necessary evil, if only to make decisions concerning the allocation of resources. It is pointless to argue that such decisions may be wrong, because they are made nonetheless. It is even more pointless to argue that testability is not relevant in assessing a theory. Unless one eschews induction altogether and favors an intuitive assessment of theories, confidence in the predictive accuracy of a theory cannot be divorced from testability. Further, testability is linked closely to the problem of limited research resources; if a theory has been formulated without concern for resources, repeated tests of it are most unlikely.

Consider two theories, both of which assert some relation between the suicide rate and other properties of social units. As such, the two theories are contenders, for both of them deal with a particular phenomenon—the suicide rate. However, it could be that one of them asserts a relation between the suicide rate and urbanization, while the other makes assertions about the relation between the suicide rate and social mobility, the homicide rate, unemployment, and average family size. If so, the *scope* of the latter theory exceeds that of the former, for it makes assertions about a greater number of properties. If a theory purports to identify or create order, then the *amount* is contingent on the theory's *scope*. Hence scope is another dimension of predictive power.

In the philosophy of science, the notion of scope would be equated with the "fertility" of a theory. However, fertility encompasses two distinct dimensions of predictive power. All theories make assertions about properties, but different types of events or things may have properties in common. Thus, the suicide rate is a property not only of countries but also of urban areas, metropolitan areas, and various status aggregates (for example, age groups), to mention only three types of social units. Now suppose that a particular theory makes an assertion about variation in the suicide rate but only among countries, while another makes assertions about the variation not only among countries but also by sex, age, and marital status. As such, the *range* of the latter theory is greater than the former, meaning that it makes assertions about more types of social units. If a theory makes assertions about only one type of social unit, then it has identified or created less order than a theory that makes assertions about two or more types of social units.

One may question the distinction between *scope* and *range*, but the difference is obvious. To illustrate, again consider two theories, one which makes an assertion about the relation between urbanization and the suicide rate, and the other which asserts a relation between social mobility and the suicide rate. Now the two theories have the same *scope*, since both assert a relation between only two properties. However, if one asserts that the relation holds only among countries, while the other asserts that the relation holds among countries, age groups, and marital statuses, then the two theories have a different *range*. Of course, the notion of range may not be of any significance in some fields. For example, psychologists commonly regard individuals as units of observation, so much so that it is taken for granted and not made explicit. But that is not so for sociology, where numerous types of units are recognized—associations, urban areas, occupational categories, countries, and so on.

Any assertion of a relation between properties implies a temporal consideration. For example, consider an assertion of an inverse relation between the homicide rate and the suicide rate. Without specification of the units of observation (for instance, countries) and the temporal quality of the relation, the assertion is incomplete. Since no temporal quality is specified, the assertion could be construed as implying that the homicide rate of a population during a period is inversely related to the suicide rate of the population *during the same period.* However, if the assertion does refer to such a cross-sectional or synchronic relation, it should be made explicit, for that type of relation is only one of several possibilities. One could assert a longitudinal or diachronic relation, for instance, that the homicide rate during one period is inversely related to the suicide rate during a subsequent period. For one reason or another a theorist may consider only one type of relation (such as synchronic) and make no assertion whatever about the other types. However, the various types are not mutually exclusive, so the theorist may make several assertions about the relation between two properties, with each assertion specifying a different temporal quality. The theory may be extended even to assertions that particular types of temporal relations between properties are closer than other types, for example, that the synchronic relation between the suicide rate and the homicide rate is closer than their diachronic relation.

When a theory comprises assertions about all types of temporal relations, it has maximum *intensity*; conversely, if the assertions are restricted to one type of temporal relation, intensity is minimal. Obviously, a temporal quality is entailed in the notion of order, but some theories comprise more assertions about that quality than do others.

Even when the intensity of a theory is minimal, the theory may be such that two or more assertions of relations between variables are derived from it. As a simple illustration, suppose that three assertions of *synchronic* relations among countries are derived from a theory as follows: (1) the suicide rate varies directly with the degree of urbanization; (2) the suicide rate varies inversely with the homicide rate; and (3) the homicide rate varies inversely with the degree of urbanization. In tests of the theory it could be that each relation is in the direction asserted and fairly close; but it is also possible (even likely) that some of the relations are closer than others. Accordingly, the theory should be such that one can derive predictions as to the *relative* magnitude of the asserted relations (for example, the inverse relation between the suicide rate and the homicide rate will be closer than the direct relation between the suicide rate and degree of urbanization). To the extent that such predictions are correct, the theory *discriminates*, and discrimination is still another aspect of predictive power.

It is generally recognized in all fields that parsimony (or simplicity) should be a criterion in assessing theories. However, in no field is there consensus as to the appropriate procedure for judging parsomony,[8] and recognition that parsimony is a matter of degree complicates the matter further. But parsimony can be treated as an aspect of predictive power, and it can be reckoned quantitatively in that context.

In each test of a theory a prediction about particular things or events is derived from one or more universal assertions of a relation between properties. Since an infinite number of predictions are potentially derivable from any theory, that consideration is not relevant in judging parsimony. However, a theory comprises more than one universal assertion. In the proposed mode of theory construction, some of the universal assertions, those designated as theorems, are derived from other universal assertions, and the theorems enter into the derivation of predictions. So parsimony is simply taken as the ratio of theorems to other universal assertions as components of the theory.

Many discussions of parsimony do not relate it to predictive power; but, since theorems represent different *types* of potentially derivable predictions, then surely the ratio in question has *something* to do with the predictive power of a theory; and it hardly does violence to the English language to call that something *parsimony*. The proposed conception is all the more defensible since there is no conventional alternative in sociology or the philosophy of science, let alone one that can be applied systematically. True, the proposed conception is limited to theories in which some of the assertions are derived, but that is hardly a serious limitation. In any case, the conception is consistent with the proposed mode of theory construction, and it is dubious to presume that there can be a procedure for assessing the parsimony of a theory regardless of its mode of construction.

Logical consistency is one conventional criterion for assessing theories that is not considered as an aspect of predictive power.[9] Consensus on that criterion is so universal and long-standing that it tends to be accepted uncritically, that is, without a concern for the rationale. Although the two are distinct, logical consistency is a necessary condition for predictive accuracy, which is to say that contradictory predictions can be derived only from a logically inconsistent theory.

Predictive Power and Dissensus in Sociology

As suggested previously, sociologists differ in their opinions as to the appropriate kind of explanation of social phenomena; for instance, a mechanistic, functional, teleological, genetic, historical, or reductionistic explanation. Adoption of predictive power as the primary criterion for assessing theories could resolve those differences, as it would apply to any theory regardless of the kind of explanation it offers. To illustrate, suppose that there are two contending theories on social stratification, one of which is considered to be functionalistic and the other genetic. Given the present dissensus in the field, some sociologists will prefer the functionalistic theory and others will prefer the genetic theory, with the difference in preference seemingly beyond argumentation. However, if sociologists should accept predictive power as the primary criterion, then consensus could be realized in assessing contending theories. But it is naive to presume that the criterion will be accepted. On the contrary, members of at least two camps or schools of thought are certain to reject it.

Functionalism.   For more than a generation a debate has raged intermittently in sociology over "functional" analysis. Today, even some of the partisans would concede that the debate has been inconclusive; and, judging from Kingsley Davis' statements (1959), it shows signs of disintegrating into a semantic squabble as to whether the term "functionalism" designates a controversial or even a distinct perspective in sociology. Davis' commentary on "functional analysis" *as a method* or strategy of research ignores the controversial question: How is the adequacy of a functional explanation to be judged? If adequacy is to be judged by predictive power, than a theory must make assertions about differences in the properties of social units or, stated another way, it must make explicit assertions about the relation between properties of social units. But there are no such assertions in "functional" theories, and their exclusion is entirely consistent with the principle of functional alternatives or equivalence.

In formulating a "functional" theory one may identify the "system consequences" of some type of institution or organization without committing himself to any prediction about the presence or absence of that institution or organization in particular social units. If it is absent in a given case, then, presumably, some other type of institution or organization performs the function in question. Accordingly, as revealed most clearly by Hempel (1959), no specific predictions flow from the typical functional explanation of a socio-cultural phenomenon; and the functionalists have not refuted Hempel on that point.[10] Consequently, there is only one interpretation: predictive power is simply not a criterion by which functionalists assess their own theories. Exactly what they regard as appropriate criteria is not clear, and commentaries on functionalism ignore the issue (for example, Martindale, 1965). In any event, it is not surprising that the debate over functionalism has been inconclusive; given divergent criteria for assessing theories, some of which are not even made explicit, there is no mechanism for resolving the issue.

Sociological subjectivism. Several properties of social units (such as the division of labor) do not represent an average of individual traits; so it is not surprising that some sociologists assert relations between properties without considering motives, perception, or values. Since such assertions are not subject to an "empathetic interpretation," they can be defended only by pointing to their predictive power. However, no evidence of predictive power will compel members of three schools of sociological thought to accept such assertions.

The three schools—phenomenology (see Schutz, 1967; Tiryakian, 1965), ethnomethodology (see Garfinkel, 1967), and Weberian sociology—differ in several respects; but the members are as one in assessing theories. Specifically, regardless of its predictive power, if a theory excludes reference to motives, perception, or values, that omission is regarded as a sufficient reason for either rejecting the theory or ignoring it.[11] Of course, the argument may be that the predictive power will be minimal if the theory ignores "meaning for the actors"; but that judgment is entirely premature and suggests a dubious contention—that the predictive power of theories can be known *a priori*.

Theories which incorporate reference to motives, perception, and values can be judged by the predictive power criterion; but the criterion cannot be utilized unless testable assertions about the relation between infinite classes of phenomena are derivable from the theory. Conceivably, one can formulate universal assertions about the relation between nonpsychological properties of social units (such as the degree of urbanization, the division of labor) and the prevalence of particular types of motives, perceptions, or values in social units. But these would not be testable without specification of the behavioral manifestation of the type of motivation, perception, or value in question. Further, if tautologies are to be avoided, the behavioral manifestation would have to be defined independently of the nonpsychological properties stipulated in the assertion. Nonetheless, at least in principle a theory which incorporates reference to dispositional phenomena can be formulated so as to derive testable predictions.

What phenomenologists, ethnomethodologists, and Weberians could do in the way of theory construction and what they actually do are quite different. Eschewing formal theory construction, they consistently fail to formulate explicit universal assertions. Rather, their theories comprise a mixture of definitions, argumentation, observations on the particular, and constant appeal to motives, perception, values, and/or "meaning." As such, it is questionable to regard their theories as other than a loosely formulated set of ideas; certainly they are not a set of explicit universal assertions. Even when such an assertion appears to be made, the theory commonly does not extend to a specification of behavioral manifestations of dispositional phenomena, let alone a procedure by which data are to be gathered and expressed numerically. For that matter, assertions about motivation, perception, values, or meaning are likely to be misinterpreted. When made in connection with a particular individual or a particular social unit, such an assertion may appear to imply a generalization; but the theorist may deny that interpretation, which is to say that the assertion is intended to be an *ad hoc* explanation. Such explanations are alien to science, but sociologists are uncritical of an *ad hoc* explanation if it is expressed in dispositional terms. They do have a basis for assessing such an *ad hoc* explanation; it is either plausible or implausible, but there is no assurance of consensus in that assessment. Indeed, a more effective way to make science personal and private cannot be imagined. But if the phenomenologists, ethnomethodologists, and Weberian sociologists would have their theories judged by some criterion other than plausibility, they certainly do not make the alternative explicit.

Even when the observations of an "interpretive" sociologist transcend the particular and the unique, it is not clear whether an assertion about infinite classes of phenomena is intended. Needless to say, if the observations are intended to apply only to *some* instances of an infinite class, the assertions cannot be falsified. For that matter, the observations may be conceptions rather than assertions and hence irrefutable. As examples, Garfinkel (1956) has made observations pertaining to "degradation ceremonies," and Goffman has done the same in connection with asylums (1961a), encounters (1961b), and stigma (1963). But a question is left unanswered: Do the observations represent generalizations about *or* conceptions of degradation ceremonies, asylums, stigmas, and encounters?[12] Garfinkel has analyzed the *idea* of a degradation ceremony and Goffman has done the same with regard to asylums, encounters, and stigmas; but their observations are not refutable assertions.

Whatever criterion of adequacy the phenomenologists and ethnomethodologists use to judge their ideas or theories, it is clearly not predictive power, and the same may be said of Max Weber's disciples. For Weberian sociologists the goal is understanding, and they evidently presume that it can be reached only by ascribing values to actors.

Values cannot be observed directly, but that is not the problem with the *Verstehen* notion. If all component terms of a theory must designate phenomena that can be observed or otherwise experienced directly, then that requirement might well preclude sociological theories altogether. Nonetheless, an assertion about motives, perception, values, or meaning is not testable unless additional assertions are made about behavioral manifestations, and those additional assertions cannot be divorced from the theory. Thus, if the degree of urbanization is asserted to be a behavioral manifestation of a particular value and average family size is asserted to be still another behavioral manifestation of the same value, then the theory leads to a derived assertion about the relation between urbanization and family size as an infinite generalization. Far from making such assertions explicit, Weber and his followers offer *ad hoc* explanations or "interpretations" of a particular thing, event, or finite universe. Thus, even if Weber had clearly designated the behavioral manifestations of the Protestant ethic (1930), such that they could be identified in any country or culture (something which he did not do), it could be argued that he asserted a relation between those manifestations and capitalism only in a particular historical context. A further complication is added by the notion that Weber made assertions about the Protestant ethic and capitalism only as "ideal types."[13] So the very way in which the theory was stated made tests of it debatable from the outset.

The question of alternatives.   As suggested in the foregoing, members of some sociological schools of thought formulate theories in such a manner that they cannot be judged in terms of predictive power, and that alone indicates a pronounced reluctance to accept the criterion. However, if consensus in the assessment of theories is to be achieved, what alternative is there to predictive power as the primary criterion? In a remarkable passage Dubin dodges the question.

> Theories of social and human behavior address themselves to two distinct goals of science: (1) prediction and (2) understanding. It will be argued that these are separate goals and that the structure of theories employed to achieve each is unique. I will not, however, conclude that they are either inconsistent or incompatible. In the usual case of theory building in behavioral sciences, understanding and prediction are not often achieved together, and it therefore becomes important to ask why. It will be concluded that each goal may be attained without reference to the other.
>
> I mean one of two things by prediction: (1) that we can foretell the value of one or more units making up a system; or (2) that we can anticipate the condition or state of a system as a whole. In both instances the focus of attention is upon an *outcome*.
>
> As I employ the term *understanding*, it has the following essential meaning: it is knowledge about the interaction of units in a system. Here attention is focused on processes of *interaction* among variables in a system (Dubin, 1969:9-10).

Dubin's statements suggest a very narrow conception of prediction, evidently limited to forecasts, projections, or extrapolations. No other writer in sociology or the philosophy of science has adopted such a narrow conception, and Dubin indulges in logomachy to make an argument. A more conventional conception would include any instance where one derives an assertion about an unknown, either a prediction or a postdiction.[14] Of course, Dubin eschews that conception, for it implies knowledge of interaction between variables, which he would relegate to "understanding." The truth is that Dubin's distinction will not bear examination, because a description of an interaction is a description of an outcome. In any case, how are we to evaluate a particular "understanding"? Significantly, Dubin does not mention criteria for assessing theories that offer "understanding" without predictions, but he should know that many scientists and philosophers (even a few sociologists) argue that prediction and understanding are related. Consider Walker's statement: "If a specific occurrence is predicted by a law that has predicted many such occurrences accurately, then the scientist says that he *understands* that occurrence" (1963:1). Now suppose that Walker and the entire tribe of positivists err in their insistence on some connection between understanding and prediction. The question thus becomes: Given a theory that offers understanding and not prediction, what criteria are to be used in assessing it? Stated another way, if two theorists have different "understandings" or offer divergent explanations of a system or some phenomenon, how is that difference to be resolved without reference to predictive power?[15] Shall we say that it is a matter of personal opinion, conviction, or taste? In any event, the issue survives Dubin's distinction.

Sjoberg and Nett (1968:289-290) have contrived an ingenious strategy to dodge the central issue in the assessment of theories. They have one school of thought making the following argument: prediction *is* explanation. By vulgarizing the positivist's argument, Sjoberg and Nett seek a following among those who insist that explanation somehow differs from or involves more than prediction. Of course it does, for we can make predictions without offering explanations, provided that the predictions are not derived (see Brodbeck, 1962:233). The central question is not the difference between explanation and prediction but rather: By what criterion shall we judge the *adequacy* of an explanation? One answer is "predictive power." Hopefully, Sjoberg and Nett will cease their equivocation and argue for an alternative.

It will not do to reply that the predictive power criterion entails accepting physics as the model of science, the usual cryptic way that sociologists face the issue. Confining our attention exclusively to criteria for assessing theories, in what way is physics a peculiar model of science? Stated another way, in what "natural" science is predictive power not the primary criterion in assessing theories? In any case, if sociologists reject predictive power, they should adopt an alternative that promises effective consensus in assessing theories. What could it be? Many sociologists are reluctant to face that question, but it will not go away.

Notes for Chapter 3

[1] See especially Weber, 1947:88-112 and 1949:113-188; Abel, 1967; Wax, 1967; and Baar, 1967.

[2] The impression conveyed by Mills' *The Sociological Imagination* (1959) is that he would dismiss formal theory construction as merely another "abstracted empiricism."

[3] Gouldner is unique among sociologists in his candid observations on the assessment of sociological theories. He states: "Some theories are simply experienced, even by experienced sociologists as *intuitively* convincing; others are not. How does this happen? . . . The theory felt to be intuitively convincing is commonly experienced as *deja vu*, as something previously known or already suspected. It is congenial because it confirms or complements an assumption already held by the respondent, but an assumption that was seen only dimly by him precisely because it was a 'background' assumption" (1970:30). Gouldner's statement is correct, but note that he does not condemn intuitive assessments of theories, nor does he speak to the implication. Given two critics with inconsistent "background assumptions," their intuitive assessments of a theory will differ. So the question is: How are such differences to be resolved? Unless sociologists can agree on *public* rather than *private* criteria for assessing theories, the field will remain a debating club. Nevertheless, Gouldner's observations are sound advice for a sociologist who aspires to become a prominent theorist. Do not be concerned with tests or predictive power; just give the audience what it wants to hear.

[4] See *Sociologist*, 9 (1933:298-322).

[5] Bierstedt's comment on Sorokin's method: "it is of the essence of subjectivity" (1937:819-820).

[6] Nagel observes that "the social sciences often produce the impression that they are a battleground for interminably warring schools of thought" (1961:448). In sociology, it is more than an impression and continual rather than often.

[7] "We choose the theory which best holds its own in competition with other theories; the one which, by natural selection, proves itself the fittest to survive" (Popper, 1965:108). The statement is noteworthy because Popper has advocated the idea that a theory can be "falsified." Were that so, then theories could be assessed in accordance with some absolute criterion. However, as we shall see in subsequent chapters, such a criterion is dubious. For the moment it will suffice to say that Popper's idea is questionable, and note that in this statement he subscribes to a relative rather than an absolute criterion for assessment of theories.

[8] For observations on divergent conceptions of parsimony or simplicity, see Popper, 1965:136, and Frank, 1956:4.

[9] Needless to say, several possible criteria for assessing theories are deliberately excluded, one of them being "plausibility." Invoking that criterion can have only one outcome—assessments of theories will be personal, with disagreements common and beyond resolution.

[10] Note that we are concerned with actual "functional" explanations or theories, not with arguments as to how they might have been stated or how they could be restated (see Nagel, 1961, and Isajiw, 1968).

[11] In keeping with a tradition, Wilson (1970) identifies "action" as the central concern of sociology. He states (again in keeping with tradition) that behavior is not action unless it is meaningful to the actor, but he then concedes that "complex social patterns" need not themselves be meaningful to the actors. However, consider his example and conclusion: "the rate of theft in a group may not be known to the members of the group; nevertheless, an individual act of theft is behavior meaningful to the members, and the rate expresses a regularity in those actions that can be taken as a phenomenon for sociological investigation" (1970:698). The conclusion is simply a bone thrown to sociologists who do not work in the tradition of interpretive sociology. Wilson may grant that macroscopic phenomena "can be taken" as subjects for theory construction, but it is difficult to imagine "interpretive" sociologists giving such a theory a serious hearing. After all, since the phenomena dealt with by the theory are not "meaningful" to the actors, how could such a theory possibly be judged adequate by the principles of interpretive sociology? Interpretive sociologists, of course, may merely ignore the theory rather than reject it, but the distinction is of no consequence.

[12] Glaser and Strauss put the question differently: "One reader has wondered in conversations with us: Does Goffman's 'Total Institutions' represent a model or a description of many if not most mental hospitals, or is it really a

description mainly of St. Elizabeth's Hospital, where Goffman did most of the field work for his paper on total institutions?" (1967:137). The reader has cause to wonder. However, Glaser and Strauss do not comment on the significance of the question. Had Goffman explicitly stated that his observations are descriptions of St. Elizabeth's, it would have been quite clear from the outset that his work is not theory. But Goffman can have it both ways; his work is theory until someone questions it with data, and then it is really a description of St. Elizabeth's. Even if it is taken as a model or theory of all "total institutions," it cannot be confuted, for Goffman does not stipulate what would be negative evidence.

[13] As used by sociologists, the notion of "ideal types" does not clarify the logical structure of a theory or further its predictive power. On the contrary, it obscures empirical assertions and prevents consensus as to what would be an appropriate test of the theory. For observations on the ineffective utilization of the notion in theory construction, see Lopreato and Alston, 1970.

[14] "Scientific explanation, prediction, and postdiction all have the same logical character" (Hempel, 1958:37).

[15] Some philosophers of science (see Kaplan, 1964:350) simply dodge such questions.

CHAPTER 4 SOME CONTEMPORARY PRACTICES

Since most sociological theories appear to be untestable, the criterion of predictive accuracy receives scant attention in assessments of them. But why are so few testable? The simplest answer is that sociologists employ a discursive mode of theory construction; however, that answer does not identify those features of the discursive mode that are inimical to tests. Rather than describe those features in the abstract, this chapter examines some contemporary practices in sociology that are antithetical to formal theory construction.

Definitions and Empirical Assertions

Most sociologists would argue that their field has not developed to a point where it is feasible to state a general theory in the form of equations. So, conventionally, the principal components of theories are statements, and that practice is not questioned.

Two major types of statements can be distinguished; instances of one type are empirical assertions, and instances of the other are definitions. Conceivably, a theory could comprise only empirical assertions, but terminological problems in sociology preclude it. In brief, the field's vocabulary is not such that sociologists agree as to the meaning of all terms, let alone how to apply them to identify particular events, things, or attributes. Consequently, when two investigators differ in their interpretation of terms in an empirical assertion, they are likely to disagree in stipulating what would be relevant tests.

Given the field's vocabulary, at least some constituent terms of a sociological theory should be defined. So the issue in sociology centers on a question: What kind of definition is appropriate? That question underlies the debate over "operationalism," and its importance cannot be exaggerated. For all practical purposes the predictive accuracy of a theory depends entirely on definitions, but in regard to theory construction there is a more immediate consideration.

The discursive mode of theory construction makes it very difficult to distinguish synthetic and analytic statements. Sociologists appreciate the distinction, but it is blurred in the typical theory by the intermingling of empirical assertions and definitions.[1] The distinction is confounded further by the inclusion of rhetoric in the statement of a theory. If one argues that empirical assertions and definitions are distinct even when not set apart, the argument reflects a profound ignorance of sociological terminology. The meaning of most sociological terms is vague, and for that reason alone it is difficult to distinguish empirical assertions from analytic statements, especially since the latter includes not only complete and express definitions but also statements variously identified as incomplete or partial definitions, negative definitions, and tautologies. All such statements stipulate a relation between words or symbols and nothing more.

Unlike empirical assertions, analytic statements are made and accepted without regard to evidence or tests; nonetheless, they assume various forms, and that diversity makes it difficult to distinguish them from empirical assertions. Consider the following: All organized groups have a communication system. Is the statement an empirical assertion? Unless the theorist offers a definition of the two key terms, "organized group" and "communication system," any answer would be conjecture.

Some illustrations.   The statements below exemplify the terminological problem under discussion. In examining each statement, the reader should ask himself: Is the statement intended to be taken as an empirical assertion or as an analytic proposition (that is, true by definition)?

Statement 1: "The withdrawal of an expected reward is not just something that releases emotional behavior; it is also a punishment, and its avoidance is accordingly a reward" (Homans, 1961:77).

Statement 2: "for conflict among individuals and groups to be kept within bounds, the roles and role clusters must be brought into appropriately complementary relations with one another" (Parsons and Shils, 1951:198).

Statement 3: "The more occupational success is related to educational achievements, which are open to all, the greater the occupational mobility" (Lipset and Zetterberg, 1964:445).

Statement 4: "To the extent that the social structure insulates the individual from having his activities known to members of his role-set, he is the less subject to competing pressures" (Merton, 1957b:114-115).

Statement 5: "To the extent that a society is stable, adaptation type I—conformity to both cultural goals and institutionalized means—is the most common and widely diffused" (Merton, 1957a:141).

Statement 6: "Power is the resource that permits an individual or group to coordinate the efforts of many others, and legitimate authority is the resource that makes possible a stable organization of such coordinated effort on a large scale" (Blau, 1964:232).

Commentary.   Even if the reader can identify each statement as an empirical assertion or as true by definition, the practice of intermingling the two types in theory construction is indefensible. Consensus in the identification of statements should be realized, but this cannot happen if identification is left to the discretion of the audience.

The distinction between analytic and synthetic statements is debatable in all fields (see Quine, 1960; Putnam, 1962), especially if, as in sociology, there are no conventional primitive terms. Further, the problem is not solved even when the theorist explicitly identifies each statement either as true by definition or as an empirical assertion. Such an identification would be indisputable only if the theorist could avoid any presumption about linguistic conventions, which cannot be done in using a natural language (see Hempel, 1952:9). Nonetheless, unless the theorist explicitly identifies each statement, that is, stipulates what is to be "taken" as an empirical assertion, the theory is ambiguous from the outset. An explicit identification can be achieved by a simple procedure— formulate empirical assertions in one part of the theory and definitions in another. That procedure would in no way alter the content of a theory.

Issues

Consensus in sociology is realized on two features of a theory: (1) it comprises more than one empirical assertion and (2) the empirical assertions are logically related. However, those two features should not be accepted uncritically, as though simply true by definition. Theorists do not make more than one empirical assertion merely to insure that their work will be labeled honorifically as a theory. If a theorist regards his initial empirical assertion as untestable, he should formulate others and eventually derive testable statements. But derivations are possible only when empirical assertions are interrelated logically. So the two features of a theory are not merely true by definition; without them, tests of the theory might be precluded.

The foregoing observations introduce some major issues in theory construction. If the purpose of multiple assertions has nothing to do with tests, then each assertion can be taken as testable; and that is precisely the argument suggested by radical operationalism. The operationalist's argument is that all terms of a theory must be empirically applicable; as such, each component assertion is testable. The proposed scheme departs from the putative tenets of operationalism; briefly, a term may be utilized effectively in theory construction even though it is not considered as empirically applicable.

Given logically interrelated assertions, additional assertions can be derived. If the derivations enter into tests, then the results are evidence for assessing the direct assertions or premises. So derivations are not an end in themselves; they are undertaken primarily to bring evidence to bear on the direct assertions. Yet in Zetterberg's mode of theory construction (1965) the goal is merely to minimize the number of premises and to maximize the number of derivations.

The goal is desirable, but nothing is gained when the derived assertions are no more testable than the premises; and that is the defect in Zetterberg's version of the axiomatic method (1965:159), by which assertions are derived without regard to the empirical applicability of their constituent terms.[2] Contemplate a set of assertions in which none of the constituent terms are regarded as empirically applicable. Should Zetterberg's scheme be used, other assertions could be derived from that set; but the derivations would not alter the empirical applicability of the terms, and hence the derivations would be no more testable than the premises. To be sure, if some terms in the premises are empirically applicable, then some of the derivations may be testable and others untestable; yet Zetterberg ignores the distinction. Assertions are derived without regard to testability; and in that sense his scheme is not an improvement on the discursive mode of theory construction.

Zetterberg's scheme is useful only if all sociological terms are taken as empirically applicable. Were that so, then the primary goal in theory construction could be to maximize derivations. Tests of derivations do generate evidence for assessing the premises, even when the premises can be tested directly (that is, without derivations); but Zetterberg's scheme is grossly unrealistic, for no one will argue that all sociological terms are empirically applicable. Moreover, "indirect" tests of premises through derivations entail difficult problems and issues. So, if all premises were directly testable, it would be most questionable to assess them inferentially, that is, through indirect tests. As stated by Northrop: "the basic assumptions of a deductively formulated science are the postulates, and it is certainly preferable, if one cannot directly verify all propositions in a science, to verify the basic propositions making up the postulates than the secondary and derived propositions making up the theorems" (1947:108).

The derivation of empirical uniformities.   As emphasized in Chapter 2, a mode of theory construction should recognize the problems and conditions of work in a field. One condition in sociology is especially important when contemplating the advanced sciences (such as physics). In those fields putative empirical uniformities often are discovered in experimental work, and one major concern in theory construction is the derivation of those uniformities. Such derivations may be construed as explanations, but in any case the theorist need not be preoccupied with testability. To illustrate, suppose the theorist commences with two putative uniformities: phenomena W and X are related in some particular way, and so are phenomena Y and Z. In attempting to derive those two uniformities, the theorist need not be concerned with the empirical applicability of the related terms, W, X, Y, and Z, not even when he derives an heretofore undetected association (for example, W and Z are related in a particular way). After all, the four terms were used in experimental work that led to the discovery of uniformities. So in a field where theorists commonly commence with putative uniformities, empirical applicability is not a major problem.

There is very little systematic evidence of a relation between any *specific* sociological variables; as such, derived assertions typically do not represent putative uniformities, and more than their predictive accuracy is questionable. The derivation itself does not provide any assurance that the terms of an assertion are empirically applicable. Hence the present emphasis on empirical applicability as a problem in theory construction is consistent with the conditions of work in sociology.

Some sociologists do construct theories on the presumption

that there are established uniformities, Homans' work (1961) being especially noteworthy in that regard. Although Homans does not glorify "grand theory," like Zetterberg he does not emphasize terminological problems in theory construction. On the contrary, he takes theory construction as having only one purpose—explanation—and hence the exclusive emphasis on derivation or "deduction."

Consistent with his views on theory, Homans writes as though there are numerous established uniformities in sociology.[3] If that were so, a theorist could ignore the problem of empirical applicability by presuming that the constituent terms of his derivations have been used by investigators in reporting uniformities. But Homans' strategy is unrealistic. For one thing, he regards the findings of one or a few isolated studies as having established a uniformity, and in so doing he conveniently ignores a painful condition of work in sociology—a few findings do not establish a uniformity. On the contrary, the relation between sociological properties varies appreciably from place to place and time to time. As for quantitative properties, there is very little evidence for regarding any relation as invariant, not even approximately so. Accordingly, if Homan's strategy in theory construction were adopted, sociologists would busily engage in the "deduction" of what are nothing more than conjectural uniformities.[4] In any case, when Homans "abstracts" uniformities from the findings of sociological investigations and states them as general "propositions," he habitually uses terms that were not employed by the investigators. Such abstractions are creative, but Homans cannot claim that the empirical applicability of the constituent terms is demonstrated by research findings, as the investigators did not use the terms in reporting their findings.

Types of Statements

The structure of a theory should be consistent with three principles: (1) the component statements are interrelated logically; (2) definitions and empirical assertions are differentiated; and (3) not all empirical assertions are of the same type. Only the third principle is exemplified in actual practice.

Sociologists recognize types of empirical assertions and use a variety of labels in that connection (hypothesis, proposition, theorem, axiom, and postulate), but there are no established conventions. Moreover, sociologists habitually use labels to identify types of empirical assertions without any rationale, and for that reason it is difficult to see what is accomplished. In any case, the statements quoted below suggest divergence rather than consensus in the use of labels.

Statements designated as hypotheses

H1: "The more a member conforms to the norms of a formal organization, the greater the likelihood that he will be promoted" (Zetterberg, 1965:104).

H2: "The greater the socioeconomic rewards derived from present American life in comparison with past Cuban ones, the greater the integration" (Portes, 1969:511).

H3: "A positive association was anticipated between an individual's most-valued source of information and the time at which he adopted the study innovations" (Becker, 1970:272).

H4: "only those actually sharing in societal power need develop consistent societal values" (Mann, 1970:435).

H5: "French policies in the Caribbean were perceived, even by enlightened Antillean leaders, as having achieved a more-or-less satisfactory spread of equality, and this satisfaction, in turn, largely 'cut the ground from under' the desire for political independence there" (Murch, 1968:554).

H6: "The size of the nuclear family is an inverse function of position in the stratification system" (Smelser, 1969:14).

H7: "work would have the most consequence for community status in the Costa Rican village, next most consequence in the Michigan village, and would be least important in the Guatemalan village" (Faunce and Smucker, 1966:393).

H8: "The larger a city, the larger the range of its occupational composition" (Schwirian and Prehn, 1962:815).

Statements designated as propositions

Pr1: "The second derived proposition is that the larger an organization is, the larger the average size of its structural components of all kinds" (Blau, 1970:207).

Pr2: "hypergamy is more common that hypogamy in the United States" (Rubin, 1968:752).

Pr3: "by combining either propositions 5 and 2, or 3 and 4 we get the sixth proposition: 6. The actual responses of others toward the individual will affect the behavior of the individual" (Kinch, 1963:482).

Pr4: "Values in the rest of the United States are opposed to the practice of segregation by the white south" (Heer, 1959:593).

Pr5: "the more a member conforms to the norms of his group, the more favorable valuations does he receive from his group" (Zetterberg, 1965:107).

Statements designated as theorems

Th1: "the primary structure of the human personality *as a system of action* is organized about the internalization of *systems* of social objects which originated as the role-units of the successive series of social systems in which the individual has come to be integrated in the course of his life history" (Parsons, 1955:54).

Th2: "The suicide rate of a population varies inversely with the degree of status integration in that population" (Gibbs, 1969:522).

Th3: "people generally *want* to do what they are supposed to do, and this is what the society needs to have done in order to continue" (Goode, 1960:485).

Statements designated as postulates

P1: "The greater the number of associates per member, the greater the division of labor" (Zetterberg, 1965:161).

P2: "there is a universal set of basic human needs which have attributes of their own which are not determined by the social structure, culture patterns, or socialization processes" (Etzioni, 1968:871).

P3: "although the development of the Puritan-Parliamentarian culture accelerated the emergence of industrial society in England, this cultural development depended on cumulative reactive interaction between its adherents and the adherents of the simultaneously developing Anglican-Royalist culture" (Israel, 1966:597).

P4: "The individual's self-concept functions to direct his be-havior" (Kinch, 1963:482).

Statements designated as axioms

A1: "An increase in the number of associates per member will produce an increase in the division of labor" (Blalock, 1969:19).

A2: "The probability of remaining in any state of nature increases as a strict monotone function of duration of prior resi-dence in that state" (McGinnis, 1968:716).

A3: "Every social pattern continues to manifest itself in con-stantly recurring social action at an unaltered rate unless some social force modifies the rate or pattern of such action" (Catton, 1966:235).

A4: "individuals who face common role obligations *can* gen-erally fulfill them" (Goode, 1960:484-485).

Commentary.   There is no conspicuous and consistent difference between the above types of statements. In particular, at least one instance of each type appears to be a generalization about an infinite universe (that is, a universal assertion). But some of the hypotheses (see H2) pertain to a finite universe, and that is also the case for some of the propositions and postulates. So only the axioms and theorems appear to be infinite generalizations; however, that form also characterizes some of the hypotheses, propositions, and postulates.

As for the idea that theorems are distinct by virtue of having been derived, Th1 and Th3 do not qualify.[5] Further, some of the propositions (see Pr1) and hypotheses (see H8) were derived, and they take the form of infinite generalizations, as do the theorems.

Only the axioms appear to be consistent in form, that is, none of them were derived and all appear to be infinite generaliza-tions. Yet there are differences. Observe that nomic terms (for example, "produces") appear in only some of the axioms, and not all of the axioms were stated in the context of a deductive scheme.

Needless to say, it would be desirable for theorists to employ the same nomenclature in typifying empirical assertions, but the illustrative statements clearly indicate that there are no established conventions.[6] However, the uncritical use of labels, rather than divergence, is the basic problem. Far from developing a typology of empirical assertions and a rationale for it, sociologists commonly use labels indiscriminately. So when a theorist identifies an empirical assertion as a postulate (or some other type), he is not likely to justify that label, and hence its purpose is obscure.

Definitions of types.   The identification of empirical assertions serves no purpose unless made in accordance with an explicit typology. Now it might appear that such a typology could be borrowed from the more advanced sciences, but it may be that no typology is suitable for all fields. In any case, when one turns to the literature of sociology or the philosophy of science, there is no conspicuous agreement in definitions of conventional labels (e.g., hypothesis, propositions). The magnitude of divergence is suggested in the definitions or partial definitions quoted below.

Definitions or partial definitions of an hypothesis

DH1: "The general form of an hypothesis is a conditional prediction about the relationship between two or more things, followed by a figurative question mark" (Dubin, 1969:7).

DH2: "propositions for which more evidence is needed are called *hypotheses*" (Zetterberg, 1965:101).

DH3: "All scientific hypotheses will be taken to be generalizations with an unlimited number of instances" (Braithwaite, 1953:14).

DH4: "a statement of the logically, and presumably temporally, prior conditions under which dependent variables may be expected to vary in certain ways" (Smelser, 1969:14).

DH5: "A hypothesis may assert that something is the case *in a given instance*, that a particular object, person, situation, or event has a certain characteristic" (Selltiz, *et al.*, 1959:35).

Definitions or partial definitions of a proposition

DPr1: "Propositions relate variates to each other" (Zetterberg, 1965:64).

DPr2: "A proposition . . . is a truth statement about a model that is fully specified in its units, laws of interaction, boundary, and system states" (Dubin, 1969:166).

DPr3: "The law of interaction states the general relationship among units, whereas the proposition predicts the specific values that one unit will have in relation to the values of another" (Dubin, 1969:178).

Definitions or partial definitions of a theorem

DTh1: "The theorems of a deductively formulated theory are all the empirical propositions in the theory other than the postulates" (Northrop, 1947:140).

DTh2: "Theorems . . . are derived by reasoning, or *deduced*, from the axioms" (Blalock, 1969:10).

Definitions or partial definitions of a postulate

DP1: "The postulates of a deductively formulated theory are those propositions which are assumed in the theory in question as logically unprovable and which are sufficient to enable one to prove, i.e., to logically deduce, the theorems." (Northrop, 1947:140)

DP2: "From the list of original propositions (inventories or matrices) a certain number are selected as *postulates*. The postulates are chosen so that all other propositions, the *theorems*, are capable of derivation from the postulates and no postulate is capable of derivation from other postulates." (Zetterberg, 1965:97)

Definitions or partial definitions of an axiom

DA1: "We may distinguish, within a theoretical system, statements belonging to various levels of universality. The statements on the highest level of universality are the axioms; statements on the lower levels can be deduced from them" (Popper, 1965:75).

DA2: "axioms may be regarded either (i) as *conventions*, or they may be regarded (ii) as empirical or scientific *hypotheses*" (Popper, 1965:72).

DA3: "a deductive science needs a set of fundamental truths —axioms—which would serve to test all the others, but would not need to be tested themselves. These axioms would be derived either from experience or from some other source. If they were derived from experience, they would be inductive, and therefore from the point of view of the deductive standard should be again tested by deduction from other truths. If not, they would be inapplicable to empirical reality, unless that reality were first made to fit them, which means that it would not be empirical reality, independent of mind, but the work of mind" (Znaniecki, 1934:219).

DA4: "Axioms are propositions that are assumed to be true" (Blalock, 1969:10).

Commentary. According to some definitions (see DH3 and DH4), an hypothesis is an infinite generalization, and viewed that way, a proposition, postulate, or axiom is a type of hypothesis.[7] But DH5 suggests that an hypothesis may take the form of an assertion about a particular thing, event, or finite universe; and certainly sociologists commonly use the word in that sense. Even if the word "hypothesis" is a generic designation of infinite generalizations, the definitions do not suggest consensus as to distinctions among propositions, postulates, and axioms. The matter is complicated further by definitions (see DPr1) that blur the distinction between propositions on the one hand and hypotheses, postulates, or axioms on the other.

Then there is the question of the distinction between an axiom and a postulate. Some of the quoted definitions suggest no distinction, and the two terms commonly are used as though interchangeable (see Blanche, 1962:9). Feigl makes the following distinction: "'Postulate' in modern science, especially in mathematics, means an assumption which serves as a premise for deduction. In contradistinction to the term 'axiom,' it does not carry the traditional connotation of self-evident truth or indubitability" (1956:27). But some writers (see DA2) question the idea that an axiom is always a self-evident truth.

The only consensus realized in the foregoing definitions is that theorems are derived statements. However, in actual practice, the term "hypothesis" often is used to designate derived statements, and there are even instances (see Hopkins, 1964:65) where "propositions" are derived.

Apart from the general lack of consensus, observe that several of the definitions are ambiguous and questionable. Thus, considering DH2, can anyone imagine a sociological proposition for which more evidence is not needed?

Since terms are components of statements, it is surprising that none of the definitions reflect that consideration. In any case, the present argument is that a typology of terms is a prerequisite for a defensible typology of theoretic statements.

Types of Terms

As already suggested, typification of empirical assertions should serve some purpose, and hence the need for a typology with a rationale. Of course, typification may clarify the structure of a theory, but that is not an end itself; the ultimate purpose should be to facilitate tests.

If only some empirical assertions of a theory enter directly into tests, then a label should distinguish those assertions. The point may appear obvious, but one conspicuous defect of the typical sociological theory is that testable and untestable assertions are not distinguishable.

Even when some constituent terms of direct assertions are not considered empirically applicable, it may be possible to derive testable statements from the theory. However, derivations are testable only when *some* terms of *some* premises are empirically applicable, and that consideration implies a distinction between two types of direct assertions: (1) those in which only some constituent terms are designated as empirically applicable and (2) those in which all terms are so designated.

The foregoing commentary suggests a fundamental point—typification of terms is necessary for typification of empirical assertions. However, the former entails more than a consideration of empirical applicability. A statement cannot assert order in the universe unless it includes the following terms: (1) one that denotes a class of objects or events, (2) one or more that designates a property or properties of that class, and (3) one that stipulates a relation either between the class and the property or between properties of the class. These three types are identified respectively as (1) unit terms, (2) substantive terms, and (3) relational terms. For the moment one illustration will suffice: Among urban areas, population size varies directly with population density. The unit term is "urban area," while "population size" and "population density" are substantive terms, and the relational term is "varies directly with." Each type of term is explained more fully in Chapter 5, and we are concerned here only with the notions as regards contemporary practices in sociology.

Unit terms.   Sociologists use various unit terms—countries, voluntary associations, age groups, occupations, and urban areas. Without such terms, a theory is scarcely intelligible, let alone testable; but in actual practice, sociological theorists are careless in the stipulation of units. Indeed, in many instances they make empirical assertions either without incorporating any unit term or using an ambiguous term. Some illustrations follow.

Statement I: "If the middle classes expand, the consensus of values increases" (Zetterberg, 1965:98).

Statement II: "A high degree of complexity varies directly with a high number of joint programs" (Aiken and Hage, 1968:915).

Statement III: "the more centralized the decision-making structure, the higher the level of outputs" (Clark, 1968:587).

Statement IV: "In general, the greater the solidarity, the narrower will be the belt of acceptable spouses that the rules of exogamy and endogamy define" (Young, 1967:593).

Statement V: "suicide varies inversely with the degree of integration of the social groups of which the individual forms a part" (Durkheim, 1951:209).

Although the foregoing statements are intelligible, they are ambiguous. The statements assert relations between properties, but in each case the class of objects or events is conjectural. For example, in statement I the unit term appears to be "middle class," but Zetterberg evidently had countries in mind when making the statement (that is, "expansion of the middle classes" and "increase of consensus in values" are treated as properties of countries).

It will not do to argue that the properties imply the class, for different classes of objects or events may be characterized by the same property. As an instance, population density is a property of territorial units in general, for instance, countries, metropolitan regions, metropolitan areas, urban areas, and cities.

Although a unit term may be implied contextually, it is far better to include a unit term in each empirical assertion. This is only a first step taken by a theorist to facilitate tests of a theory. Additionally, he should define each unit term, and the definitions should provide assurance of empirical applicability. Specifically, unless independent investigators achieve substantial agreement in applying a unit term to identify objects or events (such as populations or acts), tests of the theory will be idiosyncratic.

Sociologists rarely formulate theories without at least implying a unit term, but they often fail to define the term explicitly and carefully. Such a definition is all the more necessary when the unit term is exotic, meaning one that is seldom applied in actual research. As an example, two unit terms in Parsonian theory are "social system" and "collectivity," neither of which is used frequently in the research literature. Rather, investigators report findings on particular types of territorial units (for example, metropolitan areas), organizations, and status aggregates. Although Parsons may not consider such conventional units to be social systems or collectivities, his definitions of those terms certainly do not clarify the distinctions. Consider three statements: "A social system . . . is a system of interaction of a plurality of actors, in which the action is oriented by rules which are complexes of complementary expectations concerning roles and sanctions" (Parsons and Shils, 1951:195). "A social system having the three properties of collective goals, shared goals, and of being a single system of interaction with boundaries defined by incumbency in the roles constituting the system, will be called a *collectivity*" (Parsons and Shils, 1951:192). "A collectivity may be defined as the integration of its members with a common value system" (Parsons and Shils, 1951:192).

The terminology of the foregoing definitions is largely alien to the language of research sociologists, and for that reason alone their empirical applicability is doubtful. To be sure, a theorist is not bound by research language. He can introduce a novel term in formulating a theory; but tests are precluded unless he defines that term so that independent investigators can agree in applying it to identify objects or events. The matter is complicated further when the unit term designates a type of population; in such a case independent investigators must agree not only in the identification of instances but also in the delimitation of populations as to territorial extent and/or membership. Parsons stipulates no instructions whatever for the delimitation of social systems or collectivities. He is indifferent to the notion of empirical applicability, but the toleration of his indifference is significant, for it indicates that many sociologists only give lip service to the desirability of testable theories.

Perhaps the most common practice in the construction of sociological theories is the use of a generic unit term, such as "population," "social unit," or "social group." Since a generic unit term maximizes the range of a theory, it may appear desirable. However, the practice suggests two dubious assumptions: first, that any particular unit term (say, metropolitan region) is empirically applicable without a definition or delimitation procedure; and, second, that the theory holds equally well for all types of units.

Substantive terms. Radical operationalism notwithstanding, testable assertions can be derived from a theory even though some of its substantive terms are not considered as empirically applicable. Needless to say, a theorist enjoys an enormous advantage in using terms without regard to their empirical applicability, but that advantage is abused by sociological theorists. They habitually fail to distinguish terms as to their empirical applicability, and hence the distinction is left to the discretion of investigators. But it is idle to presume that investigators can or should distinguish testable from untestable assertions, let alone that they will agree. The distinction is not somehow "objectively" given; it is a matter of judgment, and systematic tests of a theory will be realized only when the theorist makes that judgment.

The testability of an assertion cannot be divorced from its constituent substantive terms. Obviously, if the theorist does not regard all substantive terms as empirically applicable, then he should stipulate that the assertion is untestable. Although a mode of theory construction cannot provide a criterion for judging empirical applicability, it should allow the theorist to express his judgments systematically and reveal the implications. In that connection the conventional language of theory construction in sociology is inadequate. The conventional distinction between a theoretical and an empirical language suggests that (1) the empirical applicability of a term is somehow objectively given, and (2) a term is either empirically applicable or it is not. Even ignoring the oversimplification, the dichotomy does not recognize differences among theoretical terms. To illustrate, suppose a theorist contemplates using two terms in theory construction but regards neither as empirically applicable. Even so, he may be able and willing to define one term completely, that is, such that the definiton closes the meaning of the term *in his own thinking*. Now suppose that by his own admission the other term designates a vague notion, meaning that the theorist does not regard his definition as closing the meaning of the term. Surely the difference should be recognized, but sociological theorists do not work with a typology of terms that recognizes the distinction.

For all practical purposes, the conventional language of theory construction in sociology is limited to four types of substantive terms: construct, concept, operational, and indicator. Two of the labels, "construct" and "concept," are used uncritically by sociologists, even to the point that they appear interchangeable; and hence the utility of the distinction in conventional theory construction is doubtful.

What has been said of construct and concept also applies to the other labels, "operational" and "indicator." Additionally, it could be argued that those labels are not elements in the language of theory construction, since they are used only in tests. Certainly few theorists identify any constituent terms of a theory as "operational" or "indicators"; for that matter, they seldom bother to typify any terms and never in accordance with an explicit typology. So it is not surprising that sociological theorists are so careless and uncritical in their typifications of empirical assertions, for a systematic typification of assertions requires a typology of terms. Nor is it surprising that sociological theories are amorphous; they are so largely because they have been constructed discursively, and the conventional language of theory construction offers no viable alternative.

Although sociological theorists correctly reject radical operationalism, they remain insensitive to one fundamental point— systematic tests of a theory are precluded unless the theorist identifies all substantive terms in at least one component assertion as empirically applicable. Most theorists studiously refrain from that identification, and they do so for an obvious reason. In the vast majority of sociological theories the substantive terms designate quantitative properties. Given such terms, it is inconceivable that investigators can apply them (let alone agree in their applications) unless the theorist stipulates formulas and procedures for obtaining the requisite data. But such stipulations are excluded from well-known sociological theories, contemporary or otherwise.[8] The generality of the omission is emphasized if only to recognize that Parsons is not the only sociologist who promulgates untestable theories.

Homans' work is especially noteworthy, as he creates the impression of being a staunch empiricist. Yet in many of his purportedly testable propositions, Homans uses the terms "value," "cost," or "reward" to designate a quantitative property of an activity; but he gives no specific instructions as to the measurement of such properties.[9] How, then, are investigators to test Homans' propositions in which value, cost, or reward is a constituent term? Surely they would have to use a formula and employ some procedure for gathering the requisite data, but Homans stipulates neither.

No defect of sociological theories is more glaring than the omission of formulas and procedures for obtaining data. The defect is rationalized by the myth that investigators will know what formulas and procedures are appropriate for tests of a theory. The myth suggests that formulas and procedures enter only into tests; that is, they are not components of the theory; but if this is true, it is difficult to see how an investigator knows what formulas and procedures to employ.

If it be argued that formulas and procedures somehow logically follow from the definitions of theoretic terms, one must surely wonder what rules of syntax and semantics apply. Presuming there are such rules, one wonders even more why the theorist does not derive the formulas and procedures himself. In any case, without such rules, what assurance is there that independent investigators will use the same formula and procedure? The question is embarrassing because the answer is obvious—there is none. Now consider the consequence of investigators using different formulas and procedures; the tests would be idiosyncratic, meaning that the outcome would depend on the investigator, not on the theory.

Merton's work is especially relevant in contemplating theory construction in sociology, for his advocacy of "middle-range" theory (1967:39-69) is a plea for testable theories. However, his conception of middle-range theory is so incomplete that it is of little or no consequence (see Willer, 1967:xv). Granted that a middle-range theory differs from the typical "grand" theory in that it is less inclusive, any theory is "global" in that it comprises assertions about an infinite universe. Above all, a narrow range of variables in itself does not insure testability. As a case in point, although Merton's theory of anomie and social structure (1957a) deals with a narrower range of variables than does the typical grand theory, one would be hard pressed to identify any item in the literature as a report of a test of it. But the issue is more general. Merton's conception of middle-range theory is not a mode of theory construction; and sociology will not have an abundance of testable theories until an alternative to the discursive mode of theory construction is *imposed* on theorists, which is to say that a plea for testable theories is not sufficient.

Relational terms and derivations.   The following is an illustrative list of terms used by sociologists to assert some kind of relation between properties: varies directly with, depends on, is a function of, influences, is associated with, causes, is proportionate to, is a necessary condition for, increases with, is correlated with, due to, determines, is contingent on, results in, leads to, produces, is the basis of, insures, is connected with, is organized about, partially determines, creates, changes along with, is a correlate of, is manifested in, and reflects. The sheer variety indicates that relational terms are used indiscriminately by sociologists, and the license they enjoy creates three problems.

The most obvious problem is the inclination of sociologists to employ nomic words (for example, causes, determines) as relational terms. Within sociology there is no accepted procedure to *demonstrate* that a property causes, determines, influences, another; nonetheless, so one argument goes, a nomic connection can be inferred from space-time relations. Even accepting that argument, any assertion of a nomic connection must be translated into an assertion of a *particular type* of space-time relation, and that translation gives rise to seemingly insoluble problems. Further, assertions of some types of space-time relations (for instance, invariant conjunctions) are simply alien to the conditions of work in sociology. But sociological theorists persist in using nomic words as relational terms, and they commonly do so without reference to any type of space-time relation.

Another problem is the ambiguity of certain relational terms, even those with no nomic connotations. Since the ultimate goal in theory construction is the derivation of testable assertions, relational terms should be consistent with some standard test procedure. Given a theory that deals with quantitative properties, the data that enter into tests of it take the form of numbers, usually two or more sets of values. To conduct a test, a prediction is derived from the theory as to the relation between the values, and the truth or falsity of that prediction is a matter of its consistency with a statement that describes the actual relation. The description is expressed by some conventional statistic, such as a product-moment coefficient of correlation. Accordingly, the relational terms should be such that derived predictions can be judged as either consistent or inconsistent with a conventional statistic.

The uncritical use of relational terms becomes an especially difficult problem when numerous types appear in the same theory. That practice is common in sociology, and the passages below (with relational terms italicized) are instances.

Passage 1: "Any mobility which occurs in a given social system, which is not a consequence of a change in the supply of statuses and actors *must necessarily result from* an interchange. Consequently, if we think of a simple model, for every move up there *must be* a move down. Interchange mobility *will be determined in large part by* the extent to which a given society gives members of the lower strata the means with which to compete with those who enter the social structure on a higher level. Thus the *less* emphasis which a culture places on family background as a criterion for marriage, the *more* class mobility that can occur, both up and down, through marriage. The *more* occupational success is related to educational achievements, which are open to all, the *greater* the occupational mobility" (Lipset and Zetterberg, 1964: 444-445).

Passage 2: "The amount of social energy devoted to a value *is mainly determined by* whether it is defended by full-time workers or by amateurs. One of the main advantages of full-time workers *is* their greater degree of reflection and rationality. One of the main *determinants* of what a man thinks about *is* what he gets paid for thinking about. Directly or indirectly, power-holders in institutions *get paid for* thinking about how to achieve and preserve the values and interests embodied in an institution. The *more* elaborate an argument in favor of a value, the *more* extensive the data collection on which a solution to a problem is based, the *more* explicitly alternatives are explored and evaluated, and the *longer* the time span planned for, the *more* likely is it that the analysis was done by somebody who gets paid for it. The greater rationality with which values embodied in institutions are defended and disseminated *is* one of their main advantages in competition with alternative values" (Stinchcombe, 1968:113-114).

Passage 3: "Safety-valve institutions *may serve to maintain* both the social structure and the individual's security system, but they *are* incompletely functional for both of them. They *prevent* modification of relationships to meet changing conditions and hence the satisfaction they afford the individual *can be* only partially or momentarily adjustive. The hypothesis has been suggested that the need for safety-valve institutions *increases with* the rigidity of the social structure" (Coser, 1956:155-156).

Passage 4: "The function of finding means to single goals, without any concern with the choice between goals, *is* the exclusively technical sphere. The explanation of why positions requiring great technical skill *receive* fairly high rewards is easy to see, for it is the simplest case of the rewards being so distributed as to *draw* talent and *motivate* training. Why they seldom if ever *receive* the highest reward is also clear: the imporance of technical knowledge from a societal point of view *is never so great as* the integration of goals, which *takes place on* the religious, political, and economic levels. Since the technological level *is concerned solely with* means, a purely technical position *must ultimately be subordinate to* other positions that are religious, political, or economic in character" (Davis and Moore, 1945:247).

Commentary. A theorist may use a variety of relational terms to signify that the component statements of his theory assert different kinds of relations between properties. However, that interpretation would be conjectural in analyzing the foregoing passages or a typical sociological theory. Moreover, when diverse relational terms appear in a theory, there is a problem even if each term is defined. As suggested earlier, some assertions of a theory are derived, and those derivations should be systematic; that is, the audience should be able to see how the theorist arrived at each one. Should the theorist use only one type of relational term, the audience may follow his derivations even without explicit rules (see Hopkins, 1964:51). However, when diverse relational terms are used, the derivations become personal; only the theorist knows how they were made, especially if he has used the discursive mode of theory construction.

The foregoing is not a suggestion that systematic derivations can be realized by simply limiting relational terms to one type. On the contrary, even with only one type and a careful definition of it, derivations are not truly systematic unless made in accordance with formal rules.

No feature of the discursive mode of theory construction is more telling than the omission of explicit rules of derivation, and a formal mode is incomplete without the stipulation of such rules. For example, Zetterberg's version of the axiomatic method excludes formal rules of derivation, evidently on the presumption that "ordinary language" is adequate (see Costner and Leik, 1964:820). [10] But contemporary practices in theory construction clearly suggest otherwise. In one theory after another there are passages ending with an assertion that supposedly was derived, but all too often the "logic" of the derivation is conjectural. [11] Some examples follow.

Passage I: "Usually people can adopt religious or magical beliefs to relieve their anxieties fairly 'cheaply.' Religion and magic do not have many negative effects. Hence religion and magic are fairly constant features of societies" (Stinchcombe, 1968:145).

Passage II: "For if . . . crystallization reduces status ambiguity and increases visibility, it presumably acts to clarify the interest or reward positions on the basis of which people act and form opinions. Thus, a general expectation derived from these thoughts is that the effects of status *per se* on given dependent variables should increase with crystallization" (Smith, 1969:911).

Passage III: "The proposition that resource-using communication and socialization tend to popularize institutional values implies that the less spontaneous and private socialization is, the more popular institutionalized values will be" (Stinchcombe, 1968:113).

Passage IV: "Given a middle-class—working-class structure in which conservative political behavior is normal to the middle class, and leftist political behavior is normal to the working class; 1. Associated with a status in each class are pressures to espouse political ideologies appropriate to the class. That is, if there is downward mobility, then the newcomers in the working class are likely to be subject to political resocialization into the class. 2. Men, however, are influenced by superior status, in the sense that they tend to emulate behavior that gives them a sense of superiority over others. Thus, skidders are likely to emulate the political orientation of their more prestigious class of origin, thereby mitigating the force of resocialization to which they are subjected as members of the working class. 3. Therefore, the normal political behavior of skidders is intermediate between the class of origin and the class of destination. 4. However, the efficacy of the resocialization factor varies inversely with the strength of the success ideology which: a. underscores individual responsibility for both failure and success; b. promotes the hope of class re-ascent by accentuating the reality of existing opportunities; c. asserts the reality of distributive justice. 5. It follows that: a. If skidders held a strong success ideology, they are likely to resemble their class of origin in political behavior; b. conversely, if they have an image of limited opportunities, they are likely to resemble politically their class of destination" (Lopreato and Chafetz, 1970:449-450).

Passage V: "Members whose conduct deviates from a norm present a threat to conformers, the more strongly held the norm, the greater the perceived threat. Men when threatened are apt to punish the persons who threaten them, provided they can punish with impunity. The punshiment may be quite mild, such as the withdrawal of approval, but the more strongly held is the norm, the greater the punishment is apt to be. And men are apt to act so as to avoid punishment, provided that no alternative is more rewarding. A deviate can avoid punishment by turning conformer. He may have an alternative in the approval of fellow-deviates, but the larger the number of members who concur in the norm, the fewer these alternative sources of approval are apt to be, and the greater the likelihood that a deviate will avoid punishment by conforming, rather than accept punishment and continue to deviate. It follows that a group, many of whose members agree strongly on the value of conformity to a particular norm, is apt to be one in which some other members conform, too.

Now, this argument in its crude way—and, of course, I have not spelled it out in full—is a deductive system. It consists of propostions from which the conclusion follows more or less logically from the other propositions" (Homans, 1969:83).

Passage VI: "From these assumptions—(1) that technical complexity plus one is a valid measure of the number of items to which administrative attention must be given in order to assure production flow, and (2) that five such items is the maximum effective limit of the human span of attention—it follows that: 1. Organizations having a technical complexity of five or more tend to possess at least three levels of authority" (Udy, 1961:250).

Passage VII: "If the rights and perquisites of different positions in a society must be unequal, then the society must be stratified, because that is precisely what stratification means. Social inequality is thus an unconsciously evolved device by which societies insure that the most important positions are conscientiously filled by the most qualified persons" (Davis and Moore, 1945:243).

Commentary. All of the foregoing passages illustrate how sociologists rely on the conventions of a natural language rather than formal rules of derivation. Those conventions are putative and notoriously imprecise. Accordingly, when a theorist uses them rather than formal rules, the logical structure of the theory is conjectural. As a case in point, Parsons claims to "carry deductive procedures further than is common in the social sciences" (Parsons and Shils, 1951:49). But given the amorphous character of Parsonian theory, it is difficult to detect any deductive *procedures*. As stated bluntly by Max Black: "There is very little strict deduction in Parsons' exposition" (1961:271). No better confirmation of Black's comment can be found than an instance where Parsons states a theorem and an hypothesis as follows.

Theorem: "*this* structure of personality develops, not *primarily* by a process of the modification of 'primary drives' or 'instincts,' but by a process of differentiation of a very simple internalized object-system—we feel it legitimate to postulate a *single* such object at the beginning—into progressively more complex systems" (1955:54).

Hypothesis: "the principle of differentiation is that of binary fission" (1955:54).

Parsons describes the hypothesis as "secondary" to the theorem, but it is difficult to see what that word ("secondary") could possibly mean. Nevertheless, it is left to the reader to make of it what he will. Surely the hypothesis is not derived or deduced from the theorem by any "logic," including the conventions of the English language.

But Parsons is not alone in making the logical relations among the components of a theory ambiguous; the practice is endemic in sociology. Further, sociologists persist in the belief that testable predictions can be derived systematically from the well-known theories of the field. For example, according to Stinchcombe (1968:16), Durkehim derived the following statement from his theory of egoistic suicide: "Protestants in France will have higher suicide rates than Catholics in France." Quite the contrary; in formulating his theory Durkheim took the difference between Catholic and Protestant suicide rates as a "fact," not a derived prediction. If Stinchcombe or anyone else chooses to argue that testable predictions about suicide rates can be derived from Durkheim's theory, they should contemplate a question:

Given any two particular populations, what statements by Durkheim would enable independent investigators to derive the same testable prediction about the difference between the suicide rates of those populations? The myth notwithstanding, Durkheim did not provide a basis for the systematic derivation of predictions.

Notes for Chapter 4

[1] Davis (1963) and Dahrendorf (1959:237-240) are two of the very few sociologists who have made an effort to set definitions and empirical assertions apart in the formulation of a theory.

[2] The same can be said of some other conceptions or schemes of formal theory construction (see Gross, 1959b, and Kyburg, 1968). Kyburg's scheme represents one of the few instances where a philosopher of science has formulated a mode of theory construction with a view to its application by sociologists. However, his scheme ignores differences among sociological terms as to empirical applicability. As such, assertions would be derived without regard to the empirical applicability of their constituent terms, and hence his scheme does not even recognize the central problem in constructing sociological theories.

[3] Rather than speak of established uniformities, Homans commonly refers to "propositions" and even to "tested propositions"; but such cryptic references are misleading. There are few explicit propositions about the relation between specific variables (as opposed to trite expressions about culture, social forces, and so on, or merely reports of isolated findings) in the sociological literature. Further, systematic and comparable tests of propositions are rare, with the results far from consistently positive.

[4] Despite Homans' views on strategy and his commitment to "psychological" explanations, his conception of theory (1964b) is akin to that proposed in Chapter 1, and of the major contemporary theorists only Homans appears to favor the formal over the discursive mode of theory construction. Hopefully, at least some aspects of the scheme proposed here would be to his liking.

[5] Goode cites Parsons as the source for Th3, designating it as the "theorem of institutional integration." However, although Parsons alludes to such a theorem (1951:43), he does not state it, let alone derive it. In the case of Th1, Parsons states it baldly without even a pretense of derivation.

[6] The diversity in the identifications of statements cannot be dismissed by arguing that the statements were not made as components of a "theory." Sociologists hold quite divergent conceptions of theory and, in any case, it is difficult to determine where a discursively formulated theory starts or ends.

[7] Some writers identify virtually any kind of statement as an hypothesis, including descriptions, facts, guesses, hunches, and speculation (see Stephens, 1968).

[8] One possible exception is Stouffer's theory on migration (see Galle and Taeuber, 1966). It is a mystery why sociologists tolerate the omission of formulas and procedural stipulations from theories. Certainly the practice is not tolerated in the more advanced sciences. Examine Eddington's statement: "It is never the task of the experimenter to devise the observational procedure which is the *ultimate* test of the truth of a scientific assertion. That must be indicated unambiguously in the assertion itself, having regard to the definitions of the terms employed in it" (1929:33).

[9] For elaboration on this criticism of Homans' work, see Abrahamsson, 1970.

[10] Zetterberg is not alone in presuming that formal rules of derivation are not needed. Blalock (1969) creates the impression that a causal language insures valid derivations without rules of derivation; Dubin (1969:178) refers cryptically to "a set of logical rules" (but specifies none); and Stinchcombe (1968) and Willer (1967) ignore the problem altogether.

[11] The mode of derivation may be obscure even in what purports to be a "formal" sociological theory. As one instance, Davis (1963) derives fifty-six "propositions" from balance theory without any explicit rules of derivation.

PART TWO: A LANGUAGE OF THEORY CONSTRUCTION

CHAPTER 5 THE TERMS OF THEORY CONSTRUCTION

Two major divisions of a theory are distinguished: (1) the *intrinsic* part comprises statements in the form of empirical assertions; and (2) the remainder, the *extrinsic* part, defines the terms of the intrinsic statements.[1] The distinction serves two purposes. First, logical relations are more evident when intrinsic statements are not interspersed with definitions; and, second, the separation of the two parts underscores the difference between synthetic and analytic statements.

The choice of labels for the two parts reflects recognition that a theory could comprise only empirical assertions and yet be both intelligible and testable. Such parsimony would be feasible only in fields where the meaning of all terms is unquestioned. That condition is alien to the social sciences generally and hence the importance of the extrinsic part of a theory.

Commentary on the Extrinsic Part

The proposed mode of formalization focuses primarily on intrinsic statements, which is to say that the definition of a sociological term is a substantive rather than a formal problem. However, the problem does depend on the position or role of the term in the theory. When an intrinsic term is identified as empirically applicable, then its definition must facilitate agreement between independent investigators who apply the term. If an intrinsic term denotes a quantitative property, investigators cannot apply it, much less agree in its application, unless the term designates a formula in the extrinsic part of the theory. Further, unless the theorist regards the symbols in a formula as primitive, he must offer what purports to be empirically applicable definitions of them. In either case, he should stipulate the requisite data and instructions for obtaining them. No mode of theory construction can govern the theorist's judgments pertaining to formulas and data; he must rely on his imagination and knowledge of the field's resources.

In striving for empirically applicable definitions, the end justifies the means. To say it once more, the form, type, or appearance of a definition is strictly secondary to the empirical applicability achieved. In that connection, succinct definitions of sociological terms are seldom feasible. If a one-sentence definition is offered but its constituent terms are not regarded as primitive, the theorist should define the nonprimitive terms. Thus, in the extrinsic part of the theory there may be a series of definitions, all referring to one intrinsic term.

Even though the proposed mode of formalization deals primarily with intrinsic statements, the importance of the extrinsic part beggars description. Given sociology's terminological problems, the intrinsic part of a theory alone is virtually meaningless. But the extrinsic part does more than make the intrinsic statements intelligible; it also determines testability and predictive accuracy.

The importance of the extrinsic part cannot be exaggerated if only because it is the most neglected aspect of theory construction in sociology. Its neglect is a tradition, one reflected in the naive belief that the "great theorist" makes statements and someone else (the faithful technician) can define the terms. On the contrary, should someone else define the terms, he has created rather than completed a theory.

Two Features of Intrinsic Statements

A theory is not constructed formally unless its intrinsic statements are distinguished as to type. In the proposed scheme, statements are distinguished by reference to their "position" and their constituent terms.

All theories leave at least one question unanswered. In every theory some intrinsic statements are direct assertions (premises), from which other intrinsic statements are derived. Test results may suggest that the direct assertions are true, but tests cannot reveal why they are true. As such, direct assertions are "intuitive" components of a theory (that is, the theorist cannot or does not go beyond them), but that is not the case for derived statements. So, as far as "position" is concerned, there are only two possibilities—some intrinsic statements are direct assertions and others are derived.

The "character" of intrinsic terms is much more complicated than the position of statements. Seemingly infinite types of terms could be distinguished, but the relevant possibilities are reduced by recognizing again that assertions about order necessarily include three types: unit, substantive, and relational. Further, if testable assertions are to be derived, all terms of *some* intrinsic statements must be considered as empirically applicable, and hence that notion enters into terminological distinctions.

Unit Terms

All intrinsic statements refer to properties, but a property implies a class of event or things. Consider the unintelligibility of the following utterance: are black, less than two feet long, and weigh less than ten pounds. Although the properties (color, length, and weight) are understandable, the utterance itself becomes intelligible only in reference to a class of events or things, such as: crows are black, less than two feet long, and weigh less than ten pounds. Accordingly, all intrinsic statements should include a term that designates a class of events or things as *units*.[2]

Sociological theories should be concerned primarily with *differences* among social units. As instances, polygamy is an accepted marriage form in only some jurisdictions, urbanization varies internationally, and the official suicide rate is much greater for some age groups than others. When a theory does not make assertions about differences among units, it is questionable from the outset. To illustrate, one unit term in Parsonian theory is "social system," but Parsons often refers to *the* social system, as though there is only one, and that terminology is revealing. As a "methodological essentialist" (to use Popper's term, 1950:34), Parsons appears concerned primarily not with differences among actual populations but rather with the *idea* of a social system; as such, one must surely wonder what would constitute a test of his theory. In any case, Parsonian theory is clearly contrary to Homans' argument that "sociology has to explain . . . the actual features of actual societies and not just the generalized features of a generalized society" (1964a:813).

It will not do for Parsons to argue that he is concerned with "social order" and thus he is indifferent to conventional sociological research. Social order is a quantitative property, the degree of which can be assessed only by reference to riots, rebellions, revolutions, and various types of isolated acts, all conventional subjects of sociological research. Far from being concerned with the *amount* of social order, Parsons suggests that *by definition* order is a qualitative attribute of a social system, and hence empirical questions about order are alien to his perspective.

In some fields it is conventional to assert that units of the same type are characterized by a particular attribute (for example, the specific gravity of copper), and such statements may be synthetic. So empirical assertions can be made about "sameness" as well as "difference," but in sociology monovariate assertions are a thicket of tautologies. Statements about the function of particular instituions or organizations (such as the family or the government) may appear to be true, but commonly such statements are implied by the very idea of the institution or organization in question. By contrast, assertions about differences among units are more clearly synthetic, and certainly they are much more realistic. Although social units of the same type must have at least one attribute (the definitional attribute) in common, as for *any contingent* attribute (one not true by definition), it is difficult to conceive of all units as homogeneous.[3] Data on social units of the same type (for instance, urban areas, associations) reveal differences, not homogeneity. In any case, the proposed mode of formalization is relevant only when a theorist is concerned with differences among units of the same type and wants to make assertions about the association between such differences.[4]

The empirical applicability of unit terms. The proposed scheme is not limited to particular unit terms, nor does it imply invidious distinctions. The choice of unit terms is a matter of substantive interest, and that point is emphasized because the illustrations (urban area, metropolitan area) may suggest a preoccupation with human ecology, where most unit terms designate territorially differentiated populations. But some sociological unit terms designate acts, events, or conditions rather than populations.

Whatever the unit term, the theorist must consider it as empirically applicable; in effect, he asserts that investigators can agree substantially in applying the term to identify events or things. [5] Thus, should a unit term denote a type of population, the implied assertion is that investigators can agree substantially in designating aggregates as instances.

A definition of a unit term may be such that investigators use it to typify socially recognized categories. If so, empirical applicability is a matter of agreement among investigators in identifying instances (for example, is Harvard University a socially recognized category and, if so, is it an association?). In other cases, investigators delimit a population (that is, establish the territorial limit and/or membership) in accordance with a procedure stipulated by the theorist. For example, the Los Angeles metropolitan *region* may not be considered as a socially recognized category; if it is not considered as a socially recognized category, then its limits and members (residents) would be determined by investigators working with a definition of metropolitan region (a unit term) and a delimitation procedure.[6] Absolute agreement would be realized when investigators delimit the same boundary and identify the same individuals as residents of the Los Angeles metropolitan region.

Unless the unit terms are empirically applicable, defensible tests of a theory are precluded. To illustrate, suppose the following assertion is derived from a theory: Among associations in a metropolitan area, the greater the number of members, the greater the number of organizational divisions. Unless independent investigators agree substantially in reporting the number of members and organizational divisions of each association, tests of the assertion would be idiosyncratic. But agreement in the identification of associations is a prior consideration.[7] Should investigators working independently in the same metropolitan area fail to agree on the universe of eligible associations, it would be difficult to justify using data gathered by any of them. However, even agreement in the identification of associations would not be sufficient. A unit term that designates a type of population may not be considered as empirically applicable without a procedure for delimiting territory and/or members. Accordingly, when two investigators disagree in identifying members of an association, they may further disagree as to the number of members. Should they disagree substantially, then data reported by either would be questionable. Despite the disagreement, each investigator's data could be used in a test, but the two test results might well be inconsistent.

To further the empirical applicability of a unit term, the theorist will provide an elaborate definition of it in the extrinsic part of the theory. He may assume that the unit is a socially recognized category; that is, when an instance is present, there will be a label for it in the natural language of that population. If so, his definition will begin with the identification of a *type* of socially recognized category; for example, an association is a socially recognized category of persons who consciously and deliberately coordinate some of their activities and exclude other persons from participation. Alternatively, the theorist may decide that a socially recognized category is vague and proceed to define it as well as the unit term in question ("an association"). To that end, he may commence with instructions for identifying socially recognized categories. As suggested earlier, the terms of a natural language may be taken as indicative of socially recognized categories, but only the theorist should judge the relevance of natural language in defining unit terms.

The difficulties in framing empirically applicable definitions of unit terms defy exaggeration; linguistic and other cultural differences alone are major obstacles. However, the theorist has some advantages. As for linguistic and other cultural considerations, the definition of a unit term does not imply that instances are ubiquitous. For instance, granted that a definition of an "urban area" in English may be difficult to apply in non-English speaking populations, the definition does not imply that instances are found in all populations. If the definition is empirically applicable, independent investigators will agree as to the presence or absence of urban areas in a territory, whatever the native language.

Maximum empirical applicability need not be realized by a definition of a unit term, nor must the theorist offer evidence as to the degree realized. What he asserts (as part of the theory) is *sufficient* empirical applicability. The theorist need not stipulate what is "sufficient," but he should recognize that the predictive accuracy of the theory depends, *inter alia*, on the empirical applicability of its unit terms. However, the relation between predictive accuracy and empirical applicability is contingent on the properties considered in the theory. Suppose the unit term is "associations," meaning that the intrinsic statements make assertions about properties of associations. If one of the properties pertains to members (for example, their average age), then the definition of the unit term should enable investigators to achieve sufficient agreement in identifying both associations and the members of each. But it could be that the properties are structural (for example, the number of organizational divisions of an association), which can be described without reference to traits of members.

In striving for an empirically applicable definition of a unit term, the theorist should ignore the dicta of methodologists. Again, the form or appearance of a definition is irrelevant; only the empirical applicability achieved is decisive.[8] Of course, particular definitions and procedures may come to be conventional, and the theorist can draw on the work of others. Even today the research literature indicates the unit terms and related definitions that sociologists tacitly regard as empirically applicable; and no theorist should be ignorant of that literature. But theorists are not bound by conventions. After all, only a theorist can decide to use a particular unit term, and he is free to define it as he sees fit. However, his judgment is subject to question. When investigators report that they are unable to apply the theorist's definitions, or when there is evidence of disagreement in their applications, then the theorist has failed.

Practical considerations.   The theorist does not merely assert that independent investigators *could* achieve substantial agreement in applying a definition of a unit term; rather, he asserts that the identification of instances is feasible, a consideration that touches on the resources of the field. For example, if a theory makes assertions about urban areas, data on instances of that type of unit must be secured for tests; but the resources of sociology are not such that investigators can readily delimit numerous urban areas and gather data on them. So, to further testability, the theorist should consider published data, such as census statistics. But published data are relevant only if reported for territorial units that qualify as urban areas as defined by the theorist. He need not assume that all national census agencies delimit such units, nor that the delimitations are made in strict accordance with his definition. Census units in some countries may qualify "sufficiently" as an urban area, but only the theorist can make that judgment. However, he cannot make decisions in particular cases, nor can he point to a finite class of units, such as Urbanized Areas of the United States, as suitable for tests. If the theorist must make decisions about particular data, then the tests will cease on his demise and the testability of the theory is finite.

Should a theorist formulate definitions of unit terms without anticipating problems, investigators may decide that published data cannot be used or fail to agree in their decisions as regards particular instances, and in either case empirical applicability would not be realized. All of the foregoing admonitions are contrary to sociological tradition, which clearly justifies the formulation of theories without regard to data. One may tolerate that practice; but it is then ludicrous to bemoan the paucity of testable theories.

The preceding is not a suggestion that sociologists should formulate all theories with a view to basing tests on published data. Whether one contemplates using published data depends on the theory and the field's resources. To illustrate, social psychologists are not likely to regard census statistics as relevant, and it is feasible for them to gather data. Their unit terms commonly designate individuals or small groups; the data then can be gathered without enormous expenditures. This is not so for human ecology or demography, where some unit terms designate entities so large that the cost of gathering data may be prohibitive. Of course, a human ecologist or demographer may regard census data as too unreliable for tests, but that is a judgment which only the theorist should make.

If a theorist makes no reference to published data, then in the extrinsic part of the theory he must specify a procedure for identifying and delimiting instances of a type of unit as the first step in *gathering* data. The specification should be elaborate, if only to anticipate problems, especially those stemming from linguistic and other cultural differences, that will confront investigators. When a delimitation procedure is couched in terms that are peculiar to a particular culture or historical period, tests of the theory will be finite. To illustrate, in specifying a procedure for delimiting urban areas, it would be questionable to speak of "house numbers," as the practice of numbering houses is by no means universal.

The relativity of assertions.   As suggested in Chapter 4, sociologists commonly formulate theories without stipulating a unit term. That practice ignores a crucial consideration: the predictive accuracy of any assertion is relative to the unit term. One illustration should suffice: There is a direct relation between population size and population density. The assertion may be interpreted as applying to all territorial units, but predictive accuracy is substantially less when the assertion is applied to countries rather than urban areas.

The argument is not that a theory should be restricted to one type of unit term. It is desirable to maximize the range of a theory; but when a theorist fails to specify unit terms or uses a generic designation (such as "populations" or "social units"), there is only one interpretation—that his assertions hold equally well for all types of units. However, the predictive accuracy of the theory is probably less for some types of units than others, and hence predictive accuracy could appear minimal because investigators happened to select a particular type of unit for the initial tests.

The solution is simple: restrict each intrinsic statement to one type of unit term. That requirement does not minimize the range of a theory, for should the theorist believe that a particular relation holds for several types of units, then he should treat each type separately. Thus, if the theorist is convinced that there is a direct relation between population size and density among cities, urban areas, metropolitan areas, and countries, he should make four intrinsic statements in which the same relation is asserted but for different types of units. Such distinctions eliminate ambiguities and facilitate the interpretation of tests. Given negative findings, the theorist should modify a theory rather than reject it totally. The prescribed practice makes it possible to retain some intrinsic statements and reject others, with the only difference between those rejected and those retained being the unit term.

Consider the following assertion: Among urban areas, the greater the population size, the greater the population density. As stated, the assertion could be tested by comparing urban areas in different countries, for example, Bombay and Los Angeles. However, the theorist can restrict the *context* by stipulating a universe of units; thus: Among urban areas *in a country*, the greater the population size, the greater the population density. The wording limits comparisons to urban areas *in the same country*.

The limitation of an assertion to a type of *universe* reflects more than a concern for clarity. Obviously, it is intended to maximize predictive accuracy, and the limitation could be relevant for several reasons, one being the idea that some relations hold only in a "closed system." A completely closed social system is virtually unknown, but some units approximate it more than others. Thus, so one may argue, a country more nearly approximates a closed system than does an urban area; and that argument may lead a theorist to restrict an assertion to countries. Then consider the possibility that technology is more constant among urban areas in the same country than among urban areas in different countries. That possibility may lead a theorist to limit an assertion to a particular type of universe—urban areas *within* a country. Similarly, the phrase "among industrialized countries" restricts an assertion to a particular type of universe.

The foregoing anticipates the criticism that the proposed scheme forces a theorist to assume that relations between variables are invariant. Quite the contrary; it enables the theorist to anticipate variance in a relation and qualify his assertions. The restriction of each intrinsic statement to a particular type of unit term is one such qualification, and the theorist may further qualify an assertion by a contextual stipulation (that is, a type of universe of units).

Observe that unit terms and universes are relevant in contemplating "ideal conditions" for tests of a theory. An intrinsic statement may be purely theoretical in that the relation asserted holds only in a condition never found in the "actual" world. Thus, just as perfect competition or a frictionless state may never be realized, no social unit qualifies as a completely closed system.

The proposed mode of formalization recognizes the notion of ideal conditions, but the notion should not be a crutch to defend untestable theories. Maximum predictive accuracy may never be realized because tests cannot be made under ideal conditions, but that does not make testability any less important. When the theorist invokes the notion, he must specify what would be ideal test conditions and stipulate at least two types of "actual" conditions, one of which more nearly approximates the "ideal" condition. By that route the theorist makes an assertion about *variation in a relation*, for instance, the relation between properties X and Y will be closer in a type I condition than in a type II condition. But he does not assert a "perfect" relation between X and Y in condition I; rather, he asserts that the X-Y relation will be closer in the type I condition than in type II. What the relation would be in the stipulated ideal condition (which type I only approximates) remians unknown; nonetheless, the theorist would have extended the theory's range and perhaps enhanced its predictive accuracy.

Substantive Terms

Consider four terms: lower social class, urban area, association, revolution. The terms designate units, but they are not statements or assertions. No assertion can be made without using terms that designate properties of units, that is, *substantive* terms.

Since each intrinsic statement is an empirical assertion (whether testable or untestable), the constituent substantive terms pertain to *contingent* rather than definitional attributes. Stated another way, the *definitional* attributes of a unit are specified in the extrinsic part of a theory. Thus, if an urban area is defined by reference to a minimum population density (for example, at least 1,000 residents per square mile), then that attribute cannot enter into intrinsic statements. Given that partial definition, a theorist may assert some relation between population density and other properties of urban areas (such as, the greater the population size, the greater the density), but a statement that all urban areas have a population density of more than 999 would follow from the definition and thus not be an empirical assertion.

Various kinds of assertions may be made about the contingent properties or attributes of units. Thus, the statement "social classes exist" is an assertion, and the same may be true of "all urban areas have a population density of 2,700 per square mile." In both cases, however, the empirical validity of the statement obviously depends on the definition of the unit, and the definition of urban areas could be such as to make the second statement tautological. It is most improbable that anyone would so define an urban area; nonetheless, given a statement about *one* property or attribute of social units, there is reason to suspect that it is analytic rather than synthetic.

As suggested earlier, statements about the functions of organizations or institutions are a nest of tautologies, and the same is true of what Homans (1961:11) designates as "anatomical sentences" (for instance, all organizations have communication systems). Typically, the tautology is not evident or indisputable, but only because the definition of the unit is vague or implicit. For example, by virtually any definition, population size and areal extent are properties of urban areas. Hence the statement that urban areas have a population density is clearly a tautology. By contrast, the statement that all organizations have a communication system may not appear tautological but only because the definitional properties of organizations have not been made explicit.

Even when a statement about *one* property of a unit is not a tautology, it is not likely to be informative. To say that all organizations have a communication system (or any other similar statement about *a* property of a social unit) is to say next to nothing. Only assertions that social units of some particular type share a particular *attribute* in common are informative. Thus, communication system is a property, but a particular type of communication system is an attribute. Accordingly, if one asserts that all voluntary associations are characterized by a type *X* communication system, then the statement is certainly informative, but it could be a tautology. If not a tautology, it is grossly unrealistic. It is most improbable that all instances of a type of social unit share some contingent attribute in common, especially a quantitative attribute (imagine someone discovering that all voluntary associations have the same number of members).

Monovariate statements can be excluded by a simple rule: All intrinsic statements must include at least two substantive terms. Of course, sociologists addicted to "anatomical" or "functional" statements will not comply.

Types of Substantive Terms: Constructs

Even opponents of operationalism would grant that the sociological vocabulary includes a host of vague terms, meaning that conventional definitions of them promise very little empirical applicability. Most major sociological terms denote highly abstract notions, and in attempting to define such a term a theorist may conclude that he cannot recognize and relate all properties entailed in the notion. In other words, he cannot frame a definition that (even in his own thinking) closes the meaning of the term.

But suppose the theorist finally realizes what in his opinion is a "complete" definition. Even so, if the term designates a quantitative property, it is inconceivable that investigators could apply the definition without applying a formula, let alone achieve agreement in their application. But the theorist may be unable or unwilling to stipulate a formula; or, should he stipulate one, he may realize that limited resources preclude gathering the requisite data.

To illustrate the theorist's dilemma, suppose that he attempts to formulate a theory on international differences in the degree of social stratification. If the theorist thinks conventionally, he is likely to recognize so many dimensions of social stratification (power, prestige, wealth, differential opportunities, and so on) that they cannot be related in one definition. As such, the theorist has two alternatives: (1) he may use the term to designate an undefined notion or (2) he may offer a grossly oversimplified definition, such as, social stratification refers to inequality in the distribution of rewards. The definition may suffice for some purposes, but reward is a very abstract idea and leaves many specific questions unanswered, especially the distinction between wealth and prestige. Note also that the definition ignores "differential opportunities at birth." Thus, in two countries the inequality of rewards (however defined) could be approximately the same, and yet the association between social class position (however defined) of fathers and sons could be much closer in one country than in the other. If so, are the two countries socially stratified to the same degree? The theorist's definition does not speak to the question, and he may be unable or unwilling to answer it, as often happens (see Willer and Webster, 1970:753) when a definition is not intended to close the meaning of a term.

Suppose the theorist perseveres and frames a definition that (in his opinion) closes the meaning of social stratification. Even so, he may be unable to relate all of the definitional dimensions in a formula, such that investigators can compute a value as symbolizing the degree of social stratification. However, assume that the theorist does stipulate such a formula. He may be forced to conclude that some of its constituent symbols cannot be defined so as to promise sufficient empirical applicability or that practical considerations preclude application of the formula (that is, the field's resources are so limited that requisite data cannot be gathered for a series of countries and they are not available in published form).

What has been said of social stratification is no less true for other major sociological terms, such as social integration, normative consensus, and social cohesion. In brief, theorists are not likely to regard their definitions of those terms as either complete or empirically applicable.

Proponents of "operationalism" readily appreciate the terminological problem, but the proposed mode of theory construction departs from their tenets. No attempt is made to "purify" the field's vocabulary, that is, to eliminate terms that cannot be defined so as to realize empirical applicability; nor is the intention to provide a radically new vocabulary. Neither course of action is feasible, if only because neither would be accepted and neither is desirable.

Contrary to the apparent principles of operationalism, *any* sociological term can be used effectively in theory construction, even without a definition. However, since terms differ as to empirical applicability, they must be distinguished clearly as to type, and each type plays a *different role* in theory construction. That approach differs not only from operationalism but also from the traditional mode of theory construction. No better illustration can be found than the works of Talcott Parsons. The criticism is not that Parsons uses vague and "unoperationalizable" terms; such usage is justifiable and inevitable. Rather, rarely does Parsons use a term that promises empirical applicability and he consistently fails to distinguish terms in that regard.

When a theorist regards his definition of a term as neither complete nor empirically applicable, that term is identified as a "construct." Comment on the rationale of that identification is postponed until still another type of term, "concept" is introduced.

Relativity of types of terms.   It would be incorrect to describe a construct as a term that *cannot be defined* so that the definition is complete and empirically applicable. That description would suggest that all sociological terms are in some objective sense either constructs or nonconstructs, and no such claim is made. Whether a term is designated as a construct depends *entirely* on the judgment of the theorist who uses it.[9] Whereas one theorist may identify a particular term as a construct, in using the same term and the same definition another theorist may identify it otherwise.

Just as arguments about definitions are sterile, so are debates over the typification of terms.[10] To repeat, it is the theorist's judgment that determines whether a term is identified as a construct, but his judgments are not without consequence. Should he identify all substantive terms in his theory as constructs, then by his own tacit admission the theory is untestable.

The discursive mode of theory construction does not require distinctions between types of terms, and for that reason alone it is difficult to identify component statements of a typical sociological theory as testable (presuming that any are). For that matter, the terms of a typical theory alone suggest that none of the statements are testable. While typification of terms is strictly a matter of the theorist's judgment, no one is likely to regard the major substantive terms of sociology as other than constructs. Try to imagine someone formulating what he regards to be complete and empirically applicable definitions of the following designations of quantitative properties: stratification, cohesion, differentiation, integration, consensus, and technological efficiency. In that connection, had the grand theories of sociology been stated formally, all of their component substantive terms would have been identified as constructs; and from the outset it would have been abundantly clear that the theories were not testable.

Some advantages in the use of constructs. As indicated in Chapter 2, terminology is *the* problem in theory construction. It is difficult to think of conventional definitions of major sociological terms as either complete or empirically applicable, and therein lies a dilemma. A theorist may contemplate alternatives to the conventional sociological vocabulary, but he is not likely to formulate a vocabulary which is radically different, so that all definitions would be regarded as complete and empirically applicable.

Contrary to the tenets of operationalism, it is doubtful if anyone will ever define all sociological terms in such a way that he regards each definition as complete and empirically applicable. Imagine a theorist characterizing his definition of a term that designates normative phenomena (feelings or beliefs concerning the "oughtness" of conduct) as either complete or empirically applicable; and the same may be said of social integration or social stratification, terms which designate an array of phenomena so complex and macroscopic that they defy comprehension. One may state that a particular incomprehensible phenomenon is associated in some way with another particular phenomenon that is comprehensible, but how can such a statement possibly be construed as a definition? It is an untestable assertion and, as such, a matter of theory, a consideration that is only obscured by the notion of an "operational" definition.

It is desirable for a theorist to define all substantive terms, but he need not define a construct, that is, he may use the term to denote an undefined notion. Even if he should define a construct, the definition can be brief and grossly incomplete. In either case, the theorist need not be preoccupied with three difficult questions. First, how can a definition be framed so that it recognizes and relates all conceivable dimensions of the property designated by the term? Second, how can the term be defined so that investigators could apply it and further agree independently in their application? And, third, do the field's resources permit empirical applications of the definition?

Types of Substantive Terms: Concepts

Consider three terms: degree of urbanization, resident crude birth rate, and social interaction. The argument is not that those terms designate "observable" phenomena (a distinction that is scarcely relevant for sociology) or that all sociologists would identify them as something other than constructs. Nonetheless, in defining any one of them, a sociologist is likely to regard his definition as complete; that is, it equates the term with a designated property or a relation between properties, with all other conceivable properties excluded. Thus, the writer would define the term "resident crude birth rate" as: the number of live births by human female residents of a territorial unit during a period as a ratio to the average size of the resident population in the territorial unit during that period. Now the definition is regarded as complete for one reason—the term means nothing more to the writer than what the definition stipulates. Accordingly, should the writer employ the term in theory construction, it would be used as a concept.

As already suggested, a theorist should identify a term as a concept only if he regards his definition of it as complete. One justification for the suggested definition is that sociologists use the two terms, concept and construct, so uncritically that they appear to be interchangeable.[11] The proposed distinction merely recognizes that a theorist may regard his definition of one term as complete and his definition of another as incomplete.

Although regarded as complete, it does not follow that the definition also is regarded as empirically applicable. Specifically, given a term that designates a quantitative property (or a quantitative relation between properties), that term should not be regarded as empirically applicable without the specification of a formula, and a distinction is drawn between the stipulation of a formula and the framing of a definition. To recapitulate, a concept is a substantive term defined by a theorist in such a way that he regards the definition as complete but not empirically applicable.

Although the distinction between a complete and incomplete definition is commonly recognized one way or another by theorists and philosophers (see Braithwaite, 1953:chap. III; Pap, 1962:6-57; Gorskij, 1970:328-331; Homans, 1964b:954; Feigl, *et al.*, 1958:37-224; and Northrop, 1947:141), it is difficult to describe, especially if the distinction is taken as somehow objectively given. Certainly the objective character of the distinction is not made clear in the philosophy of science literature, and no pretense is made here of specifying that character.[12] On the contrary, the distinction is regarded as subjective, that is, a matter of the theorist's opinion. Nonetheless, the distinction can be related to the notion of empirical applicability, even though neither a construct nor a concept is regarded as empirically applicable. When it comes to quantitative properties, only formulas are considered to be empirically applicable. However, it is the definition of a concept (not a construct) that in some way leads a theorist to articulate a formula. The argument is not that the formula logically follows from the definition of a concept; nonetheless, it is difficult to see how a theorist can arrive at a formula without what he takes to be a complete definition of the term (for instance, resident crude birth rate) corresponding to the formula.

An issue.   Although the identification of a term is strictly a theorist's judgment, an issue is entailed in the idea that concepts are not empirically applicable. Insthat connection note again that a definition is empirically applicable only to the extent that independent observers agree in its application to particular events or things. If a concept pertains to a qualitative property, then its definition may be empirically applicable; but the situation is quite different when it comes to quantitative properties. A theorist may regard his definition of a quantitative property as complete, but independent investigators are not likely to agree in reporting a quantitative attribute of a social unit (the resident crude birth rate) unless they employ the same formula, and the definition of a concept does not stipulate a formula.

Returning to a previous consideration, one may argue that a formula somehow logically follows from a definition. That argument is examined at length in Chapter 6, and at this point a few general observations will suffice.

A term may be used in conjunction with a particular formula to the point that the connection between the two appears as somehow logically necessary, but that is never the case. Next consider the idea that a formula somehow follows from a definition. For example, the foregoing definition of "resident crude birth rate" may *suggest* the following formula: RCBR = $[(B/3)/P]$ 1,000, where B is the number of live births by resident females of a territorial unit during a three-year period, and P is population size (that is, total residents) at some point in the second year. But there is no assurance that all investigators would use that formula, unless so instructed in the extrinsic part of the theory, and such instructions are distinct from definitions.

Types of Substantive Terms: Referentials

A referential is an intrinsic term that designates a formula in the extrinsic part of a theory. Unlike other intrinsic terms, a referential appears as a capitalized acronym (for example, RRCBR, meaning referential of the resident crude birth rate), the purpose being to signify that the meaning of the term is technical and relative to a particular theory. As such, a referential designates nothing more than a formula, and in that sense the two have the same meaning.

Some issues. Each formula in the extrinsic part of a theory is a *referential formula*, and that identification is stressed because there is a tendency to think of formulas apart from any particular theory. One may articulate a formula (or measure) without constructing a theory, or use a formula with no particular theory in mind, and the same formula may be used in several theories. Nonetheless, the term *referential formula* signifies that the formula is relative to the theory in which it is employed. More specifically, the meaning and "validity" of a referential formula entails reference to a particular theory.[13] One illustration should suffice.

Suppose a theorist uses a novel referential formula in constructing a theory. Critics may think that the meaning and validity of the referential formula can be assessed apart from the theory; but the present argument is quite to the contrary. The justification of a referential formula is entirely a matter of theoretical context. If the theory proves to have substantial predictive accuracy, that alone justifies the component referential formulas; indeed, other than rhetoric and intuitive conviction, predictive accuracy is the only justification. To be sure, when the predictive accuracy of a theory appears negligible, that alone does not prove that a particular referential formula is somehow invalid. Negligible predictive accuracy could be due to other referential formulas in the same theory; or it could be due to something unrelated to any referential formula. So, insofar as one considers the notion at all, the validity of any referential formula is a matter of uncertainty. There is no way the uncertainty can be resolved, but that does not negate the more general consideration. Again, insofar as the notion is entertained at all, the validity of a referential formula cannot be assessed independently of a particular theory.

Although a particular referential formula can be used in more than one theory, that possibility touches on an issue. Suppose that a theorist employs a novel referential formula, "X," in constructing theory "Y." Now suppose that tests of "Y" reveal impressive predictive accuracy, and for that reason a second theorist employs referential formula "X" in constructing another theory, "Z," so that the two theories share one referential in common. But suppose that tests of theory "Z" reveal negligible predictive accuracy. Would those tests be relevant in assessing the meaning or validity of the referential formula *as used in theory "Y"*? Not at all. The results suggest that the second theorist may have made a mistake in "borrowing" referential formula "X," but that possibility has absolutely no bearing on theory "Y" or the meaning and validity of the referential formula in the context of "Y."

Enamored with the philosopy of science, some sociologists (for instance, Dumont and Wilson, 1967) insist that the terms of a theory be "significant," and they are prepared to accept or reject a theory by that criterion. The notion of "significance" is vague, and it can be made intelligible only when translated as follows: a component term of a theory is significant to the extent that it has been utilized in accepted theories.[14] Far from questioning the interchangeability of terms among theories, the criterion of significance demands it.

As a criterion for assessing theories, the notion of significance is unrealistic. When a theorist borrows terms from accepted theories, that practice in no way insures that his theory will have any predictive accuracy. Moreover, given a theory with demonstrated predictive accuracy, the ultimate folly would be to reject it merely because the component terms were not borrowed from accepted theories. Such a practice is a luxury that sociology can ill afford, and it is difficult to see how the "significance" criterion is applicable (if at all) in any but the most advanced sciences. If one insists that a theory utilize "significant" terms, and significance is judged by "accepted" theories, how can the first theory possibly receive a favorable hearing? The question is all the more relevant for sociology, where for all practical purposes there are no accepted theories.

In replying to the foregoing argument, critics may protest that the ultimate goal of any science is a condition in which some component terms of any two theories are interchangeable. Granting the desirability of that condition, it will not be realized by merely invoking pronouncements by philosophers of science. Those pronouncements are irrelevant when alien to the conditions of work in a field (in this case, one with no accepted theories), and hence the proposed scheme disregards the notion of significance.

Referential formulas and measures.   The term "referential" or "referential formula" is preferred to "measure" for several reasons, two of which are considered at this point. As a matter of common usage, the word "measure" is ambiguous; it may designate a formula, a technique, or a particular value (an actual number). By contrast, there is no ambiguity in the case of a referential; it designates a referential formula by which values can be computed, and each such value is designated as a *referent*. Then there is an even more important consideration. When a referential formula is designated as *the* measure or even *a* measure of some phenomenon, critics are likely to presume that its "validity" is not a question of theoretical context. That presumption is rejected, but sociologists are not likely to abandon it. Hence the term "referential formula" is employed to emphasize that such a formula should not be judged apart from the predictive accuracy of the theory in which it is employed.

Empirical applicability.   In stipulating a referential formula the theorist asserts that it is empirically applicable. The empirical applicability of referential formulas determines not only the testability of a theory but also in part its predictive accuracy. Consequently, empirical applicability is a central consideration and for that reason an extensive commentary on the notion is in order.

For illustrative purposes, suppose that a theorist uses the term "occupational differentiation" as a concept and defines it as: a function of the number of occupations in a population and the evenness of the distribution of members of that population among the occupational categories. That definition closes the meaning of the term in the mind of the theorist; that is, in his thinking all dimensions of occupational differentiation are recognized by the definition Nonetheless, the theorist would recognize that the term designates a quantitative property; as such, investigators cannot report the *amount* of occupational differentiation in a particular population without applying some formula. However, it would be naive to suppose that the definition will suggest the same formula to each investigator, which is to say again that the definition of concepts and the stipulation of referential formulas are distinct steps.

Suppose a theorist makes an intrinsic statement that links the concept occupational differentiation to a referential, ARCOD, and stipulates the following formula:

$$ARCOD = (1-\Sigma X^2/(\Sigma X)^2])/[1-(1/N)],$$

where X is the number of members of a population in a given occupational category and N is the number of occupational categories in that population. We are not concerned with the reasons that the theorist might give for the formula, so it will suffice to point out that by no rule of syntax does the foregoing definition of the concept *imply* that formula and *only* that formula. Accordingly, the formula represents a judgment by the theorist, and that judgment is based on imagination, intuition, or experience rather than logic, conventional or otherwise.

Regardless of how the theorist arrived at the ARCOD referential formula, his use of it in theory construction entails an assertion—that the formula is empirically applicable. That assertion reflects a series of judgments by the theorist, commencing with an assessment of the field's resources. If limited resources preclude application of the formula to instances of the unit term, the theory's testability is minimal. For example, suppose that the ARCOD is a component referential of a theory in which the unit term is "country". Since tests would require international application of the referential formulas, the time and cost required to gather occupational data by field studies (surveys) would be astronomical.

Having estimated the field's resources, the theorist may decide that it is not feasible to gather data for tests of the theory; if so, he should contemplate using published data on occupations (such as census statistics). The choice between the two alternatives, gathering data or using published data, is governed intially by one consideration—furthering the testability of the theory. However, the choice determines the explication of the referential formula. If empirical applicability is to be achieved, it is not sufficient that investigators obtain data. They must also know how to apply each referential formula; otherwise, the referents they compute for a particular country (for example, Switzerland, 1970) may diverge even though each investigator has applied the same referential formula to the same data. Accordingly, the theorist should do more than stipulate a referential formula and define its constituent symbols. He also should describe the procedure for applying the formula, including illustrative applications. But description of procedure depends on the kind of data, that is, whether they have been gathered or obtained in published form. Each kind presents special problems, all of which should be anticipated in the explication of the referential formula.

The foregoing does not mean that a theorist must defend his assessment of the field's resources, or even that the assessment must be systematic. It is not even necessary that the theorist have a precise idea of the time and cost required to apply each referential formula in a test. The overriding consideration is not the resources required for *one* test. The goal is to maximize testability and, as such, the greater the number of tests, the better; but no one can estimate the resources required for infinite tests. Nonetheless, if the theorist senses that limited resources preclude extensive applications of a referential formula, it would be pointless to incorporate that formula in a theory. Finally, although the *potential* testability of a theory is inherently conjectural, the theorist's judgment has consequences. If investigators report that requisite data cannot be obtained, they have questioned the theorist's judgment.

Empirical applicability, reliability, and comparability. If practical considerations could be ignored, the theorist should prefer to gather data for tests. Published data limit alternatives in the stipulation of referential formulas, and a theorist is likely to have more confidence in the reliability and comparability of data gathered at his instruction. In short, by stipulating the use of published data in tests, the theorist maximizes testability but at the expense of using questionable data; conversely, if he stipulates that data are to be gathered, testability may be minimized. The dilemma is inescapable; it is simply one agony of theory construction. In any case, agonized or not, the theorist must decide.

As recognized in Chapter 2, published data are not suitable for tests of all sociological theories, especially those pertaining to social psychology, small groups, collective behavior, social movements, and the sociology of knowledge. Moreover, whatever the area, even if relevant published data are available, the theorist may doubt their reliability and comparability so much that he declines to stipulate their use. Nonetheless, sociologists who formulate "macro" theories are not competent without knowledge of published data. The time and cost required to gather data on large social units are so great that, typically, a "macro" theory will remain untested unless published data are utilized.

Rarely if ever will a theorist find "ideal" published data, equivalent to that which would be gathered if resources permitted. But just as only the theorist can judge the relevance of published data, so only he can judge the discrepancy between published data and his conception of ideal data. For example, suppose the theorist contemplates application of the ARCOD referential formula to census data. A competent theorist will recognize that census agencies do not tabulate data on *specific* occupations, which number in the thousands. Rather, some specific occupations are grouped into categories (typically less than 500), and census data are reported only for the categories. Now if practical consideration could be ignored, the theorists would stipulate that data be gathered on specific occupations in each country; as such, the data would be different from corresponding occupational data in a census report. However, only the theorists can judge the consequences of that discrepancy.

The testability of a theory should be infinite; hence, a theorist should not stipulate that a *particular* body of data (such as census statistics on Australia, 1961) be used in a test, for the theory would become untestable once those particular tests have been conducted. Throughout his life, of course, the theorist could designate "suitable" data for tests of the theory, but the tests would be personal and cease on his death. If possibilities for tests are to be infinite, the theorist must speak from the grave; and, to that end, he designates a kind of published data (such as census statistics on occupations), which amounts to an assertion that such data have been and will continue to be published.

In addition to stipulating formulas, the theorist should give detailed instructions as to how the data are to be gathered or otherwise obtained. Those instructions amount to an assertion that such data will be reliable and comparable, but the "truth" of that assertion is forever unknowable. As for comparability, there is only one consideration: sets of data are comparable if gathered by the same procedure. However, given any body of data, all one can say is that the data *ostensibly* were gathered in accordance with a particular procedure, and that qualification admits uncertainty.

As for the reliability of data, doubts are inescapable. To illustrate, suppose a theorist stipulates that ARCOD is to be used in tests and provides detailed instructions for gathering data on occupations. Now suppose that an investigator reports the following occupational data on a particular population: farmers, 330; proprietors, 5; sales clerks, 3; physicians, 2; nurses, 3; ministers, 2; farm laborers, 54. How could one possibly know that the data are "reliable"? There is no way that it can be known with certainty. If another set of occupational data should be gathered subsequently by a second investigator, any discrepancy could be attributed to events between the first and second investigation. But suppose that the investigations were conducted simultaneously (though independently) and the two sets of data agree. Would that outcome prove that either set is reliable? Not at all, for it is conceivable that still a third investigator would have reported divergent data. So the occupations and the number in each cannot be known with certainty, and the same is true of any data, sociological or otherwise. To be sure, when data reported by independent investigators do agree, their credibility is enhanced; but, regardless of the number of investigators or congruence in their reports, doubts never are eliminated. For that matter, confidence in data through "repeated observations" has limited application in sociology, where practical considerations often preclude comparisons of independent sets of data.

The empirical applicability of a referential formula should be assessed in two steps. In the first step the theorist instructs two or more investigators to gather sets of data (independently) by the procedure stipulated in the explication of the referential formula. Should the sets of data diverge substantially, the theorist should revise the procedure or his explication of it. In the second step, the theorist instructs two or more investigators to apply the referential formula to the same *body* of data; and, if the computed referents diverge substantially, the explication of the referential formula should be revised.

Unless a theorist has confidence in the empirical applicability of the referential formulas, he will be haunted by doubts in the interpretation of tests. Should tests be undertaken without preliminary research on empirical applicability and the results indicate negligible predictive accuracy, the theorist would not know whether the results are due to insufficient empirical applicability or to some other feature of the theory. Even if he suspects insufficient empirical applicability, he would not know whether the insufficiency is due to the procedure for obtaining data or to an inadequate explication of the formulas. But no assessment of empirical applicability, however extensive and thorough, can eliminate uncertainty, nor can it be done without an agonizing judgment by the theorist. Regardless of the procedure for gathering data and the explication of a referential formula, some divergence in the referents computed by independent investigators is a virtual certainty. Only the theorist can judge what is "excessive" divergence, and it is not a judgment without consequence. To the extent that he tolerates divergence, the theorist settles for less than maximum predictive accuracy.

The empirical applicability of a referential formula cannot be known with certainty. Regardless of agreement in the referents computed by independent investigators, such comparisons are always finite, that is, limited to particular investigators and particular units. One cannot know what results would have obtained had other investigators gathered the data and/or applied the referential formula, nor what the results would have been had the same investigators computed referents for other units. So the dilemma of induction alone precludes certainty in the assessment of empirical applicability. Accordingly, however compelling the evidence may be, the theorist can only assert that a referential formula is sufficiently empirically applicable, and that assertion is part of the theory.

As we have seen, the idea of empirical applicability touches on two conventional notions pertaining to data—reliability and comparability. Both notions are crucial in the case of published data, such as official reports on crime, census figures, and vital statistics. The credibility of such data is debatable, and a brief consideration of census reports will reveal the major complexities. The only truly satisfactory way to pass judgment on the reliability of census data is to arrange (beforehand) for an unofficial and independent survey concurrent with the official enumeration. But agreement in the two sets of returns (official and unofficial) would not be conclusive evidence that either set is reliable, nor could an assessment in one particular case (for example, the U. S. census, 1970) be generalized to all censuses, past and future. The problem is, of course, largely academic, for if resources permitted extensive assessments of published data, the same resources could be used to gather data.

There is only one solution: however intuitive his assessment may be, the theorists must judge the reliability of published data. If he stipulates the use of a particular *kind* of published data, he asserts that the data are sufficiently reliable. Viewed that way, who else but the theorist should make the judgment? Certainly, the judgment is difficult, but the theorist enjoys one advantage— he need only assert that a particular kind of data is *sufficiently* reliable.

A theorist may qualify his designation of published data in an attempt to maximize reliability; for instance, he may exclude census data prior to the twentieth century and/or prior to the third national census for any country. Any such qualification reduces the testability of the theory, but reliance on published data restricts testability anyway, for the theorist does not assert that published data are available for all units. However, any qualification is justified if it does not necessarily result in finite tests. Nonetheless, regardless of the qualification, a theorist can only assert that published data of a particular kind are sufficiently reliable, and the truth of that assertion cannot be known with certainty. Even if an instance of a particular kind of published data should appear incredible, that alone would not refute the assertion that such published data *in general* are sufficiently reliable.

The commentary on published data applies in much the same way to data *gathered* in accordance with the theorist's instructions. Briefly, when stipulating a procedure for gathering data, the theorist believes that the data so gathered will be sufficiently reliable, but the belief is actually only an assertion. However, as regards comparability, there is a distinct difference between published data and that gathered as instructed by the theorist. When two sets of data have been gathered in accordance with a specific procedure, they are ostensibly comparable. Hence, in stipulating a procedure for gathering data, the theorist has done all that can be done to insure comparability. Published data, on the other hand, may have not been gathered and compiled, even ostensibly, by exactly the same procedure. To illustrate, in contemplating international census data on occupations, the theorist should recognize that census practices vary among countries. If he considers the comparability as insufficient, then he will decline to stipulate that a referential formula be applied to that kind of data; but only the theorist can judge "sufficient" comparability. His judgment may be largely intuitive, and his burden is made even greater by the necessity to consider future comparability. No one can readily question his judgment (it is an assertion about something unknowable), but the judgment partially determines the predictive accuracy of the theory.

Empirical applicability and the choice of referential formulas.   Some of the foregoing statements may erroneously suggest that a theorist selects referential formulas and then considers data. A theorist may have a referential formula in mind at the outset, but he must be prepared to alter it in light of judgments about data. So there is a constant interplay between thinking about referential formulas and judgments pertaining to data, especially when contemplating published data. For instance, he is likely to find that requisite data have not been published. If so, unless the theorist is determined to add one more untestable theory to sociology's overstocked supply, he will abandon the referential formula and seek an alternative. Even if requisite data are published, the theorist may doubt their comparability and reliability to the point that he abandons the referential formula.

In addition to his concern with reliability, the theorist should devise referential formulas with a view to enhancing the comparability of published data. As an illustration, reconsider the ARCOD formula. It may appear is inconsistent with the definition of occupational differentiation, for ARCOD referents reflect only the uniformity of distribution, not the *number* of occupations. Why the discrepancy? The answer is revealed by considering an alternative:

$$IROD = N((1-[\Sigma X^2/(\Sigma X)^2])/[1-(1/N)],$$

Where N is the number of occupations and X the number of individuals in a given occupation. IROD is an abbreviation for *ideal referential of occupational differentiation*, ideal in that IROD referents reflect both distribution and number of occupations. Accordingly, given the resources to gather occupational data, IROD would be preferable to ARCOD. Of course, there is still another alternative—apply IROD to census data—and that alternative may appear to be all the more desirable since the number of occupational categories tabulated in census reports varies internationally. However, the variation probably reflects incomparable data. Specifically, as observed earlier, most census agencies tabulate data on occupational categories rather than specific occupations, and the number of categories varies internationally only because of arbitrary and divergent census practices. Hence application of IROD to census data would be questionable, and ARCOD is preferred precisely because it does not reflect the number of occupational categories. There is no assurance that greater predictive accuracy is realized if ARCOD is used rather than IROD (as applied to census occupations), but the choice illustrates the kind of judgments required of a theorist.

Choosing among alternative referential formulas is essentially no different when the theorist stipulates that data are to be gathered for tests. The theorist can consider a range of alternative referential formulas, but resources are never unlimited. Consequently, even if the theorist has an ideal referential formula in mind, once he commences to stipulate a procedure for gathering data he may realize that (1) practical considerations preclude application of the procedure and/or (2) the procedure is not likely to produce sufficiently reliable data. In either case, an alternative procedure and/or referential formula should be sought. Although the referential formula and the procedure for gathering the requisite data are distinct, abandoning one procedure for another may require alteration of the formula itself.

Ordinarily, the theorist will choose among alternative referential formulas, but if he cannot choose (that is, anticipate greater predictive accuracy in the use of one rather than the other), all alternatives can be incorporated in the theory. Thus, rather than choose between ARCOD and IROD, both referentials could be linked to the concept occupational differentiation in the intrinsic part of the theory. When two or more referentials are linked to one concept, the theory is still testable, and the possibility of negative results in *all* tests is not precluded.

The stipulation of alternative referential formulas is quite different from a common practice in sociology, whereby theorists specify no formulas or measures, leaving it to the investigators to discover the "appropriate indicators." If "indicators" are appropriate only when tests are positive, the logical possibility of negative evidence for the theory is precluded. The practice is alien to the proposed mode of theory construction. When a theorist stipulates alternative referentials (that is, when he links more than one referential to a concept), he asserts in effect that some or all of the alternatives will yield a sufficient degree of predictive accuracy. If one yields much greater predictive accuracy than the others, the theorist can revise the theory by eliminating alternative referentials. He would do so by induction rather than deduction, but surely induction can play a constructive role in the formulation or revision of theories (see especially Znaniecki, 1934).

Empirical applicability and the notion of concordance.

According to one common conception, a test is simply a matter of determining the consistency or inconsistency of the theory with "facts." The conception is an oversimplification in that several crucial questions are ignored. What is a fact? How is the *relevance* of a particular fact determined? Who makes the decision as to relevance? How is the consistency of a relevant fact and the theory determined? Who makes that determination? In what sense (if any) are there doubts about facts?

If tests are to be public rather than private, the central notion is not experience. Individuals may report their experiences as "facts," and others may be willing to accept their reports as "factual." Such reports make the tests public; but, nonetheless, the very notion of "reports of experience" is debatable. We shall never know in any direct sense what others experience, for reports of experience cannot be equated with experience itself. Indeed, if science is a public enterprise, then reports of experience should be regarded as assertions or allegations. Accordingly, a theory is consistent or inconsistent not with "facts" but rather, with assertions. Stated another way, a theory is consistent or inconsistent with facts only insofar as one regards assertions as facts in themselves.

The three notions—referential, referential formula, and referent—are entirely consistent with the commentary on "facts" and assertions. A referent (that is, a particular numerical value) may be taken as "fact," but it actually represents an assertion. The assertion is that a particular referential formula has been applied to an instance of the class of things or events designated by the unit term. Thus, if a RRCBR referent of 16.2 is reported for the United States over the years 1968-1972, there are two assertions: (1) that a particular body of data has been gathered or otherwise obtained in accordance with a stipulated procedure and (2) the RRCBR formula has been applied to those data. Since no one has or ever will experience a resident crude birth rate as sense data, how can one be certain that the value of 16.2 is correct, valid, and reliable? The answer is simple—no one can, which is to say that doubts about referents are inevitable.

Of course, one may have fewer doubts about some referents than others, and one goal of science is to reduce such doubts by "repeated observations" on the same unit. The notion of empirical applicability is entirely consistent with the strategy of repeated observations. In stipulating a referential formula and instructions for obtaining requisite data, the theorist asserts that should two independent investigators apply the referential formula to a particular unit, the two referents will agree substantially.

If two individuals apply a particular referential formula to each of a series of populations, there are several possible outcomes, three of which are illustrated in Table 5-1. For illustrative simplicity, only two investigators and six units (populations) are shown, but subsequent comments are not contingent on the number of investigators or populations. In actual research, teams of investigators might be involved rather than individuals, and each referent might be computed by different individuals or teams; but none of those considerations are relevant.

In an actual situation the theorists would hope for outcome "A" in Table 5-1. Observe that there is substantial agreement (as such things go in sociology) between the two sets of referents, one reported by investigator I and the other by investigator II. Accordingly, given a theory in which IROD is a constituent referential, tests results would not differ appreciably regardless of which set of referents (I or II in outcome "A") is used. However, the significance of an outcome like "A" is not merely consistency in two particular tests. Agreement between referents in one instance furthers confidence in the *general* empirical applicability of the referential formula.

Table 5-1
HYPOTHETICAL IROD REFERENTS

| | Outcomes (A,B,C,) and Investigators (I,II) | | | | | |
| | A | | B | | C | |
Populations	I	II	I	II	I	II
P-1	754	749	754	85	754	370
P-2	342	348	342	160	342	167
P-3	1,284	1,300	1,284	50	1,284	638
P-4	47	45	47	1,400	47	24
P-5	427	427	427	140	427	217

Tests of a theory should be public and impersonal, that is, the results should not be contingent on particular investigators. But should the referents diverge substantially, as in outcome "B," the test results could not be consistent. Moreover, in such a case the theorist cannot somehow know which referents are reliable. Should test results be taken as the criterion (the referents are reliable if the test is positive), then the predictive accuracy of the theory could never be minimal. Accordingly, given an outcome resembling "B" in Table 5-1, the theorists would have no alternative but to conduct two tests, one based on referents I and the other on referents II. Both tests could not be positive. The relation between the two sets of referents is inverse; consequently, if one set of referents is directly related to another variable (as predicted by the theory), that relation will not hold between the same variable and the other set. But outcome "B" would have far greater implications than two inconsistent test results; it would suggest that future tests will be markedly inconsistent. Accordingly, given an outcome like "B," the theorist should consider altering the referential formula so as to realize greater empirical applicability.[15]

While the theorist hopes for an outcome like "A," he may settle for less—specifically, an outcome like "C." In any case, the contrast introduces a terminological consideration. Up to this point empirical applicability has been described as a matter of agreement between referents, but there are two kinds of agreement—absolute and relative. Outcome "A" approximates maximum *absolute* agreement between sets of referents (each pair of referents differs very little), and outcome "C" illustrates substantial *relative* agreement. In the case of "C," while there are great differences between the referents, the two sets are directly correlated.[16] Consequently, if a theory is to be judged (at least initially) by predictions of *ordinal* differences between units rather than absolute values, even in outcome "C" test results would not differ appreciably for the two sets of referents. So, the theorist hopes to realize a substantial *concordance* between sets of referents, that is, *either* absolute *or* relative agreement.

The notion of "relative agreement" is still another reason why a referential formula is not identified as a "measure." Rightly or wrongly, the common conception of a measure emphasizes precision, which is traditionally reckoned by absolute rather than relative agreement. But a theorist may be willing to use a referential formula even if only substantial relative agreement among referents is realized. He should never be truly content with less than absolute agreement; but he is justified (at least temporarily) in settling for less, especially if he is unable to modify the referential formula, his explication of it, or instructions pertaining to data.

While substantial absolute agreement commonly is required in the physical sciences, in sociology the conditions of work and especially the quality of data are quite different. As such, were absolute agreement demanded, it could have a crippling effect on theory construction. Certainly any theory is better than none, even if the constituent referential formulas promise only relative agreement. However, a theorist cannot settle for relative agreement, even temporarily, unless he is content with prediction of ordinal differences (that is, a referent for a particular unit will be greater than the referent for another particular unit, without regard to the magnitude of the difference).

The idea of ordinal prediction is controversial, for some sociologists demand that a theory generate predictions about the absolute magnitude of values (see Blalock, 1964). That demand is simply alien to the present condition of work in the field, and it is an excellent way to reject theories prematurely. If a theory generates even fairly accurate ordinal predictions, it is better than no theory or an untestable theory.

Regardless of the kind of concordance sought, an assessment of empirical applicability is complicated and fraught with uncertainties. When data have been gathered and substantial concordance realized, the theorist has reason to be satisfied with his explication of the referential formula and to have confidence in the comparability and reliability of the data. But if the referential formula has been applied to published data, substantial concordance *does not* substantiate the theorist's assertion that such data are sufficiently reliable and comparable. Nonetheless, the theorist should be concerned with empirical applicability even in the case of published data. If substantial concordance is not realized, there are three possibilities: (1) the investigators have not applied the referential formula to the same published data and/or (2) they have not applied the formula in the same way (that is, errors have been made in applying the formula). In either case, insufficient concordance between referents serves notice that the theorist should modify the identification of requisite data and/or revise his explication of the referential formula.

Regardless of the kind or degree of concordance realized, note again that a theorist can never be certain about empirical applicability. Specifically, an outcome resembling "A" or "C" in Table 5-1 is always relative to particular investigators and populations. Of course, when resources permit, each test can be based on alternative sets of referents, meaning that both the predictive accuracy of a theory and the empirical applicability of the referential formulas are assessed simultaneously. But not even absolute agreement between sets of referents is conclusive evidence of reliable data. For that matter, research resources are never unlimited, and it may be necessary to accept any set of referents as credible without evidence of concordance. Such acceptance is consistent with the procedure of the most advanced sciences, where the use of a particular method furthers the credibility of data without repeated observations.

While limited resources may preclude assessment of empirical applicability, without it the theorist is likely to err when interpreting test results. Negative tests may be due solely to insufficient empirical applicability; if so, the theorist is likely to make one of three errors. He may (1) reject the theory as a whole, (2) modify components of the theory other than the referential formulas, or (3) modify the wrong referential formula.

The dilemma of induction is not the only complication in assessing empirical applicability. Rarely, if ever, will maximum concordance be realized, and only the theorist should judge what is sufficient. He will recognize a relation between concordance and predictive accuracy, but the relation is not a simple one. Maximum concordance does not assure maximum predictive accuracy; rather, the former is only a necessary condition for the latter. Moreover, even if the theorist has a precise idea of the minimum acceptable level of predictive accuracy, he may be uncertain as to exactly how much concordance is a necessary condition for that level. Nevertheless, judge he must.

Additional considerations. Concordance is only one of four aspects of empirical applicability. The other three are feasibility, intelligibility, and universality.

Unless the theorist is sensitive to research resources, investigators may report that they cannot apply his referential formulas. When a theorist stipulates application of a referential formula to a particular kind of published data, application of the formula is *feasible* only if such data actually are published. Further, should the data cease to be published, sooner or later investigators will report (in effect) that application of the referential formula is no longer feasible.

The theorist's judgment of resources is no less critical in stipulating a procedure for gathering data. If he overestimates the field's resources, investigators will report that practical considerations preclude gathering the requisite data. But resources are not the only consideration. The theorist may misjudge the conditions of work in his field. Taking an extreme example, should the theorist stipulate that a count by direct observation be made of criminal acts in social units, it would be impossible for investigators to comply. Obviously, individuals commonly attempt to conceal criminal acts, and regardless of resources or the definition of a criminal act, an "observational" procedure is not feasible.

Feasibility is a matter of degree, meaning the relative ease with which the referential formula can be applied; as such, it is likely to be reflected in the number of tests. Specifically, if tests of a theory are rare, the theorist's judgment of resources and/or conditions of work is questionable. Tests also reflect prevailing interests within the field, something beyond the theorist's control. Nonetheless, if he is practical and wishes to maximize the testability of his theories, he will recognize that science is a human enterprise governed to some extent by interests.

Even if a theorist does not misjudge research resources, an ambigous exposition of a referential formula negates its empirical applicability; that is, investigators will report that they cannot understand the formula and/or related procedure for obtaining requisite data. In other words, they find the referential formula to be *unintelligible.*

Before considering another criterion of empirical applicability, a brief digression is necessary. Far from being concerned with feasibility, intelligibility, and concordance, the typical sociological theorist does not stipulate referential formulas or "measures" in a theory, not even a theory that deals with quantitative properties. The only possible rationale for that practice (short of a candid admission of a lack of concern with testability) is the belief that investigators can formulate the "operational" definitions necessary for tests of a theory. Since sociologists are likely to equate referential formulas with operational definitions and construe the present emphasis on empirical applicability as only another plea for operationalism, a comment on the subject is in order.

Suppose that a theorist uses the term "upper-class families" but offers no definition of it. Then suppose further that in testing the theory an investigator defines the term by reference to a "disposable income in the previous calendar year of over 30,000

United States dollars." Similar definitions are frequently employed in sociological research and typically designated as "operational definitions."

In actual practice, the use of so-called operational definitions in tests is contrary to the principle of empirical applicability. When an investigator must define component terms of a theory to conduct a test, the definitions do not make the terms empirically applicable *in the context of the theory*. There is no assurance that other investigators will employ the same definitions and, if their definitions diverge, the terms in question are not empirically applicable. To be sure, should one investigator define the term "upper-class family" by reference to income in United States dollars and another define the same term by reference to occupations, then the extension of those two definitions to a particular population (for example a metropolitan area) might result in the identification of the same families as "upper-class." But that possibility is remote, and to the extent that families do not qualify as upper-class under both definitions, the term is not empirically applicable.

Whatever the principles of operationalism may be, the notion of an operational definition has not facilitated tests of sociological theories. On the contrary, theorists use the notion as a license to formulate untestable theories and to rationalize it by the tacit argument that someone else will somehow discover "appropriate" operational definitions of the constituent terms.[17] That practice is flatly rejected. When someone other than the theorists must define the intrinsic terms to conduct tests, the theory is incomplete. So if only to discourage a common practice in sociology, referential formulas are not identified as "operational definitions." To say it again, referential formulas are components of a theory, not additions to it made by someone other than the theorist.

As commonly used by sociologists, so-called operational definitions are contrary to one particular criterion of empirical applicability—universality. If a definition is applicable only in a particular social unit or culture, it lacks universality, and that parochial quality minimizes the testability of theories in which the definition is employed. The point is that so called operational definitions in the sociological literature typically promise very little in the way of universal application, and that is true whether they are formulated by theorists or investigators.

Universality is important in stipulating referential formulas, for it determines testability, but testability may be infinite even though the constituent referential formulas are not universally applicable. To illustrate, suppose the following definition of "upper-class families" is used in a theory: families with a disposable income of 30,000 or more United States dollars during the calendar year of 1970. Given that definition, tests of the theory would be finite; however, without the temporal limitation (1970) testability could be infinite, even though tests are limited to social units in which income is reckoned in United States dollars. Now the theorist may argue otherwise (that is, he may deny the limitation), and in a sense any definition without temporal limits is universally applicable. But should the definition be applied outside the United States, then investigators will report there are few if any "upper-class families"; and if tests of the theory are undertaken in countries outside the United States, they are likely to reveal minimal predictive accuracy. Further, even if tests are limited to the United States, the predictive accuracy of the theory in the distant future would be questionable, for the income structure of the country could change to the point where there are few if any "upper-class families" (as defined) or where virtually every family qualifies.

The foregoing should make it clear that it is difficult to assess the universality of a referential formula or any definition, and arguments in a particular case are sterile. When the theorist insists that a particular referential formula is empirically applicable and universally so, then it is his judgment. But the soundness of his judgment is likely to be reflected in the predictive accuracy of his theory. If a theorist is genuinely concerned with maximizing predictive accuracy, he will readily entertain any suggestion that a referential formula is not universally applicable; indeed, he should be willing to make the limit explicit, for instance to populations where family income is reckoned in United States dollars. For that matter, a theory can be limited to a particular social unit, such as the United States. There could be such a theory, provided that no temporal limitations are stipulated (only for 1970) and the unit term is any country known as the United States of America. Of course, limiting a theory to a particular social unit is questionable, for the limitation does not insure predictive accuracy of the theory in future tests.

Having recognized the complexities entailed in assessing universality, the previous condemnation of so-called operational definitions is now qualified. Even though limited to a particular social unit, a definition can be used in theory construction provided it does not eliminate (in principle) the possibility of infinite tests. However, in using such a definition, the theorist should entertain doubts about the predictive accuracy of the theory in the distant future.

However the constituent symbols of a referential formula be defined, its universality is difficult to reckon. For example, when a theorist stipulates that a referential formula should be applied to census data, he has limited applicability to countries in which censuses are taken. So testability is reduced, but the amount would be difficult to estimate.

Finally, purely linguistic considerations do not necessarily limit universality. When a theory is stated in the English language, it does not follow that the referential formulas are applicable only in English-speaking countries. Even though translation of terms would be required, substantial concordance could be realized in any population, but it may not be realized if the theorist uses terms in one language that have no equivalent in others. However, a translation of terms is necessary only when the theorist makes reference to the natural language of a population.

As stated in Chapter 2, cultural and historical relativity creates a difficult terminological problem for sociology. One solution is what Blalock (1969) and others designate as "auxiliary" theories. The idea is that for each theory there are auxiliary theories, and the latter are used in tests of the former. Thus, to illustrate, suppose that the following phrase is used in a definition in the extrinsic part of a theory: "disposable income in the previous calendar year of more than 30,000 United States dollars *or its equivalent.*" The last three words expand universality enormously, but they raise a difficult question: How is equivalent income to be reckoned? An answer would take the form of an auxiliary theory, for equivalence is surely not a matter of "definition." Nonetheless, the notion of an auxiliary theory is rejected, as it represents a tacit approval of traditional practices in sociology. If the notion is not rejected, theorists will abuse it in precisely the way they have abused the notion of an operational definition or an indicator, that is, they will construct untestable theories and leave it to others to formulate "appropriate" auxiliary theories.[18] Finally, one may think of a referential formula (including the procedure for gathering or obtaining data) as an auxiliary theory; but the identification is misleading, for a referential formula is part of a theory, not "auxiliary," and the theory would be incomplete without it.

The achievement of empirical applicability.   In stipulating a referential formula, the theorist does more than assert empirical applicability. His explication of the formula is an attempt at a self-fulfilling prophecy; that is, he makes every effort to insure that the assertion of empirical applicability will be substantiated.

In designating a kind of published data, except for illustrative purposes, the theorists should not refer to any particular social unit, year, or publication. Rather, he should describe the data in general terms, and the description must be such that independent investigators will agree in identifying the relevant data for a particular social unit (such as Canada, 1961). To illustrate, in the case of the ARCOD referential formula, the theorist would stipulate (in the extrinsic part of the theory) that the formula applies to census data on occupations. But there are various kinds of such data (for example, in one series of reports occupation may be tabulated by age, sex, income, marital status, or education), and the theorist should specify that only data on the number of individuals in each occupational category are relevant.

Given explications of the ARCOD and other referential formulas in the extrinsic part of a theory, an investigator could select a tentative universe of countries for a test, for instance, all countries in which a national census was taken during the period 1969-1971. Several countries might be subsequently excluded, as the census reports or other publications for some countries will not include the data required for the application of the ARCOD referential formula and all other constituent formulas. However, if the requisite published data are adequately described and investigators have access to the same publications for the same period (1969-1971), they should agree in their selection of particular units (countries) for tests of the theory. While agreement in selection is desirable, it is not absolutely essential; but substantial concordance in the referents computed by independent investigators for a particular unit would be essential. After all, if they have applied the same referential formula to the same units (such as the United States, 1960; Canada, 1961; Sweden, 1960), substantial concordance should be realized. Otherwise, some investigators have erred in applying the formula and/or they have not applied it to the same data for each unit.

A detailed explication of a referential formula can reduce errors in the computational procedure, but no explication assures that the investigators will apply a formula to the same published data for a particular social unit. The primary reason is that similar data may appear in two or more tables, sometimes in different volumes. For example, in reporting occupational data, national census agencies often include at least two tables in a publication, one for a short list of "gross" occupational categories and the other for a more detailed list. The theorist would prefer to use data on the *same* detailed list for each country, but census data on occupations are not internationally comparable; that is, the occupational titles vary from country to country. Of course, the theorist could provide a list of detailed occupations and stipulate that census data are to be used only if tabulated for that list; yet that restriction would drastically reduce testability. So the theorist is likely to accept some degree of incomparability. He must, however, confront the question: Given the possibility of two or more similar tables in a publication, how can the instructions be worded so that investigators will select the same table? The question is difficult primarily because the theorist cannot specify any particular number of occupations, let alone particular occupational titles, without being arbitrary and reducing testability.

The theorist's instructions should not only lead investigators to select the same table as a source for occupational data on a particular unit but also to anticipate questions as to alternative uses of the data. Some examples follow. Should the ARCOD formula be applied to data only for males? Must data on individuals below a certain age be excluded? If so, what is the minimum age level? Should data on the occupations of the unemployed be excluded or included? What should be done with such categories as "occupation unknown," "housewife," "retired," and "all other occupations"? The theorist speaks to such questions with a view to minimizing the decisions of investigators and to enhancing the comparability of published data without seriously reducing testability. One possibility in the case of ARCOD is to stipulate some minimum age, but such judgments are not obvious. For example, if occupational data are tabulated for individuals above six years of age in one country but the minimum is thirteen in another, both ages could be "realistic" even though not comparable; that is, in one country it could be that very few individuals below the age of fourteen have an occupation. So, conceivably, the theorists may decide not to stipulate a uniform minimum age for all countries.

Still other aspects of comparability are by no means conspicuous. For one thing, census practices pertaining to the classification of women as to economic activity vary internationally, and with that variation in mind the theorists may stipulate that census data be restricted to males.

As the foregoing suggests, a theorist must make numerous decisions in designating published data, and each decision could affect the predictive accuracy of his theory. Nonetheless, however uncertain and reluctant the theorist may be, judge he must; for the requisite decisions are not "technical" considerations that can be left to investigators.

The theorist's only advantage is that he can stipulate more than one kind of data for any referential formula. To illustrate, he could stipulate that $ARCOD_1$ referents are to be based on census data on males, while $ARCOD_2$ referents are to be based on census data for both sexes; so the referential formula would be the same in both cases, but the data would be different. The stipulation of alternative data in no way assures positive test results; that is, predictive accuracy could be minimal in all tests, whether based on $ARCOD_1$ or $ARCOD_2$ referents. Nonetheless, unlike the typical sociological theory, there would be no uncertainty as to relevant evidence. In the illustrative case, each test would be based on either $ARCOD_1$ or $ARCOD_2$. Of course, investigators should conduct two tests if reasible—one based on $ARCOD_1$ referents and the other on $ARCOD_2$ referents. After a series of such "multiple" tests, it may be obvious that substantial predictive accuracy is realized only for a particular kind of data; if so, the theorist would revise the theory and stipulate only one version of the referential formula. He would do so by induction but, again, is justified in the formulation or modification of a theory.

If data are to be gathered for tests, the explication of referential formulas is complicated enormously. The theorist must have confidence in his definition of the unit term and the related procedure for delimitation. Obviously, data gathered by independent investigators are not likely to be congruent unless the investigators agree in the identification and delimitation of units, but agreement is only a necessary condition for congruent data. To achieve congruence the theorist must stipulate a step-by-step procedure for gathering data once a unit has been delimited. The stipulation will reflect the theorist's assessment of research resources, but not even unlimited resources would solve all problems. Should the theorist describe the procedure in terms that are peculiar to a particular social unit, he reduces the universality of application. Further, unless each procedural step is described in detail, investigators will use their own discretion, and to that extent congruent data become problematical.

As a brief illustration, consider the IROD referential formula. Suppose the unit term is "sustenance associations in a country," meaning associations in which all members earn at least some of their livelihood as members (companies or corporations would qualify as instances). The extrinsic part of the theory would define a sustenance association and stipulate criteria of membership. Further, if practical considerations preclude a comparison of all sustenance associations in each country, the theorist should stipulate a sampling procedure.

The theorist's treatment of the unit term is preliminary to his specification of the IROD formula and stipulation of a procedure for gathering data. The specification may be brief, but the stipulation of a procedure must be detailed. Initially, the theorist should decide whether investigators are to identify the occupation of each member; if not, a sampling procedure must be stipulated. The next consideration would be the identification of a member's occupation; to that end, the theorist should provide an elaborate definition of an occupation, but he should not presume that a definition alone is sufficient. There are various alternatives in the way of identifying an individual's occupation, such as questioning the individual, inspecting records maintained by the association, asking other members of the association to identify the individual's occupation, or observing his activities. In choosing among alternatives, the theorist should consider both feasibility and universality. To illustrate, the identification of occupations by observation may be judged as too time consuming; and, as for universality, it is improbable that all sustenance associations maintain occupational records. In any case, whatever the choice among alternatives, the procedure should be described in detail. Rather than merely stipulate that each individual is to identify his own occupation, the theorist should state the specific questions to be asked of each individual. Moreover, he should anticipate problems (for example, individuals who may have two occupations) and formulate special rules to resolve them.

In stipulating a procedure for gathering data, the theorist has advantages in the way of resources and alternatives. As for resources, he is free to draw on the heritage of the field, such as conventional techniques for sampling. If such a technique is incorporated in the stipulated procedure, the theorist need only refer to a publication where the technique is treated at length. However, only the theorist should judge the universality and feasibility of alternative sampling techniques, and he is free to depart from conventions.

When a theorist cannot decide between alternatives at a particular procedural step, such as the choice between asking individuals to identify their occupation or arriving at the identification by inspection of records, he can leave the choice to investigators, assuming that independent investigators will agree in their choices or that the difference between alternatives is inconsequential. Finally, if his doubts cannot be resolved, the theorist may stipulate that alternative kinds of data be gathered and that the referential formula be applied to each kind. Thus, the theorist could stipulate that two IROD referents be computed for each sustenance association; in tests each set of referents would be considered separately.

Temporal Quantifiers of Substantive Terms

Any substantive term (construct, concept, or referential) designates a property, but the designation is incomplete without reference to time. To illustrate, suppose a theorist identifies the "division of labor" as a construct. When he speaks of the division of labor in a social unit, the reference pertains either to the property at a point in time or to change. Accordingly, to avoid ambiguity, no substantive term should be used in theory construction without stipulating the temporal dimension.

Change in a property can be reckoned absolutely or proportionately, and in either case some period is implied. Hence, taking the "division of labor" as a term, we have the following possibilities: absolute change in the division of labor over T_{0-5}, proportionate change in the division of labor over T_{0-5}, absolute change in the division of labor over T_{0-10}, proportionate change in the division of labor over T_{0-10}. Each phrase qualifies the term "division of labor," and the entire phrase constitutes a substantive term, for without the qualification reference to the division of labor in a social unit would be ambiguous.

Note that each temporal quantifier (for instance, T_{0-5}) stipulates the length of a period. The numbers themselves (0-5) would refer to a unit of time (days, months, years) as stipulated in the extrinsic part of the theory. In the present illustrations the units are thought of as years, but T_{0-5} and T_{0-10} are only two of an infinite series of temporal quantifiers (for example, T_{0-1}, T_{0-12}) that a theorist could consider.

No temporal quantifier refers to an historical period, that is, T_{0-5} would not refer to any particular years, such as 1960-1965 or 1965-1970. Rather, a temporal quantifier designates an interval of or point in time. For example, if the substantive term is "absolute increase in ARCOD over T_{0-10}," the interval could be 1960-1970 for one unit, 1961-1971 for another, and 1955-1965 for still another. So any ten-year interval is an instance of T_{0-10}.

Temporal quantifiers are especially important in the case of referentials, as it would be ambiguous to stipulate that investigators are to compute change in ARCOD without specifying an interval. Thus, if the temporal quantifier of ARCOD is T_{0-5}, investigators would apply the referential formula to two sets of occupational data on each unit, with the census dates representing (approximately) a five-year interval. In the case of New Zealand, for example, if the census year 1966 is taken as T_0, then 1971 (another census year for that country) would be T_5; but in the case of Japan the census year 1960 could be taken as T_0 and 1965 as T_5. So even though the temporal quantifier is the same for all countries, it need not be the same historical period for all countries. For that matter, although unconventional, the same country could be included more than once in a test. Again assuming that the temporal quantifier is T_{0-5}, in the case of New Zealand several sets of census years would qualify, such as 1951-1956, 1956-1961, 1961-1966, and 1966-1971. So New Zealand could be included at least four times in one test, which is to say treated like four different countries. However, conventionally, each unit is included only once in a test, and the years are approximately the same for all units. Thus, for example, in one test, T_{0-5} for all countries would be some five-year interval during 1965-1975; but the years would not have to be the same for all countries.

If a theorist does not make assertions about absolute or proportionate change, then there are three possibilities in the way of temporal quantifiers, illustrated as follows: T_0, $_{-2}T_2$, or $T_{0,5}$. Again the numbers designate units of time (hours, days, years), and the three possibilities are only illustrative; they could be T_0; $_{-3}T_3$; or $T_{0,1}$. The temporal quantifier T_0 signifies some point in time, but that point need not and will not be the same for all units or all tests. To illustrate, if the temporal quantifier of ARCOD is T_0, then it simply disignates the census year of any country for example, (T_0 for the United States could be 1950, 1960, 1970, and so on).

Census data ordinarily are gathered during some brief interval (for example, April 1-2, 1970), and the same is true of survey data. However, some events are registered continuously (as in the case of vital statistics), so that all of them taken together are thought of as having occurred not at a point in time but during a period (say the calendar year of 1960 or the period 1960-1964). If a referential formula applies to data for a period, the temporal quantifier could be $_0 T_0$ (a period of one year), $_{-2} T_2$ (a period of five years centering on the year of T_0), or $_{-4} T_5$ (a period of ten years). Thus, when the temporal quantifier of RRCBR (referential of the resident crude birth rate, is $_{-2} T_2$, it signifies that the referential formula applies to the number of births *during* a five-year period centering on the year of T_0. As such, in applying the RRCBR formula, the years could be 1968-1972 (inclusive), 1961-1965, or any other five-year period.

Finally, if the unit term is "within a country over time," then the temporal quantifier could be $T_{0,5}$ or $T_{0,10}$. Such a quantifier signifies a property at each of two *or more* points in time. To illustrate, when the temporal quantifier of ARCOD is $T_{0,10}$, it signifies that the ARCOD referential formula should be applied to census data on occupations at two or more points in time with an interval of ten years. In the case of the United States, for example, if 1930 is taken as T_0, then ARCOD referents are to be computed for 1930, 1940, 1950, 1960, and 1970. The illustrations pertain to census data, which are reckoned in years; but temporal quantifiers can designate any unit of time.

The only difference between a temporal quantifier of the form T_x and one of the form $_x T_y$ is that the latter designates a period of time and the former designates a point in time. Thus, suppose the complete substantive term is "absolute increase in RRCBR over $_{-2} T_2 - _8 T_{12}$." As such, the quantifier stipulates that the RRCBR referential formula is to be applied to the number of births during a five-year period and again to the number of births during a subsequent five-year period, with the interval between the two periods being five years. So the two periods could be 1958-1962 and 1968-1972. In such a case T_0 is some point during 1960, and all of the other years follow from that coordinate. Then consider the case where the unit term is "within a country over time" and the substantive term is "RRCBR at $_{-1} T_1, _4 T_6$." The quantifier stipulates that the RRCBR referential formula is to be applied to the number of births during two or more periods, each of three years with a two-year interval between the periods. Accordingly, two periods for a particular unit could be 1959-1961 and 1964-1966.

Both the notational system and the illustrative temporal quantifiers are consistent with the kinds of data commonly used by sociologists. Nonetheless, there are some issues. As suggested previously, a substantive term is incomplete without a temporal quantifier, and that is true for constructs, concepts, or referentials. Temporal quantifiers for constructs and concepts may appear paradoxical, for those two types of terms are not considered as empirically applicable. However, the derivation of statements is more systematic if a temporal quantifier is a part of each substantive term.

The idea that a temporal quantifier must be stipulated for each referential is more controversial, especially since theorists in sociology never make such stipulations. One may argue that the theorist should not be required to make such a judgment.[19] If so, who is to make it? Should the judgments be left to investigators, concordance in the referents are not likely to be realized. For example, if two investigators apply the RRCBR referential formula to a unit without instructions as to the length of the period, the referents they report could differ appreciably. Above all, it is idle to presume that the temporal dimension is of no consequence in assessing the predictive accuracy of a theory. To illustrate, suppose the theorist asserts that for any country an increase in the ARCOD referent is followed by a decrease in the RRCBR referent. Without instructions to the contrary, an investigator could select a country and compute four referents: ARCOD, 1960; ARCOD, 1970; RRCBR, 1959-1961; and RRCBR, 1969-1971. Suppose that ARCOD is greater for 1970 than 1960 but the 1959-1961 RRCBR referent is not lower than the 1969-1971 referent. If so, a theorist cannot dismiss the finding by arguing that the "wrong" intervals were considered. After all, his assertion did not stipulate any particular interval, leaving the choice to the investigator. Now suppose that another investigator computes an ARCOD referent for 1940 and 1950 but then computes an RRCBR referent for 1960 and 1970. One may think that surely the theorist did not mean that an increase in ARCOD will be followed by a decrease in RRCBR twenty years later. But since the theorist did not specify temporal quantifiers, there is no way to know what he meant. The general point is that leaving the stipulation of temporal quantifiers to others makes tests idiosyncratic and jeopardizes the predictive accuracy of the theory.

Relational Terms

Each intrinsic statement asserts an empirical relation between properties of units; hence the statement must include words that denote some particular kind of association. Such words in the proposed scheme are identified as a *relational term*.

Given the present quality of sociological data, all that should be expected of a theory is predictions of ordinal differences between units (a numerical value for one particular unit will be greater than for another). Accordingly, a relational term need only signify the *direction* of the association between properties. With that end in mind, only two relational terms are needed: "greater, greater" and "greater, less."

By way of illustration, consider two contradictory intrinsic statements: (1) Among urban areas in a country, the greater X at T_0, the greater Y at T_0; and (2) Among urban areas in a country, the greater X at T_0, the less Y at T_0. In both statements, X and Y designate properties, and the temporal quantifier (T_0) stipulates that X and Y are to be considered at the same points in time. However, the relational terms differ, which means that the statements assert a different association between X and Y.

Continuing the explication, suppose that X and Y referents have been computed for each of two urban areas, identified as I and II. Suppose further that the X referent for urban area I is greater than the X referent for urban area II. As such, intrinsic statement "1" would lead to the prediction that the Y referent for urban area I is greater than the Y referent for urban area II. But intrinsic statement "2" would lead to an opposite prediction—that the Y referent for urban area I is less than the Y referent for urban area II. Stated in terms of conventional statistics, intrinsic statement "1" anticipates a positive coefficient of correlation between the X and Y referents, while intrinsic statement "2" anticipates a negative coefficient.

An issue.   The proposed relational terms are such that no intrinsic statement permits a prediction as to the absolute magnitude of a referent. To illustrate, suppose the following assertion is derived from a theory: Among countries, the greater ARCOD at T_0, the less the RRCBR during $_{-2}T_2$. Given knowledge of an ARCOD referent for a country, the assertion would not generate a prediction as to the *magnitude* of the RRCBR referent for that country. However, it does permit another kind of prediction: If the ARCOD referent for that one country is greater than the ARCOD referent for another country, then the RRCBR referent for the first country will be less than the RRCBR referent for the second country. Stated more abstractly, the assertion generates predictions about ordinal differences but not about the amount of difference between any two units.

Observe that the relational terms are symmetrical, that is, "the greater X, the less Y," is equivalent to "the greater Y, the less X." The symmetry is controversial. Blalock (1969) and Costner and Leik (1964) reject symmetrical relational terms and demand that theories be stated in a causal language (for instance, X causes Y or Y causes X). Since they make their argument largely in reference to the "logic" of derivation or deduction, the issue is considered later. For the moment, a brief comment will suffice. If a causal language is used in the statement of a theory, it creates problems and issues that Blalock and Costner do not resolve. Then consider the use of conventional statistics (rank-order coefficient of correlation, product-moment coefficient of correlation, chi-square) in tests of a theory. Those techniques do not reveal a "causal relation." What, then, is gained, as Blalock and Costner would have it, by stating a theory in a causal language when that language is alien to the conventional techniques for testing theories? At least the proposed relational terms are consistent with those techniques.

Blalock and Costner are free to use a causal language in promulgating their own theories, but it remains to be seen how they will translate that language into relational terms that are empirically applicable and defend the translation.[20] Certainly they are not likely to defend any translation as long as their conception of causation is incomplete, and some critics will reject their idea that a causal relation can be inferred from synchronic correlations. Accordingly, as the matter now stands, their advocation of a causal language is most debatable.

Notes for Chapter 5

[1] The distinction is close to that recognized by Campbell (1953:290) when he spoke of one group of statements as the "hypotheses" of a theory and another group as the "dictionary."

[2] Dubin (1969) considers *units* to be properties of things rather than things themselves. The difference may appear unimportant, but unless the component assertions of a theory include some term to designate classes of things or events, the assertions are ambiguous. Dubin does not recognize the role of such a term in theory construction. For example, he says: "Gresham's law is made up of three kinds of units—good money, bad money, and a money market" (1969:88). In the present scheme, the "money market" would be a *unit term*, while "good money" and "bad money" would be *substantive terms* (that is, they designate properties of a money market).

[3] The argument is not that units (things or events) have an "essence" that exists independently of their attributes. On the contrary, a thing or event is a configuration of attributes, but that characterization does not preclude a distinction between definitional and contingent properties.

[4] The enphasis on units of the same type precludes assertions of the following form: Any Ux has a greater Z than any Uy. In such an assertion Ux designates one type of unit (for example, an urban area), Uy designates another type of unit (say, a country), and Z designates some quantitative property (population density). In any case, the proposed mode of theory construction is to be used only when the theorist's assertions refer to a quantitative difference among units of the same type. For further commentary on the other kind of assertion (that is, a difference between units that are not of the same type), see Zetterberg, 1965:94-96.

[5] One could argue that "gene" is a unit term for genetics even though it is not empirically applicable. The counterargument is that the unit term is plant or animal rather than gene. However, such arguments presume that a language of theory construction must transcend particular scientific fields, and that presumption is denied.

[6] Note the distinction between empirical applicability and "observability." Investigators never see a "metropolitan region" (or a socially recognized category of individuals for that matter); nonetheless, they may achieve substantial agreement in the delimitation and designation of territorial units as metropolitan regions.

[7] The entire unit term in the assertion is "among associations in a metropolitan area." So the identification and delimitation of a metropolitan area is the first consideration. To simplify illustrations, the presumption is that investigators have been able to agree in their delimitation of a particular metropolitan area.

[8] In defining a unit term the theorist is identifying or constructing a type of population, act, event, or condition; as such, his concern is very much in the mainstream of sociological tradition (see McKinney, 1970 and 1966). However, the sociological literature on typologies or "social types" pays scant attention to empirical applicability. The present emphasis is not a tacit denial that the relation between quantitative properties depends on the definition of units, and predictive accuracy of a theory may be enhanced by altering the definition of a unit term. The theorist may employ the strategy of "analytic induction" (see Znaniecki, 1934) even though the theory deals with quantitative properties. Indeed, some of the objections to analytic induction (Robinson, 1951) are not relevant when the theory deals with quantitative properties. So it is admitted readily that the definition of unit terms is a matter of "theory." Nonetheless, whatever the initial definition of a unit term or whatever the subsequent alteration of that definition, the immediate concern of the theorist is with empirical applicability.

[9] By contrast, Willer and Webster (1970) suggest that sociological terms can be identified as constructs by some absolute, objective criterion.

[10] Some critics may not object that a term itself is never a construct (or a concept), the argument being that those words refer to a precept or conception. This poses no problem. If one chooses to regard the definition of the term as the construct (or the concept), it makes no difference, for the definition is given in the extrinsic part of the theory. As a matter of syntax, what is true of the definition (that is, the conception or notion) is true of the term, and the terms are identified as to type (for example, construct) for a very simple reason—definitions are not literal constituents of empirical assertions. For further commentary on the subject see Caws, 1965:chaps. 5 and 10; Zetterberg, 1965:32, and Brodbeck, 1968.

[11] Willer and Webster (1970) have attempted such a distinction. Their treatment suggests that *concepts* designate (or stand for) "observable" phenomena, while constructs designate "unobservables." As indicated in Chapter 2, the notion of "obsrvability" is scarcely relevant in considering the semantics of sociological terminology. Examples of concepts used by Willer and Webster to the contrary, one does not "observe" an officer or a clerk. The terms may be empirically applicable, but it does not follow that they designate "observables."

[12] Of course, the very use of a term (whether defined or undefined) in a theory assigns some meaning to it (see Northrop, 1947:141), but no one would argue that such meaning amounts to a complete definition. Furthermore, whatever the "relational" meaning of a construct, a theorist may elect

to offer a definition of it in the extrinsic part of the theory. He will not regard the definition as complete, but nonetheless the definition may be distinct from the "relational" meaning. To illustrate, when a theorist asserts that X varies inversely with Y, then the assertion itself assigns some meaning to X (it is something that supposedly varies inversely with Y), but he could give still another incomplete definition of X in the extrinsic part of the theory without any reference to Y. The point is stressed because it could be argued that there is an objective criterion of an incomplete definition: if a term is defined only "implicitly" (that is, by its position in a theory), then the definition is incomplete. That is so, but it does not follow that all "explicit" definitions (those in the extrinsic part of a theory) are complete. So, in brief, there is no objective distinction between a complete explicit definition and an incomplete explicit definition.

[13] The point is lost on Blalock, who attempts to distinguish between a logical or theoretical conception of validity and an "empirical approach to the problem of validity." Consider his statement: "In the empirical usage, validity may refer to the degree to which a given index can predict to some outside criterion" (1968a:13). Blalock ignores the crucial question: Why would a given "index" be expected to be associated with one particular "outside criterion" rather than another? The expectation in any instance implies a theory, but Blalock would have us believe that the choice of "outside" criterion is "technical" or simply "empirical." What has been said of Blalock's treatment of the notion of validity applies to virtually all treatments in the sociological literature (see Zetterberg, 1965:114-123; Selltiz, et al., 1959: 154-166; Dubin, 1969:206-210; Phillips, 1966:160). In all such treatments validity is analyzed as though independent of theory construction. If it be objected that "external" or "empirical" validity is a matter of theory but not "logical" or "internal" validity, the latter conception is not defensible. It suggests that there are rules of syntax by which one can *derive* a formula, measure, operation, or procedure from a definition of a construct or concept. The existence of such rules is denied. Moreover, no theorist is likely to stipulate any formula, measure, procedure, or operation unless he is convinced that it is valid (logically or internally), and debates in any given instance will be sterile, for validity in the sense indicated is clearly a matter of opinion. Above all, it would be ridiculous to reject a theory with impressive predictive accuracy because of someone's opinion about logical or internal validity. For the latest round of sterile debates on validity, see Deutscher, 1969; Ajzen, et al., 1970; and Lastrucci, 1970.

[14] The conception is suggested by Bergmann's statements (1958:50) on the significance of concepts.

[15] In altering a referential formula the theorist may revise the formula itself, change his explication of the formula, and/or stipulate a different procedure for obtaining data.

[16] The notion of relative agreement can be designated as "correlation" (see Robinson, 1957).

[17] Without exception, what is said in this chapter of operational definitions also applies to the notion of "indicators."

[18] "Given a main body of theory, anyone wishing to test this theory may then construct an auxiliary theory containing a whole set of additional assumptions, many of which will be inherently untestable. This auxiliary theory will be specific to the research design, population studied, and measuring instruments used. For example, in one population it may be reasonable to ignore a particular set of disturbing influences that would have to be explicitly considered in another population" (Blalock, 1968a:25). Blalock's strategy would not only free the theorist of any responsibility for constructing a testable theory, but also make the tests (by auxiliary theories) idiosyncratic. Further, the theorist can deny negative evidence by claiming that the "auxiliary theory" was not appropriate.

[19] "Tolerance of ambiguity is as important for creativity in science as it is anywhere else" (Kaplan, 1964:71). It is statements such as Kaplan's that sanction the indifference of sociological theorists to empirical applicability and testability. Kaplan conveniently ignores the point that "ambiguity" should be tolerated only in defining certain constituent terms of a theory (constructs in particular) but not all terms.

[20] Despite his condemnation of symmetrical relational terms in theory construction, Blalock has used such terms in stating "theoretical propositions." Indeed, he occasionally uses the very relational terms proposed here, that is, greater . . . greater or greater . . . less; but he says: "Causal asymmetry is intended in the wording of these propositions" (1967:48). One must surely wonder why Blalock does not simply use the word "causes." In any case, if the words "greater . . . greater" denote causal asymmetry, Blalock cannot use conventional statistical measures of association (for example, *rho, tau, r*) to test his propositions, as those measures do not reveal an asymmetric relation, let alone causation. Further, since Blalock uses a bewildering variety of relational terms (such as, fewer . . . lower, are most likely to be), the kind of relation asserted in several of his propositions isconjectural. Finally, if a causal language facilitates derivations, it is curious that Blalock (1967) formulates ninety-seven propositions but does not derive an assertion from any two of the propositions.

CHAPTER 6 TYPES OF INTRINSIC STATEMENTS

The intrinsic part of theory comprises assertions of empirical relations; as such, statements rather than terms are the principal components of a theory, for a term is not an assertion.[1] However, the typology of terms in Chapter 5 is a step toward a typology of intrinsic statements.

In any theory only some of the intrinsic statements enter into tests. All such statements are derived rather than direct assertions, but that distinction is not sufficient. In the proposed mode of formalization a derived statement does not enter directly into tests unless all of its constituent substantive terms are referentials, so it is actually the constituent substantive terms that typify a statement, whether derived or direct. With three types of substantive terms (constructs, concepts, and referentials), there are five types of intrinsic statements: axioms, postulates, propositions, transformational statements, and theorems.

Axioms

An axiom is a direct intrinsic statement in which the substantive terms are constructs. Note especially that typification of a statement follows from the theorist's identification of substantive terms, which is entirely relative to a particular theory. So no claim is made that some component statements of extant sociological theories are axioms, for those theories were not constructed in accordance with the proposed scheme. There are a few instances where sociological theorists have identified statements as axioms, but typically the rationale for the identification is obscure; in any case, what some sociologists think of as an axiom may have no connection with the present definition.

Although the typification of a statement depends on the theorist's judgment, he is not free to identify statements whimsically. Specifically, unless he has identified its constituent substantive terms as constructs, he cannot regard a statement as an axiom.[2] Further, if the theorist identifies all intrinsic statements as axioms, he tacitly admits that the theory is untestable. Other statements can be derived from a *set* of axioms, but in a sense a derived statement would be another axiom, for its constituent substantive terms would be constructs.[3] The goal in theory construction is the derivation of testable assertions, and nothing is gained by deriving statements that link constructs.

An illustration.   Suppose a theorist considers an assertion suggested by Durkheim—that the division of labor varies inversely with normative consensus. Division of labor and normative consensus are substantive terms, but it does not follow that both are constructs. That identification depends on the judgment of the theorist who uses Durkheim's observations; however, no one is likely to argue that Durkheim provided a complete and empirically applicable definition of either term. Accordingly, unless the theorist can formulate such definitions, he should identify the assertion as an axiom; and it would be incomplete without stipulating a unit term, a relational term, and temporal quantifiers. So, in accordance with the proposed mode of formalization, the theorist could commence the intrinsic part of his theory with the following axiom, A1: Among countries, the greater the "division of labor at T_0," the less the "normative consensus at T_0."

An extensive commentary on the illustrative axiom is in order. As in all intrinsic statements, the initial phrase is the unit term. One may wonder why the unit term designates countries, but the theorist need not defend his choice, and it would be pointless to criticize the theory on that ground. Nonetheless, the theorist would surely have some reason for designating countries as units. Following Durkheim, he could have used the word "society," but an empirically applicable definition of it would be difficult. So, in effect, the theorist has equated a country with a society, and he has done so to formulate a testable theory. Even when contemplating an axiom, the theorist should recognize that it will enter into the derivation of other statements, whose testability depends in part on unit terms, relational terms, and temporal quantifiers. As one consideration, if the theorist contemplates using published data to test the theory, the unit term in the axioms should be countries rather than societies. After all, data are published for countries, not "societies."

When only one type of unit term appears in a theory, then its range (as an aspect of predictive power) is minimal. Accordingly, in thinking of a relation between properties, a theorist should consider a variety of unit terms. However, he may conclude for one reason or another, that a relation between two properties holds only for one type of unit. As for the illustrative axiom (A1), the theorist may conclude that the relation between the division of labor and normative consensus holds only for countries. Even if he considers other types of units (such as metropolitan areas) as well, it could be that adequate data for tests can be secured only for countries. The importance of that consideration canot be exaggerated, for sociological theorists habitually formulate theories without ostensible concern for testability. The proposed mode of formalization breaks with that tradition; specifically, in each step the theorist should strive to maximize testability. Even though axioms (or any direct intrinsic statements) are not testable, their constituent unit terms and relational terms determine the testability of derived statements.

Critics of a theory may question the theorist's judgment by arguing that (1) the asserted relations hold for other types of units and (2) adequate data are available for other types of units. But testability and predictive accuracy pertain to the assertions actually made, not those that could have been made, and it would be ludicrous to reject a theory because various assertions have been excluded. If a critic thinks that additional assertions should be made, he is free to make them, but in so doing he will have altered the theory, and his judgment becomes subject to question. Of course, if the theorist restricted the units to one type because of practical considerations (the availability of data in particular) the theory should be expanded as the conditions of work in the field change for the better. But no one would seriously argue that a theory should be rejected because its range is expandable. In any case, the expansion of a theory changes it and requires scarcely less judgment than the initial version.

Returning to the illustration, should the theorist or someone else expand the range of the theory, additional axioms would be formulated, but axiom 1 (A1) would remain unchanged. Specifically, if it is thought that the asserted inverse relation between the division of labor and normative consensus also holds among metropolitan areas, then another axiom would be formulated in which the unit term is metropolitan areas rather than countries.

The partial repetition of intrinsic statements may be questioned on the ground that it is cumbersome and restricts the generality of the theory. As for generality, just the opposite is true; the partial repetition of intrinsic statements expands predictive range. Of course, the theorist may stop short of making the same assertion for all types of units, but no one is likely to reject a theory simply because its range is less than maximum. For that matter, intrinsic statements that imply an unlimited range are suspect. If no unit term is stipulated in axioms and other intrinsic statements, from the outset investigators will have doubts in attempting tests of the theory. Such an omission implies that the asserted relations hold for all types of units, but that implication taxes credulity. Furthermore, it is most unlikely that sociological investigators can formulate empirically applicable definitions of all types of units, let alone agree in their definitions.

As the foregoing suggests, it is indefensible to exclude unit terms from intrinsic statements. The same may be said of another idea—that the theorist should employ a generic designation of units, such as "among populations." Certainly a theorist would not employ a generic designation unless he thinks that the asserted relations hold for all types of units. No matter what his opinion, he should identify each type of unit by partial repetition of each intrinsic statement; otherwise, if the unit term in each intrinsic statement is "populations," it is as though no unit terms have been stipulated.

As for the complaint that a partial repetition of intrinsic statements is cumbersome, the explicitness achieved is more than enough compensation. Moreover, partial repetition facilitates the interpretation of tests. If no unit terms are stipulated or if the theorist employs a generic designation, the predictive accuracy of the theory is presumed to be the same regardless of the units; as such, tests could lead to the premature rejection of the theory. For example, suppose that no unit terms are specified or that the unit term is generic (for instance, social units), and suppose further that initial tests are based on metropolitan areas. As such, if minimal predictive accuracy is indicated by the tests, the temptation will be to reject the theory, but predictive accuracy could have been much greater for countries. However, that possibility may not be recognized unless the theorist stipulated countries as the unit term in one set of intrinsic statements and metropolitan areas in another. Finally, when each intrinsic statement comprises only one type of unit term, modification of the theory in light of test results is facilitated. Suppose that countries is the unit term in one set of intrinsic statements and that set is repeated for metropolitan areas. Now suppose that predictive accuracy is minimal when tests are based on metropolitan areas but not when they are based on countries. If so, the theory can be modified readily by rejecting those axioms and other intrinsic statements in which the unit term is metropolitan area.

Relational terms and temporal quantifiers in axioms.   For illustrative purposes, suppose that the theorist who formulated axiom A1 *believed* that the division of labor *causes* normative dissensus. Would it follow, therefore, that the theorist should have used the word "causes" as the relational term in the axiom? Blalock and Costner notwithstanding, the use of cause as a relational term introduces irresolvable issues and creates insoluble problems. Consider a "causal" version of the axiom: Among countries, the division of labor causes normative dissensus. Now even if the two substantive terms designate observable phenomena, no one ever will observe the division of labor causing normative dissensus. As Blalock himself admits: "All we can observe is a change in X followed by a change in Y" (1964:19). So the word "causes" cannot be used to derive testable predictions unless it is translated into the language of space-time relations, and that translation introduces all of the irresolvable issues examined in Chapter 2. To restate them again briefly: some sociologists do not accept the idea that causation can be equated with space-time relations, while those who argue to the contrary do not agree in their specifications of a particular type of space-time relation as the manifestation of causation. So the question is: Why employ "cause" as a relational term when it must be translated and when that translation inevitably leads to insoluble problems and sterile debates? It will not do to argue that the translation can be delayed until tests of the theory are undertaken, for that does not make the translation any less difficult or debatable.

Note at this point that the relational term (greater . . . less) in axiom A1 also requires translation, but it can be translated readily through the language of conventional statistics, which is not the case for "causes." Moreover, it is pointless to argue that the relational term somehow makes axiom A1 meaningless or unscientific. Such an argument reflects ignorance of the increasing avoidance of causal language in the most advanced sciences and the unquestioned existence of numerous "noncausal" laws in those sciences. The general point is that the relational term in axiom A1 does stipulate some kind of order, and to demand more is to invite anarchy in the assessment of theories.

Returning to the idea that the division of labor "causes"

normative dissensus, examine still another axiom: Among countries, the greater the "absolute increase in the division of labor during T_{0-10}," the greater the "proportionate decrease in normative consensus during T_{10-20}." Although that axiom is more consistent with commonly held conceptions of causation, some critics would reject it unless the theorist formulates an additional axiom that denies a close relation between change in normative consensus during T_{0-10} and change in the division of labor during T_{10-20}. However, suppose that tests of the theory are consistent with the first axiom and not with the second. Are we to abandon the theory simply because of opinion about causation? After all, the first axiom does identify order in the universe, so what could possibly be gained by rejecting it?

Now suppose the theorist merely asserts that an increase in the division of labor is "followed by" a decrease in normative consensus. Without temporal quantifiers, the assertion is vacuously true, and " $T_{0-10} \ldots T_{10-20}$ " are only two of infinite possibilities. Indeed, there are so many possibilities that the theorist may be unable or unwilling to specify a particular time lag. If so, he can consider stipulating another kind of temporal relation. In particular, rather than attempt to stipulate "lag" temporal quantifiers (for example, $T_{0-10} \ldots T_{10-20}$), he may stipulate synchronic temporal quantifiers as in the initial version of axiom A1 (that is, $T_0 \ldots T_0$). The two types of relations between the variables may be different, but that is simply one of the numerous judgments that a theorist must make. In the illustrative case at hand, presuming that the theorist is unable or unwilling to stipulate "lag" quantifiers, he has no choice but to stipulate some other *kind* of quantifier or abandon the theory altogether. If he does stipulate some other kind, the critics may object (as they never tire of doing) that the theory does not assert causation. That objection conveniently ignores the debate over the very notion of causation. Above all, even if the notion be accepted, a "noncausal" theory that identifies order is better than no theory at all.

Should a theorist stipulate "lag" quantifiers, it may reduce testability. One unfortunate conditon of work in sociology is the difficulty of obtaining comparable longitudinal data either in published form or by gathering them through field studies. Gathering such data is a costly undertaking, so much so that the feasibility of referential formulas is a major question. Moreover, the reliability of longitudinal data, whether gathered or obtained in published form, is a crucial consideration. When a referential formula is applied to cross-sectional data (for instance, the United States, 1970, and India, 1971), the referents commonly differ substantially (the referent for one unit may be more than twice that of another). In such a case the theorist may have doubts about the reliability of the data, but he can make an argument: if the referential formulas had been applied to perfectly reliable data, the rank-order of the units probably would not be altered substantially. So, in a sense, he can settle for *relative* agreement in the referents.

The situation is quite different when considering change, for variation in referents over a short time (five or ten years) may be minute. If so, reliability of the data becomes a critical question, and the theorist may have such doubts that he will refrain from stipulating the use of longitudinal data in tests. Note that the theorist's judgment is entirely relative; he may believe that data gathered by some particular procedure are sufficiently reliable for cross-sectional comparisons but not for longitudinal comparisons, and the distinction is all the more critical for published data.

One may think of data reliability as divorced from the temporal quantifiers in axioms, but that is only appearance. Again, even in formulating axioms, the theorist should be concerned with testability. In particular, the temporal quantifiers of the axioms may appear in the assertions derived from the theory, and in that sense the axioms determine the kinds of data required for tests.

Another axiom.   Since the substantive terms of an axiom are not empirically applicable, no tests of the assertion can be made. With that consideration in mind, the theorist makes additional intrinsic statements (the ultimate goal being the derivation of testable assertions).

Reconsider axiom A1. Having made that assertion, the theorist's immediate goal is to make additional assertions that link the constructs, division of labor and normative consensus, to concepts. Again drawing on Durkheim's work, the theorist could link normative consensus and the suicide rate, with the latter term identified as a concept. He may not be able to link the division of labor to a concept, but another assertion is suggested by a definition: The division of labor is a function of differences among individuals in their sustenance activities *and* the related exchange of goods and services. Given that definition, the theorist could assert a relation between the division of labor and differences among individuals in their sustenance activities. Appearances to the contrary, the assertion is not a tautology, for in effect the theorist presumes that the two components of the division of labor vary directly, which is not true by definition. Conceivably, even if individuals do not differ as to sustenance activities, they could still exchange goods and services. It would be anomalous but hardly a "logical" impossibility.

Just as the theorist could assert a relation between the division of labor and differences among individuals in their sustenance activities, so could he assert a relation between the "division of labor" and the "exchange of goods and services." However, he may prefer the former assertion in anticipation of a subsequent intrinsic statement that links the term "differences among individuals in their sustenance activities" and a concept. Nonetheless, he may identify the term as another construct, thereby signifying that he does not regard his definition as complete or empirically applicable. So the second assertion is another axiom, A2: Among countries, the greater the "division of labor at T_0," the greater the "differences among individuals in their sustenance activities at T_0."

All comments on axiom A1 also apply to A2, but a causal language would be even more questionable in the case of A2. Although axiom A2 is not a tautology, "differences among individuals in their sustenance activities" is a component of the division of labor, and hence a causal relation is not implied. Even though the axiom is interpreted as implying a relation between "differences among individuals in their sustenance activities" and the "exchange of goods and services," the theorist may not think of it as causal, certainly not "one-way" causation. Rather, he may think of the relation as reciprocal causation, which is quite different from the causation contemplated in axiom A1. So the crucial point is that even if a causal language is employed to state a theory, it does not follow that the same relational term (such as "causes") can be used in each intrinsic statement. However, if the relational terms vary from one intrinsic statement to the next (for example, "causes" in one, "reciprocally causes" in another, and "function of" in still another), systematic derivations are precluded. Such is not the case for the proposed relational terms, as they are used in all intrinsic statements.

Postulates

A postulate is a direct intrinsic statement in which the substantive terms are constructs and concepts. No claim is made that sociologists use the word postulate in that way; rather, as suggested in Chapter 4, they use it uncritically. Certainly there is no conventional distinction between an axiom and a postulate, and in actual usage the two labels appear interchangeable.

An illustrative instance. The ultimate goal in theory construction is the derivation of statements in which the substantive terms are referentials, for only that kind of statement can enter into a test. However, a theorist should not stipulate a referential formula unless he regards his conception of the corresponding phenomenon as complete, and that principle excludes an intrinsic statement in which a construct and a referential are linked. Rather, constructs are linked only with concepts and other constructs.

Returning to axioms A1 and A2, the theorist could make additional statements in which each construct is linked to a concept. However, suppose that he is unable or unwilling to link the "division of labor" to a concept. If so, he must make statements in which the other two constructs are linked to concepts; otherwise, axioms A1 and A2 cannot enter into the derivation of assertions.

As already suggested, the theorist may interpret Durkheim's observations as implying an inverse relation between normative consensus and the suicide rate, but he cannot use Durkheim's terminology to express the assertion, for Durkheim did not articulate his theory formally. Consequently, the theorist must make the assertion explicit and in so doing stipulate unit terms, relational terms, and temporal quantifiers, as follows in postulate 1 P1):

Among countries, the greater the "normative consensus at T_0," the less the 'suicide rate during $_{-2}T_2$.'

Postulates differ from axioms in that they always include a concept, which appears in single quotation marks to distinguish them from constructs. However, all previous comments pertaining to axioms also apply to postulates. Summarizing briefly, the choice of unit terms, relational terms, and temporal quantifiers should reflect a concern with testability.

Note especially that the temporal quantifier of "normative consensus" is the same in axiom A1 and postulate P1. Without that consistency, the two statements would not share a construct in common (note that, the temporal quantifier is part of the substantive term), and the systematic derivation of other statements would be precluded. Further, all temporal quantifiers in axiom A1 and postulate P1 are cross-sectional (change in a property is not designated). However, consistency in temporal quantifiers is not necessary when a new substantive term is introduced. In addition to postulate P1, or as an alternative to it, the theorist could have made an assertion something like the following: Among countries, the greater the "normative consensus at T_0," the greater the 'absolute decrease in the suicide rate over $_{-1}T_1 - {}_4T_6$.' That assertion would not exclude postulate P1, meaning that the same construct may appear in several postulates, with the only difference being the temporal quantifiers of the concepts. When such postulates are formulated, they increase the intensity of a theory (the greater the variety of temporal quantifiers, the greater the intensity). However, the theorist should not stipulate a variety of temporal quantifiers if he doubts the relations asserted or if he anticipates that sufficiently reliable data cannot be obtained for tests. In the illustrative case at hand, the simplifying presumption is that the theorist declines to stipulate additional postulates with different temporal quantifiers.

One may object that Durkheim's observations do not imply postulate P1, but such an objection only perpetuates exegetical sociology. We shall never know what Durkheim really meant, and arguments on the subject are sterile. Obviously, apart from what Durkheim may have meant, no theorist would incorporate P1 in a theory unless he has some confidence in it as an assertion. So any postulate reflects the theorist's judgment, and its "source" is irrelevant. Invoking Marx, Weber, Durkheim, or Pareto makes a theory more impressive to some sociologists, but it cannot possibly enhance predictive accuracy.

Another illustrative postulate.   Although the identification of terms is strictly the theorist's judgment, identifying "differences among individuals in their sustenance activities" as a construct would not be difficult to understand, for a theorist is not likely to regard his definition of the term as complete. Suppose that in the extrinsic part he defines the key substantive words, sustenance activity, as follows: an expenditure of energy in the production of some good or service or in the pursuit of food. The empirical applicability of the definition is questionable, if only because the word "services" is subject to all manner of divergent interpretations, especially when an attempt is made to apply it cross-culturally. Furthermore, without using the same formula, investigators are not likely to agree in describing the amount of differences among individuals as to sustenance activities, but the theorist may be unable or unwilling to stipulate a formula. Even if he could, there is no scheme for classifying sustenance activities, let alone one that can be applied cross-culturally, and no scheme could be applied to countries without enormous research resources.

In identifying a term as a construct, the theorist recognizes that the phenomenon designated by the term can be inown or analyzed only inferentially. In this case, he could reason that differences among individuals in their sustenance activities are *reflected* in occupational distinctions as elements in the natural language of the population. For example, if only certain individuals "cut human hair," some term identifies them as a category. Extending the reasoning, as differences among individuals increase, so will the occupational distinctions recognized by members of the population. So the theorist summarizes his reasoning in another postulate, P2: Among countries, the greater the "differences among individuals in their sustenance activities at T_0," the greater the 'occupational differentiation at T_0.'

This second postulate, though no more testable than the first, is not a tautology. Conceivably, there could be vast differences among individuals in their sustenance activities and yet those differences are not socially recognized, that is, occupational distinctions are minimal or entirely absent. Moreover, the theorist's conception of "social recognition" is strictly linguistic. Specifically, even if members of a population are cognizant of differences in sustenance activities, that recognition may not be reflected in the natural language of that population by the existence of occupational terms; but the assertion in P2 is to the contrary.

Propositions

A proposition is a direct intrinsic statement in which the substantive terms are concepts.[4] By way of elaboration, reconsider the illustrative axioms and postulates (A1, A2, P1, and P2). Given certain rules of derivation, the four statements imply a direct relation between the 'suicide rate during $_{-2}T_2$' and 'occupational differentiation at T_0.' However, the theorist could have arrived at the same relation by intuition or induction rather than *formal* derivation. If so, he would have commenced his theory with the following proposition, Pr1: Among countries, the greater the 'occupational differentiation at T_0,' the greater the 'suicide rate during $_{-2}T_2$.'

In the context of formal theory construction, axioms or postulates *may* preclude propositions. None of the three types of statements are testable by themselves; but as *direct* intrinsic statements they enter into the derivation of other statements, which in turn enter into tests. However, if what would otherwise be a proposition can be derived from axioms or postulates, the derivation would be superfluous. In the illustrative case, since Pr1 could be derived from A1, A2, P1, and P2, the derivation itself would accomplish no purpose as far as subsequent derivations are concerned. Thus, if statement X can be derived in part from Pr1, then X could be derived in part from A1, A2, P1, and P2.

The foregoing does not mean that propositions are excluded from formal theory construction. A theory may commence with propositions, and it need not include any axioms or postulates. Moreover, axioms and postulates do not always preclude propositions. For example, should a theorist make a statement in which the 'suicide rate during $_{-2}T_1$' or 'occupational differentiation at

T_0' is linked to another concept, that proposition would not be superfluous, because it could not be derived from A1, A2, P1, and P2.

An issue concerning propositions.   Suppose two theorists assert the same relation between two properties, with the only difference being that one of them arrived at the assertion by formal derivation and the other by intuition or induction. Now the question is: What possible difference would the source of the two assertions have on their predictive accuracy (or validity or empirical truth)? Obviously, it could have none, for the assertions are the same. That consideration should be a sobering thought for critics who demand psychological justification of assertions or a "logic of discovery." Those critics will have little use for a proposition, as the theorist may not be able to account for it. However, what the critics ignore is that the "source" or "accountability" of a proposition does not determine its predictive accuracy. Indeed, if a justification is demanded, there can be no theories, since there must be some direct assertions which in any theory are "unaccountable."

Still another school of critics are prone to question propositions, but they do so by emphasis on the "logical" character of explanation. As Homans would have it, explanation is equated with a "logical" derivation; and that equivalance may lead one to question the explanatory adequacy of propositions, for by definition propositions are not derived, certainly not formally. However, extension of the argument would deny the explanatory adequacy of any theory, for surely there are direct assertions in any theory, that is, empirical premises. Above all, the argument blurs the distinction between logical and empirical truth. Specifically, the logical basis of its derivation in no way insures the predictive accuracy of an assertion. As put by Ayer: "For it is characteristic of empirical propostions that their validity is not purely formal. To say that a geometrical proposition, or a system of geometrical propositions, is false is to say that it is self-contradictory. But an empirical proposition, or a system of empirical propositions, ay be free from contradiction, and still be false" (1946:90). Ayer actually understates the case, for an assertion may be true even though its derivation is contrary to the rules of one logic or another.

Just as some critics may question propositions, operationalists prefer them to axioms and postulates. The argument is that if a proposition can be derived from axioms and postulates, then the latter, not the former, are superfluous. Indeed, the extreme position would be to dismiss the axioms and the postulates as fictions.[5] The argument is not accepted, but the reason has nothing to do with the notion of explanation; instead, the objection is pragmatic. If axioms and postulates are excluded in preference for propositions, then much of the heritage of sociology would be rejected, since many sociological terms can be used only in axioms and postulates. Moreover, fictions or not, axioms and postulates enable a theorist to derive assertions that might not have been realized by intuition or induction.[6] When assertions are derived from other assertions, it is difficult to see how the premises can be dismissed as fictions.

Transformational Statements

A transformational statement is a direct intrinsic statement in which the substantive terms are a concept and a referential. Insofar as its component constructs and concepts designate quantitative phenomena, a theory is not testable without transformational statements. Here we realize why so few sociological theories are testable. In the "grand" tradition so ably perpetuated by Talcott Parsons, sociological theorists make assertions about quantitative phenomenal but habitually fail to stipulate referential formulas. By contrast, operationalists in sociology are preoccupied with formulas (or "measures"), but without transformational statements no formula or measure insures systematic tests of a theory.[7] The notions of "operational definition" and "indicator" are akin to referential and referential formulas, but they are not substitutes for a transformational statement.

Two illustrations. Even though both concepts in postulates P1 and P2 designate quantitative phenomena, they are not considered as empirically applicable. Although temporal quantifiers are specified, unless investigators use the same formula, they are most unlikely to agree in reporting either the suicide rate or the amount of occupational differentiation for a social unit. For that matter, even the use of the same formula will not result in agreement unless the investigators apply it to the same published data or to data gathered in accordance with a specified procedure. As stressed repeatedly in Chapter 5, the stipulation of formulas and requisite data is a distinct step in theory construction. Neither stipulation is realized by a definition of a concept, and the theorist should not presume that a definition will somehow suggest the same formula and procedure to all investigators. Regardless of conventions in the field, the formula and the procedure for securing data must be explicit and distinct from the defintion of the concept.

The 'suicide rate during $_{-2}T_2$' could be defined in the extrinsic part of the illustrative theory as follows: the ratio of members of a population who have taken their lives at some time during a five-year period to the average size of that population during the period. Having formulated that definition, the theorist may designate RFYSR as a referential, and stipulate the following referential formula in the extrinsic part of the theory: RFYSR = $[(S/5)/P]\,100,000$, where S is the number of suicides during a five-year period and P is the size of the population at some point in the third year. In addition to stipulating the formula, the theorist would specify the requisite data. If it is to be applied to published data, the theorist should specify the *kind* of published data. Should he make no reference to published data, then the procedure for gathering the requisite data should be described in great detail.

However thorough the theorist's explication of a referential formula may be, the formula cannot be linked to the intrinsic part of the theory without the following transformational statement, T1: Among countries, the greater the 'suicide rate during $_{-2}T_2$,' the greater the RFYSR for $_{-2}T_2$. With that statement the theorist completes a series of assertions that relate a construct ("normative consensus at T_0"), a concept (suicide rate during $_{-2}T_2$'), and a referential (RFYSR).

As suggested in Chapter 5, the theorist may consider a variety of alternatives in liking the concept 'occupational differentiation at T_0' and a referential. For reasons considered in that chapter, suppose he designates the referential as ARCOD and stipulates the referential formula as follows:

$$ARCOD = (1-[\Sigma X^2 /(\Sigma X)^2])/[1-(1/N)]$$

where X is the number of individuals in each occupational category and N is the number of categories. Now the theorist is not likely to consider the referential formulas as empirically applicable without the use of census data, so he would specify the *kind* of census data in detail.

Having specified the referential formula, the theorist would then link it with the concept in the following transformational statement, T2: Among countries, the greater the 'occupational differentiation at T_0,' the greater the ARCOD at T_0. By that statement he completes a series of assertions that relate a construct ("differences among individuals in their sustenance activities at T_0"), a concept ('occupational differentiation at T_0'), and a referential (ARCOD).

The question of justification and rules of correspondence. Far from being content to evaluate a theory by reference to its predictive power, some sociologists insist that a theorist "justify" his theory. The demand for justification is directed especially at those steps where the theorist specifies a formula or procedure (see Chambliss and Steele, 1966:526). The critics who make such a demand never clearly indicate what would constitute justification of a theory or any particular step in its construction. Surely a theory must be logically consistent, but even so there is no one exclusive kind of logic, and hence consistency is entirely a matter of the particular rules of logic that a theorist sees fit to use. If those rules must be justified, an infinite regression ensues. Consequently, there can be only two grounds for criticizing the "logic" of a theory: the theorist has not made his rules of logic explicit or he has constructed the theory contrary to his own rules.

Those critics who speak of "justification" evidently demand more than logical consistency, which is to say that substantive considerations are issues; but, again, it is not clear what would constitute justification. If general principles are in question, some possibilities are suggested, but all are controversial. As an instance, if one argues that "unproven" assumptions are not acceptable in theory construction, the principle is a contradiction in terms; that is, if an assumption could be "proven," it would not be an assumption. Moreover, the dilemma of induction precludes "proving" any universal assumption (that is, one that pertains to an infinite universe), and only those assumptions are relevant in theory construction.

Considering other possibilities, if "justification" pertains to testability or evidence in support of the theory, the demand is nothing more than an insistence that theories be assessed in terms of predictive power, especially accuracy. However, if by evidence in support of a theory, one means "general observations" or consistency with other theories, why even attempt systematic tests? That question touches on the central issue. If tests of a theory reveal impressive predictive accuracy, it would be ridiculous to reject the theory because it cannot be "justified" without a reference to predictive power. So the conclusion—any justification of a theory independently of tests is merely opinion, and the demand for justification encourages sterile arguments.

Rather than speak of justification of a theory, critics may invoke the notion of "rules of correspondence." The ambiguity of the notion alone makes it questionable, but there are sociologists (see Wilson and Dumont, 1968) who would make the notion a central consideration in theory construction.

Some commentaries suggest the following conception of a rule of correspondence: any statement by which a constituent term of a theory is equated with or linked to a formula, procedure, operation, or "observational" term. As such, either one of the two transformational statements (T1 and T2) could be construed as rules of correspondence. Only theorists who are indifferent to testability would question the desirability of rules of correspondence, so conceived.

Still another conception is controversial. The conception denies that a rule of correspondence is a particular statement by which a constituent term of a theory is equated with or linked to a formula, measure, procedure, or operation; on the contrary, all such statements must be made in *accordance* with some general rule of correspondence. So conceived, rules of correspondence would transcend any particular theory.

Although advocates of the "transcendent" conception speak

of rules of correspondence, they never stipulate any definite rules.[8] But the proposed mode of formalization represents a rejection of the very idea of transcendent rules of correspondence. In linking a concept and a referential formula, the theorist is guided by intuition, imagination, knowledge of procedures for securing data, and practical considerations pertaining to research resources. Briefly, in stipulating referential formulas the theorists must make judgments, and the idea that such judgments must be or can be made in accordance with rules that transcend particular theories is a myth.

Some commentaries on rules of correspondence suggest that a definition of a term implies a formula, procedure, or operation. Definitions are suggestive, but there are no rules by which one can "derive" a formula, procedure, or operation from a definition, and no rules that assure agreement among theorists. On the contrary, given the same term and the same definition, two theorists are likely to stipulate different formulas, procedures, or operations. Further, a referential formula may appear inconsistent with a definition. Thus, the RFYSR formula stipulates population size (P) at some point during the third year of a five-year period, but the definition of the 'suicide rate during $_{-2}T_2$' makes reference to "average" population size. What the theorist has done is assume a fairly close relation between P and average population size. The latter is unknown and unknowable; but, the assumption is nonetheless empirical, not a matter of syntax or semantics. For that matter, given the same definition of the concept, another theorist might stipulate a different referential formula, for example,

$$RFYSR = (S/5)/[P_1 + P_2)/2],$$

where P_1 is the population size at some point in the first year and P_2 is the population size at some point in the fifth year.

Apart from the question of the relation between RFYSR and the definition of the 'suicide rate during $_{-2}T_2$,' the latter does not stipulate how the requisite data are to be obtained. Accordingly, should the theorist designate official publications (census reports and vital statistics) as the source of data, he has stipulated something that is neither given in nor implied by the definition of the concept, and the same is true if he stipulates a procedure for gathering data. How could such stipulations possibly follow from a rule of correspondence that transcends any particular theory?

Turning to the ARCOD formula, suppose the theorist defines 'occupational differentiation at T_0' as: a function at any point in time of the number of occupational categories and the uniformity of distribution of individuals among the categories. Note especially that the ARCOD formula does not reflect the number of occupations. On the contrary, having decided to use census data on occupations, the theorist may question the credibility of international variation in the number of "census" occupations, and hence the ARCOD formula is stipulated because it does not necessarily reflect the number of occupations.

In each social unit at any point in time there is a "true" amount of occupational differentiation, but the definition of the concept gives only the meaning of occupational differentiation, not directions as to how it is to be determined in a given instance. Moreover, the theorist need not claim that his referential formula and related procedure reveal the "true" amount of occupational differentiation in any unit. Rather, he may presume that a referent computed by application of the stipulated formula to the designated kind of data will *approximate* the "true" amount. The validity of that presumption can never be demonstrated, of course, because the true amount of occupational differentiation (that which the concept denotes) is unknown and unknowable. So, even if the theorist should think of his referential formula as ideal, it is nonetheless a presumption. Certainly no rule of correspondence can validate the presumption, and for that reason alone one must surely wonder what purpose a rule would serve.

Extending the argument, suppose there were rules of correspondence such that a referential formula can be "derived" from a definition of a concept. The rules would not insure the feasibility of the formula or eliminate the need for a theorist to make a judgment in that regard. Moreover, even if research resources were unlimited, some procedure for gathering or otherwise obtaining data must be stipulated. It is grossly unrealistic to suppose that, regardless of the theory, procedures for securing data can be derived from rules of correspondence.

At this point observe again that a transformational statement asserts a close relation between (1) the referents computed by the application of the designated referential formula and (2) the "true" amount of some phenomenon (as denoted by the concept). That relation will not hold if the data are unreliable, and hence the theorist must be concerned with research procedure. However, the matter is complicated by realization that a transformational statement could be false even though the data are absolutely reliable and no errors were made in applying the referential formula. If that could be known, then the conclusion would be obvious—the theorist has stipulated the "wrong" referential formula—but no rule of correspondence can insure the "right" referential formula.

Advocates of correspondence rules may argue that the definition of a concept should extend to the stipulation of both a referential formula and a procedure for securing the requisite data; but such a definition would be contrary to the nature of a transformational statement. It is an assertion (albeit untestable) and not a definition. Further, the extension of the argument leads to the declaration that a definition of the term "age of a living human being" is incomplete unless it stipulates a procedure for determining age in any given instance. The declaration presumes that there is or can be only one procedure for the determination of age; yet the possibilities are virtually infinite, and hence the stipulation of a particular procedure in a definition is debatable. There is a still more compelling consideration—the argument presumes that the age of a living human being can be known with certainty by some procedure. On the contrary, even if considered in years rather than seconds, the "true" age of some individuals cannot be known with certainty. As such, all that can be said of any figure is that it represents an approximation of "true" age, but such a statement is an assertion and not true by definition.

The notion of correspondence rules persists despite grave doubts about even the meaning of such rules. Consider Nagel's comment:

> though theoretical concepts may be articulated with a high degree of precision, rules of correspondence coordinate them with experimental ideas that are far less definite. The haziness that surrounds such correspondence rules is inevitable, since experimental ideas do not have the sharp countours

that theoretical notions possess. This is the primary reason why it is not possible to formalize with much precision the rules (or habits) for establishing a correspondence between theoretical and experimental ideas. If we ask, therefore, what formal pattern is exhibited by correspondence rules it is difficult to give a straightforward answer. (1961:100)

Surely Nagel's observations raise doubts about rules of correspondence, but evidently he would insist that the notion play a role in theory construction. There is no paradox or issue if by a rule of correspondence Nagel means *any* stipulation of a formula, procedure, or operation, and not a rule that transcends particular theories. But note the crucial difference: a theory is not constructed in *accordance* with some general rule or rules of correspondence; rather, rules of correspondence are *components* of a theory, and a particular rule (such as a referential formula) may be peculiar to one theory. Indeed, since theories differ as to substance, how could they all possibly encompass the same rules of correspondence? In that connection, observe that Nagel questions even the idea that there is or can be one logical form for all rules of correspondence.

What has been said of Nagel's observations also apply to Torgerson's statement: "there seems to be virtually an unlimited number of ways in which such rules of correspondence can be devised" (1958:8). The comment is correct, and there is no mystery. If rules of correspondence exist at all, they are nothing more than components of particular theories, and they differ simply because theories differ. So, again, the idea that there is or can be general rules of correspondence which transcend any particular theory is a myth,[9] and a destructive one, since it suggests that theories somehow can be manufactured. To the contrary, a theory and all of its parts are created by the imagination and judgment of a theorist. Philosophers of science and methodologists may aspire to referee the game, but they are only spectators.

Recapitulation.   Transformational statements are not made in accordance with some general rules of correspondence. A transformational statement (along with the referential formula) may be regarded as a rule of correspondence in itself, but such identification is unessential and misleading, for it suggests that a transformational statement is purely a logical step and not the product of a theorist's imagination and judgment.

A transformational statement is not a purely logical step and for a very simple reason—it is always an empirical assertion. The theorist thinks of the property designated by a concept as distinct from the values computed by application of a referential formula. To be sure, he presumes that the values "reflect" the property, but uncertainty about the relation is inevitable. Even if the theorist could be certain that the referential formula has been applied correctly to absolutely reliable and comparable data, he would still have doubts about the transformational relation, for he can never be sure that another formula would not more accurately reflect the property identified by the concept. To the extent that tests of a theory indicate maximum predictive accuracy, the theorist may think of the results as somehow "justifying" the transformational statements, but that is the only sense in which they can be justified.

Transformational statements, operational definitions, and indicators. As suggested in Chapter 5, sociologists are likely to think of a referential formula as an operational definition or indicator of a concept. Of course, the choice of words is of little importance, but those two words are not used in connection with transformational statements, and some comments on their exclusion is in order.

After several decades the notion of an operational definition remains vague and controversial, especially in sociology. There is no effective consensus as to what constitutes an operational definition and sociologists typically use the word uncritically. In any case, the identification of a referential formula as an operational definition would serve no purpose, and there is an argument against the identification. If a referential formula is identified as an operational definition this suggests that the relation between the formula and the phenomenon designated by the concept is true by definition. That is not the case. Again, a transformational statement is an empirical assertion.[10] Finally, consider the following expression as a substitute for transformational statement 2: The ARCOD formula is the operational definition of 'occupational differentiation at T_0 .' The expression resembles those commonly made in putative tests of sociological theories, but it is alien to formal theory construction, and it would be so even if the theorist stipulated the referential formula and incorporated the expression in his construction of a theory. Such expressions are not substitutes for a transformational statement because they terminate the systematic use of relational terms, and that termination would preclude the formal derivation of assertions. In brief, when used by theorists, the conventional "operational" terminology is conducive to ambiguous theories.

Rather than think of a referential formula as an operational definition, some sociologists may identify it as an indicator. The two are not equated in the proposed mode of theory construction, and the arguments against it parallel those made in rejecting the identification of a referential formula as an operational definition. However, the notion of indicator is perhaps even more ambiguous than operational definition. As used in the sociological literature, an indicator may be a term, a formula or measure, an operation, or a particular datum.[11] Writers on theory construction in sociology have used the word "indicator" extensively, but they certainly have not clarified the notion (see Blalock, 1969; Dubin, 1969; and Zetterberg, 1965).

All problems and issues pertaining to operational definitions or indicators can be expressed in two questions. First, what is the relation between an operational definition or indicator and a theory? Sociologists who write on theory contruction have not provided a direct and intelligible answer to the question, and without an answer the use of either notion in the formulation or test of a theory is indefensible.[12] Second, who designates the appropriate operational definitions or indicators for tests of a theory? If the designation is left to investigators, there is no assurance that they will agree, not even in tests for the same population.[13] Without agreement, tests conducted by independent investigators are idiosyncratic in that the results depend on the investigator, not the theory. Further, the theorist or other advocates of the theory can always dismiss negative evidence by arguing that the investigator selected "inappropriate" indicators.

Theorems

A theorem is a formally derived intrinsic statement in which the constituent substantive terms are referentials. Theorems are also distinctive in that they represent the final step in theory construction; that is, the intrinsic part of a theory ends with the derivation of one or more theorems.

The rule of derivation. Theorems in the proposed mode of formalization are derived by the "sign rule," according to which the direction of the relation between any two referentials is given by the cumulative product of the intervening relational terms. [14] As the first step in the application of the sign rule, relational terms are translated as follows: "greater . . . greater" denotes a direct or positive relation and the sign is $(+)$, while "greater . . . less" denotes an inverse or negative relation and the sign is $(-)$. The sign of each illustrative intrinsic statement in this chapter is as follows: axiom 1 is $(-)$; axiom 2 is $(+)$; postulate 1 is $(-)$; postulate 2 is $(+)$; transformational statement 1 is $(+)$; and transformational statement 2 is $(+)$. The cumulative product of the signs is $(+)$, and hence the sign of the relation between RFYSR and ARCOD is $(+)$. Accordingly, translating the sign into the prescribed relational terms, we have the only theorem that can be derived from the illustrative theory: Among countries, the greater the ARCOD at T_0, the greater the RFYSR for $_{-2}T_2$.

The procedure followed in the illustrative derivation is shown in Table 6-1. One of the two transformational statements is entered in the first row and the other in the last row, and all other statements are ordered so that one of the constituent substantive terms of each statement appears in the row above and the other appears in the row below. That order shows the *syntactical* relation between the two referentials, RFYSR and ARCOD.

In column 4 the sign of the relational term in each statement is given, and the cumulative product of those signs are shown in column 5. The first cumulative product (second row of the table) is the product of the two signs in the first and second row of column 4: $[(+) (-) = (-)]$. From that point onward, the cumulative product in each row is $(X)(Y)$, where X is the cumulative

Table 6-1
ILLUSTRATION OF THE SIGN RULE

Illustrative Intrinsic Statement	Second Substantive Term	First Substantive Term	Sign	Cumulative Product of Signs
Col. 1	Col. 2	Col. 3	Col. 4	Col. 5
Transformational statement 1	RFYSR	Suicide rate	(+)	
Postulate 1	Suicide rate	Normative consensus	(−)	(−)
Axiom 1	Normative consensus	Division of labor	(−)	(+)
Axiom 2	Differences among individuals in their sustenance activities	Division of labor	(+)	(+)
Postulate 2	Occupational differentiation	Differences among individuals in their sustenance activities	(+)	(+)
Transformational statement 2	ARCOD	occupational differentiation	(+)	(+)

product in the row above and *Y* is the sign (column 4) in the given row. Thus, the last cumulative product (bottom row of column 5) is (+), and that is the sign of the relation between RFYSR and ARCOD.

Some issues. Methodologists in sociology and philosophers of science concern themselves with "logical validity" of derivations, and that concern gives rise to an issue. Needless to say, derivations should be logically valid, but it does not follow that there is only one "logic" of derivations.[15] To repeat, logic is simply any set of rules for relating symbols, and hence any derivation in accordance with a rule or set of rules *is* logically valid. Any contrary argument leads to an infinite regress in the form of a sterile debate over the rules of logic.

Whatever the logic employed, it in no way insures the empirical validity of a derivation, a point lost on Blalock and Costner, whose observations persistently blur the distinction between logical and empirical validity.[16] Specifically, they question the use of the sign rule in making derivations, ostensibly because it does not insure the empirical validity of the derivation. Indeed it does not, but Blalock and Costner ignore the more general consideration—no rule or logic of derivation insures empirical validity. There could be rules of derivation that make the empirical validity of derivations problematical, such as: If one or more of the direct intrinsic statements asserts a positive relation, then all theorems are to be derived by the sign rule with negative signs excluded from the cumulative product. One may object that the rule is "illogical," but the objection actually has to do with the probable empirical validity of theorems, and as such it merely reflects a quaint notion about logic. However, the argument is not that the empirical validity of a theorem is unrelated to the logic of its derivation; rather, it is that no logic of derivation insures the empirical validity of theorems.

The independence of logical and empirical validity of a theory stems from the direct intrinsic statements. Their empirical validity has nothing to do with logic, but the theorist presumes that the empirical validity of derived theorems is contingent on that of the direct intrinsic statements. In that connection, critics of the proposed mode of derivation may charge that it entails the fallacy of affirming the consequent.[17] It does if the intrinsic statements are judged as *true*; but some sociologists are blissfully unaware that any deductive theory, even those in the most advanced fields, cannot be taken as verified without committing the fallacy.[18] So the proposed mode of theory construction is not alien to at least some scientific fields.

Unlike analytic statements, the direct intrinsic statements are not true by definition; that is, they are empirical assertions, but they cannot be tested directly. If tests are consistent with the theorems, then the direct intrinsic statements *could be* true, but a more definite conclusion would not be justified. However, if tests conclusively falsify the theorems, then the direct intrinsic statements, *taken as a set*, are regarded with justification as false. Now it might appear that if the direct intrinsic statements are not regarded as false, then they must be true, the argument being that they are either true or false; but the direct intrinsic statements could be false even though the theorems are true. Conclusive evidence that the theorems are true verifies the direct intrinsic statements only if the theorems are derivable from the intrinsic statements and no others; but, since the "uniqueness" of a derivation cannot be demonstrated, conclusive verification of the direct intrinsic statements is precluded.

Whereas uncertainty as to the truth of the direct intrinsic statements is inevitable, conclusive evidence that the theorems are false is conclusive evidence that the direct intrinsic statements, *taken as a set*, are false.[19] So, in a sense, through derivations one may falsify a theory but never verify it. If that is taken as a limitation peculiar to deductive theory and the proposed scheme in particular, the reader is reminded that no universal generalization which asserts a relation between properties, not even one that can be tested directly (without derivations),[20] can be verified conclusively. The dilemma of induction alone precludes it.

The rules of derivation should be such that if the direct intrinsic statements are empirically valid, so are the theorems. However, again it does not follow that there is or can be only one defensible mode of derivation (see Sadovskij, 1970). What is a defensible derivation depends on the assumptions about the relations asserted in the direct intrinsic statements. The direct statements are assumed to be empirically valid, but it is necessary to articulate more specific assumptions about the asserted relations. Those more specific assumptions are not revealed by the prescribed relational terms (greater . . . greater or greater . . . less), and it is doubtful that any relational terms can convey them.

In the proposed mode of theory construction all relations asserted in the direct intrinsic statements are assumed to be sufficiently close, meaning that at least the direction of an association between values can be predicted correctly in the form of a derivation. The idea can be illustrated by coefficients of correlation. Suppose that we have three sets of values, designated as X, Y, and Z, such that each of three or more units is assigned an X value, a Y value, and a Z value. Now suppose that the product-moment coefficient of correlation between X and Y (that is, r_{XY}) is +.90 and

r_{yz} is +.85. Given those two coefficients, the prediction would be that for the same values the sign of r_{xz} will be positive (+).

No one would question the mathematical basis for predicting the sign of a coefficient of correlation in such a situation (where the known coefficients are above a certain magnitude), and hence it is curious that two critics would make the following statement: "Nothing in our present discussion can be construed as validating the sign rule for symmetric propositions" (Costner and Leik, 1964:824). The statement is all the more curious since Costner and Leik elsewhere (1964:821) admit that a sign-rule derivation is valid if it is assumed the relations that enter into the derivation are close.

Should critics argue that sign-rule derivations are invalid unless it is known that the intrinsic relations are close, this would exclude "unproven" assumptions from theories, a principle patently alien to the advanced sciences. The point is that the relations asserted by the intrinsic statements are unknown and unknowable; were it otherwise, the statements would be tested directly, that is, without deriving theorems.

Critics may insist that no relation between sociological properties is close, but that argument ignores three considerations. First, it introduces a substantive question that has nothing to do with the *logical* validity of the sign rule. Second, it cannot be argued that "measurement error" alone precludes close intrinsic relations, for those relations are not measured. And, third, alternatives to the assumption of a close relation are more questionable than the assumption itself.

The third consideration calls for additional comment. Needless to say, a belief that a relation is close does not make it so, but that belief can be altered in light of test results and an attempt made to qualify or modify the theory. So if a theorist suspects that a relation is not close and cannot qualify an assertion about the relation, he should either exclude the assertion or forego promulgation of the theory until it can be stated in a qualified form.

Above all, nothing whatever can be said for the way that sociologists typically confront the problem, that is, by invoking the notion of *ceteris paribus* in asserting a relation (see Blalock,

1967:48). The qualification is meaningless. There is no reason whatever to suppose that "other things" are ever equal, especially for an observational science, where known or unknown factors cannot be "equalized" by randomization. So the *ceteris paribus* qualification in sociology is merely a confession of ignorance and, far from serving any constructive purpose, it only leads to the formulation of untestable theories.

What has been said of the *ceteris paribus* qualification applies in much the same way to the idea of "conditional" intrinsic statements. Direct component assertions of scientific theories are not stated in the conditional form, and it is a puzzle why some philosophers of science devote so much attention to that form. Consider a restatement of an illustrative axiom: If among countries the greater the division of labor, the less the normative consensus. The statement borders on the unintelligible, and it is not even an assertion. To be sure, in deriving theorems the theorists think in terms of "if . . . then," that is, if the direct intrinsic statements are true, then the theorems are true. But without a belief that the direct intrinsic statements *are* true (not *if* they are true), there is no basis whatever for expecting the theorems to be true. So, whatever role the conditional form of statements may play in tests of a theory, it is alien to intrinsic statements.

The sign rule can be used to derive theorems without assuming that the intrinsic relations are close, meaning that there are alternative assumptions (see Costner and Leik, 1964), but they are alien to the ultimate goal of maximizing predictive accuracy. Even if one feels more confident in such an assumption and even if it is valid (that is, the relations are not close), the intrinsic statements will have to be modified in some way to increase predictive accuracy.[21] So it is pointless to assume a condition that is alien to the ultimate goal in theory construction.

The real problem in assuming close intrinsic relations is the temptation to presume that test results validate the assumption, a presumption that does entail the fallacy of affirming the consequent. However, it is recognized that the assumption of close intrinsic relations can never be verified, but that is not a limitation peculiar to the prescribed mode of derivation. Regardless of either the mode of derivation or of test results, uncertainty is inevitable;

and a demand for certainty (see Maris, 1969:51) is naive and counterproductive.

Causation and derivations. As an alternative to derivations through application of the sign rule to symmetric relational terms, Blalock and Costner argue that the statements of a theory should assert causation. Their argument is curious in several respects. A theorist may actually think of the relation asserted in a postulate or axiom other than asymmetric causation. Are all such assertions to be excluded from theory construction merely to placate methodologists who do not take Hume seriously? It seems a high price to pay. Then consider transformational statements. Is a theorist to assert that the suicide rate or occupational differentiation causes referential formulas or causes investigators to compute values? The idea is ludicrous. But once it is admitted that some of the component assertions need not be or cannot be "causal," then there is a major problem. If a variety of relational terms are used in formulating a theory (for example, "causes" in one direct intrinsic statement, "varies directly with" in another), it is extremely difficult to derive theorems systematically.

But suppose the word "causes" could be incorporated in all direct intrinsic statements. Blalock and Costner write as though the mere use of the word somehow "logically" validates derivations, but it is difficult to see how that could be. Certainly the word "causes" is not a substitute for explicit rules of derivation. In that connection, if the word "causes" is the relational term in direct intrinsic statements, the rules of derivation may be such that the word appears in the theorems.[22] So in the present illustrative case, the theorem might be: Among countries, ARCOD at T_0 causes RFYSR for $_{-2} T_2$. Even if one formula or value "causing" another is not ludicrous, there is still another controversy. Unless one argues that causation can be observed, then any "causal" theorem would have to be restated so that the relational term denotes a particular type of space-time conjunction, and that translation entails all the problems and issues considered in Chapter 2.

Another conception of derivations. Each direct intrinsic statement is logically independent of the other statement; that is, no direct intrinsic statement is derived from other direct intrinsic statements. [23] Moreover, at least three direct intrinsic statements enter into the derivation of any theorem. As such, the procedure is contrary to the common belief in sociology that the only mode of derivation or deduction is the conventional logic of classes or classic syllogism. That mode is largely alien to the advanced sciences, [24] and it has definite limitations in sociology. [25] For one thing, universal generalizations of the form "all X's are Y's" are alien to the field, where an invariant relation between qualitative attributes is virtually unknown. Even if assertions about qualitative attributes are expressed in probablistic form, the utility of the form is limited, primarily because most major sociological terms designate quantitative properties. [26] Accordingly, the proposed mode of derivation is based on a logic of quantitative relations, not the conventional logic of classes or classic syllogism.

In the advanced sciences the language of quantitative relations is mathematics, and the components of a theory are equations. But given doubts about the reliability of data and the present state of research techniques, it is premature to state sociological theories in the form of equations. In most instances sociologists can only aspire to predict ordinal differences, and hence the proposed scheme emphasizes the *direction* of relations. Granted that the emphasis is alien to what theories "should be," it is consistent with the subject matter of sociology and the conditions of work in the field.

Temporal Quantifiers and Space-Time Relations

As explained in Chapter 5, a temporal quantifier is an element in each substantive term, and in that context the notion is simple. However, an intrinsic statement contains more than one temporal quantifier, and each combination represents a *type* of space-time relation between properties. Those types require explication apart from the notion of temporal quantifiers, and the subject is important for at least two reasons. First, several types of space-time relations receive no attention whatever in sociological theory or research. Second, the variety of *types* of space-time relations asserted in a theory is an aspect of its predictive power.

Table 6-2
A TYPOLOGY OF SPATIO-TEMPORAL RELATIONS BETWEEN TWO QUANTITATIVE VARIABLES*

Variables and Quantifiers**		Number of Units	Description of Variable***								
			NCX			ACX			RCX		
X	Y		NCY	ACY	RCY	NCY	ACY	RCY	NCY	ACY	RCY
$T_{0,5}$	$T_{0,5}$	One	L-1								
$T_{0,5}$	$T_{5,10}$	One	L-IIa								
$T_{5,10}$	$T_{0,5}$	One	L-IIb								
$T_{0,5}$	$T_{0,5}$	One		L-IIIa1	L-IIIa2						
$T_{0,5}$	$T_{0,5}$	One				L-IIIb1			L-IIIb2		
$T_{0,5}$	$T_{5,10}$	One		L-IVb1	L-IVb2						
$T_{5,10}$	$T_{0,5}$	One				L-IVa1			L-IVa2		
$T_{0,5}$	$T_{0,5}$	One					L-V1	L-V2		L-V3	L-V4
$T_{0,5}$	$T_{5,10}$	One					L-VIa1	L-VIa2		L-VIa3	L-VIa4
$T_{5,10}$	$T_{0,5}$	One					L-VIb1	L-VIb2		L-VIb3	L-VIb4
T_0	T_0	Two or more	C-I								
T_0	T_5	Two or more	C-IIa								
T_5	T_0	Two or more	C-IIb								
T_0	$T_{0,5}$	Two or more		C-IIIa1	C-IIIa2						
$T_{0,5}$	T_0	Two or more				C-IIIb1			C-IIIb2		
T_5	$T_{0,5}$	Two or more		C-IVb1	C-IVb2						
$T_{0,5}$	T_5	Two or more				C-IVa1			C-IVa2		
$T_{0,5}$	$T_{0,5}$	Two or more					C-V1	C-V2		C-V3	C-V4
$T_{0,5}$	$T_{5,10}$	Two or more					C-VIa1	C-VIa2		C-VIa3	C-VIa4
$T_{5,10}$	$T_{0,5}$	Two or more					C-VIb1	C-VIb2		C-VIb3	C-VIb4

*Blank (unlabeled) cells are logical null classes.

**T_0 designates some point in time. Thus, if T_0 is taken as 1960, then $T_{0,5}$ designates 1960, 1965; and T_{0-5} is 1960-1965. Any temporal series can be extended to more than two points in time or more than two intervals. Quinquennial intervals serve as *illustrations* but neither the unit of time (months, years) nor the interval determines the type of relation.

***NCX = change not expressed in X; ACX = change in X expressed absolutely; RCX = change in X expressed proportionately; NCY = change not expressed in Y; ACY = change in Y expressed absolutely; RCY = change in Y expressed proportionately. If the interval is T_{0-5}, then *absolute* change in a variable is its value at T_5 (for example, 1970) minus its value at T_0 (for example, 1965). For the same interval, proportionate change would be the ratio of the value at T_5 to the value at T_0.

Types of space-time relations between quantitative variables.

Neither the explication of temporal quantifiers nor the illustrative theory even suggest the variety of possible space-time relations (or correlations) between quantitative variables.[27] As shown in Table 6-2, there are no less than forty-six types.

Temporal quantifiers are shown in the first two columns, and the first consideration is purely notational. Each quantifier is an instance of a type, with quinquennial integers used for illustrative purposes. Thus, the first type is of the form $T_{x,y}$, and $T_{0,5}$ is only one of infinite instances of that type ($T_{0,13}$ would be another). Note also that even though the illustrations are in terms of years, the numbers in the first two columns could be any unit of time—hours, days, and so on.

One form of temporal quantifier is not included in Table 6-2. The form is $_xT_y$, and an instance of it would be $_{-2}T_2$, denoting a period of five years. That form is excluded because there is essentially no difference between $_xT_y$ and T_z; the former represents a period of time and the latter is thought of as a point in time. Thus, $_{-3}T_3$ (a seven year period centering on T_0) is considered to be the equivalent in form to T_0 (a point in time). Similarly, $T_{0,5}$ denotes *two or more* points in time at five-year intervals, and the counterpart is $_{-2}T_2$, $_3T_7$. To give a more concrete illustration: the population size of a country in 1960, 1965, 1970 would be an instance of $T_{0,5}$; and the average annual birth rate of a country during 1958-1962, 1963-1967, and 1968-1972 would be an instance of $_{-2}T_2$, $_3T_7$. Now consider T_{0-5}, an instance of which would be the following series of numbers pertaining to the population size of a country: 1960 size minus 1955 size, 1965-1960, and 1970-1965. Similarly, an instance of $_{-2}T_2 -_3 T_7$ could be figures pertaining to the average annual crude birth rate of a country: 1963-1967 rate minus the 1958-1962 rate and 1968-1972 rate minus the 1963-1967 rate.

The third column of Table 6-2 pertains to the spatial basis of a relation between variables, and there are two possibilities. In a *longitudinal* comparison values for what is putatively the same unit are compared at two or more points or periods (for instance) the population size of the United States, 1950, 1960, and 1970).

The other possibility, a *cross-sectional* comparison, considers one value for each of two or more units (the population size of the United States, 1970; Japan, 1965; and New Zealand, 1961).

Even ignoring the difference between absolute and proportionate change, there are more types of space-time relations than recognized by the conventional distinctions—synchronic versus diachronic and longitudinal versus cross-sectional. So the words longitudinal and cross-sectional have been used to identify the spatial basis (major types L and C in Table 6-2), with temporal distinctions treated as subtypes. Those subtypes are akin to the conventional distinction between synchronic and diachronic, but the variety of subtypes is far greater than the distinction suggests.

Although the distinction between cross-sectional and longitudinal is reflected in the temporal quantifiers, it also should be reflected in the unit terms. Thus, if countries are the units and the intrinsic statement asserts a cross-sectional relation, then the unit term would be "among countries." In asserting a longitudinal relation the unit term would be "within a country over time."

The distinction between absolute and proportionate change in Table 6-2 reflects one consideration: given two coefficients of correlation, one that expresses the relation between absolute changes in two variables and the other expressing the relation between proportionate changes, the two coefficients are mathematically independent within broad limits. Accordingly, the theorist should not assume that the predictive accuracy of the theory will be the same regardless of the type of change considered. If he has no basis for anticipating that the relation between variables is closer for one type of change than the other, he should forego promulgation of the theory until he has undertaken exploratory research and made a choice inductively.

The intensity of a theory.   Maximum intensity would be realized only in a theory that makes assertion about each of the forty-six types of space-time relations between component variables (substantive terms). Of course, such a theory is simply beyond contemporary sociology, perhaps for several generations, but the idea is not nearly as complicated or far-fetched as it might appear.

In formulating a direct intrinsic statement, the theorist

neither asserts nor implies that all types of space-time relations between the properties are close. Indeed, he may believe that only one type is close and, if so, he will limit the theory to assertions about that type by his stipulation of temporal quantifiers. However, in subsequent *auxiliary* statements he may assert that the relation between variables stipulated in the theory is closer than any other type. Such an auxiliary statement would realize maximum intensity, for in effect it would be an assertion about all types of space-time relations.

Observe that in making an auxiliary statement the theorist is not required to predict the exact magnitude of any type of space-time relation; rather, he may be content to limit his assertion to a prediction of ordinal differences in the magnitude of the various types of relations. If he is uncertain, his auxiliary statements can be limited to certain types of space-time relations, as for example: Among the same countries, rho (ARCOD, T) T_0—RFYSR, $_{-2}$ T_2) is greater than rho (ARCOD, T_5—RFYSR, $_{-2}$ T_2). Finally, even if the theorist is unable or unwilling to make auxiliary statements, the theory at least asserts some order despite minimal intensity.

The foremost difficulty in maximizing intensity may be purely practical. As stressed repeatedly, in stipulating referential formulas the theorist must indicate how the requisite data are to be gathered or obtained in published form; but all such judgments are relative to the temporal quantifiers. Specifically, the theorist may conclude that the designated kinds of data can be used to test assertions of some types of space-time relations but not others; if so, he will limit the direct intrinsic statements to assertions about certain types of space-time relations, perhaps only one type, and it may be necessary to forego auxiliary statements altogether. Subsequently, as the resources of the field improve, the theorist can expand the theory by making assertions about other types of space-time relations. However, from the outset the theorist may have several alternatives. He may believe that several types of space-time relations between the variables are sufficiently close; if so, he can choose among the alternatives in accordance with his judgments pertaining to data. Those judgments may lead him to restrict his theory to one type of space-time relation.

Intensity and causation.   The typology of space-time relations in Table 6-2 will strike many sociologists as exotic and overly complicated, but the reason for the reaction is significant. No less than 90 percent of the coefficients of correlation reported in the sociological literature are of type C-I, and hence sociologists on the whole do not recognize the variety of correlations or the implications.[28] Consider the trite and cryptic expression: correlation does not demonstrate causation. The expression implies that causation can be demonstrated, but it is silent as to what would constitute demonstration. Moreover, it implies that there is only one type of correlation or that no type is any more relevant than others in contemplating causation. Either implication is indefensible. Table 6-2 should make it clear that there is more than one type of correlation and, given commonly held conceptions of causation, some types are more relevant than others. Specifically, as suggested in Chapter 2, one common conception can be expressed as follows: if a change in variable X is always followed by a change in variable Y but the reverse is not true, then X is the cause of Y. That conception is debatable, but some types of space-time relations are more consistent with it than are other types. For example, consider a type C-I relation and any C-VI type.

The general idea should be obvious. If a theorist thinks of the relation between two properties as causal, his conception of the relation can be translated by temporal quantifiers. One virtue of that strategy is that the theorist need not defend his translation or even invoke the notion of causation and thereby instigate a sterile debate. Of course, one may argue that a causal relation cannot be translated by temporal quantifiers, but if so one must surely wonder what possible purpose the notion can serve in theory construction. If the notion has any significance at all, then causation is at least more nearly reflected in one type of space-time relation than others. Viewed that way, then thinking in causal terms could maximize the predictive accuracy of a theory. To illustrate, suppose the theorist believes X to be the cause of Y and further believes (at least in this case) that causation is reflected by a close type C-VIa1 relation and a close type L-VIa1 relation. If so, with a view to maximizing predictive accuracy of his theory, he would stipulate the temporal quantifiers corresponding to the two types. The theorist could even go further by making auxiliary statements in which he asserts that those two types of relations will be closer than the other types. It would be especially important for the theorist to assert that a type C-VIa1 relation is closer than a type C-VIb1 and that the same is true of type L-VIa1 and L-VIb1 (see Pelz and Andrews, 1964). Auxiliary statements are all the more important when the quality of sociological data is considered. The quality makes unity coefficients of correlation most improbable, so the absolute magnitude of coefficients may be less relevant as a causal criterion than the relative magnitude of different types of correlations.

The proposed strategy is radically different from Blalock's causal models. He proposes that causation be inferred from type C-I correlations alone, a type that is alien to commonly held conceptions of causation, since it does not express the relation between changes in variables. Blalock may argue that a type C-VI correlation can be inferred from a type C-I, but that argument is dubious. Even given the same units and the same data, a type C-VI correlation is within very broad limits mathematically independent of a type C-I correlation. Of course, despite the mathematical independence (again at least within very broad limits), Blalock may argue that *empirically* the different types of correlation are closely related, and if he chooses to make that argument then the issue is joined and can be resolved by research findings. Given the same units and data, there will be instances where a type C-VI correlation is of much greater magnitude than a type C-I and even instances where the reverse is true.

What has been said of Blalock's causal models applies also to "path analysis," the latest fad to infest sociology. Both *techniques* presume that causation can be inferred from C-I correlations, and both are used as though they are somehow a substitute for theory.

The Structure of a Theory

A theory is not merely a collection of assertions. The intrinsic statements are interrelated syntactically, and the structure can be represented pictographically as in Figure 6-1, which depicts the illustrative theory.

Intrinsic statements. Although the structure of a theory

Figure 6-1
STRUCTURE OF THE ILLUSTRATIVE THEORY*

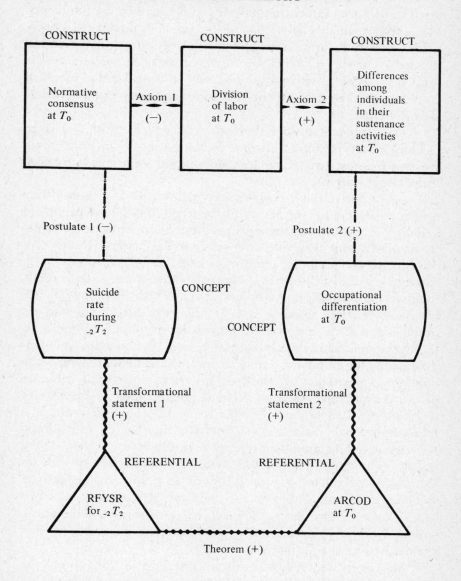

*The symbol "+" signifies an assertion of a direct relation between variables, and "−" signifies an assertion of an inverse relation.

is most clearly revealed by a pictogram, some but not all structural differences among theories can be described by reference to types. Theories differ in structure primarily with regard to the number of each type of direct intrinsic statement (axioms, postulates, propositions, and transformational statements), ignoring unit terms and temporal quantifiers. Thus, two direct intrinsic statements of the same type (such as two postulates) may be considered as identical even though their constituent unit terms and temporal quantifiers are different.

The structure of the illustrative theory is simple. It is an instance of a type 2-2-0-2 theory, meaning a theory in which there are two axioms, two postulates, no propositions, and two transformational statements.

All of the initial direct intrinsic statements of a theory may be propositions; if so, it could be type 0-0-2-3, meaning no axioms, no postulates, two propositions, and three transformational statements. However, only rarely will a theory include propositions as well as axioms and postulates; therefore, unless designated otherwise, the identification of a theory as to type omits reference to propositions. So if a theory is identified as type 1-2, the identification signifies one axiom, two postulates, and no propositions. That abbreviated designation also omits reference to the number of transformational statements. The omission is defensible when the number of transformational statements equals the number of postulates. When there are no postulates, the number of transformational statements is one greater than the number of propositions if each concept is linked to only one referential in a transformational statement. If not, the number of transformational statements should be included in the identification of the theory as to type. Thus, 0-0-2-2 would denote a theory with no axioms, no postulates, two propositions, and two transformational statements. It would signify also that one of the concepts is not linked to a referential.[29] But 2-2, an abbreviated designation, would signify two axioms, two postulates, no propositions, and two transformational statements.

No reference need be made in a designation of type to the number of theorems, as that number can be inferred from the number of transformational statements. The formula is:

$$Th = T(T-1)/2$$

where *Th* is the number of theorems and *T* is the number of transformational statements. In the way of further explication, consider the abbreviated designation of the illustrative theory as 2-2. The designation indicates that there are two transformational statements, and hence there is only one theorem.

Finally, *N* signifies one or more statements of a particular type. Thus *N*-2-2-2 would signify one or more axioms, two postulates, two propositions, and two transformational statements.

The order of a theory.   Still another feature of a theory is the number of substantive terms in the intrinsic statements, a distinction that has to do with the "order" of a theory. A lengthy discussion of the order of theories is undertaken in Chapter 7, and for the present a few explanatory comments will suffice. A first-order theory is one in which the intrinsic statements include only one substantive term, with the form of each statement as an assertion being: All *U*'s are *X*. In that form *U* is a unit term (for example, voluntary associations) and *X* is a construct, concept, or referential. Thus, the assertion "all voluntary associations have eight members" would be a component of a first-order theory.[30] Such a statement asserts that all units are homogeneous as regards some attribute or property; thus, they are likely to be tautologies rather than genuine empirical assertions. In any case, as suggested in Chapter 5, such an assertion is largely alien to sociology, and hence a first-order theory is recognized as only a logical possibility.

As the initial step in theory construction, the theorist is likely to formulate a second-order theory, in which all intrinsic statements link two substantive terms in either of the following forms of an assertion: the greater *X*, the greater *Y*; or the greater *X*, the less *Y*. In subsequent steps he may extend the assertions to a third-order or fourth-order theory. In the case of a third-order theory some of the intrinsic statements link a substantive term and a *relational* substantive term. As to the form of such an assertion, there are two possibilities: (1) the greater *W*, the greater the direct relation between *X* and *Y*; or (2) the greater *W*, the less direct the relation between *X* and *Y*. Finally, a fourth-order theory is one in which some of the direct intrinsic statments link two relational substantive terms in either of the following forms: (1) the greater the direct relation between V and W, the greater the direct relation between X and Y; or (2) the greater the direct relation between V and W, the less direct the relation between X and Y.

In a more complete identification of a theory as to type, a Roman numeral denotes the order. Accordingly, the illustrative theory is type II-2-2-0-2, meaning a theory of the second-order, with two axioms, two postulates, no propositions, and two transformational statements. Since most of the subsequent observations and illustrations relate to second-order theories, the type identification of theories is simplified by the following rule: if the order is not designated, the theory is second-order.

Divisions of a theory. Both the range and intensity of the illustrative theory is minimal. Had more than one type of unit term and/or more than one type of space-time relation entered into the formulation of the theory, it would be more realistic by virtue of being much more complicated.

Reconsider axiom A1. Depending on the theorist's sophistication and judgments pertaining to data, at least three other "versions" of that axiom could have been incorporated in the theory.

Axiom A1b: Among countries, the greater the "absolute increase in the division of labor during T_{0-20}," the less the "absolute increase in normative consensus during T_{10-30}."

Axiom A1c: Among metropolitan areas within a country, the greater the "division of labor at T_0," the less the "normative consensus at T_0."

Axiom A1d: Among metropolitan areas within a country, the greater the "absolute increase in the division of labor during T_{0-20}," the less the "absolute increase in normative consensus during T_{10-30}."

Now these axioms, like the original version (which would become axiom A1a), are only illustrative, but they do suggest some important considerations. The additional axioms would increase both the range and the intensity of the theory, for each axiom differs from the others as regards the unit term and/or the type of space-time relation asserted. Note especially that the intensity of a theory would be furthered without making auxiliary statements (another consideration ignored in the overly simplified illustration).

Had the theorist commenced with four "versions" of the axiom, he would have attempted to formulate four versions of each postulate and each transformational statement. Given four versions of each direct intrinsic statement, the outcome would be at least four theorems, with each theorem derived from a "division" of direct intrinsic statements.

The two notions, versions and divisions of intrinsic statements, introduce some complexities. Two intrinsic statements are considered as versions if their constituent *key substantive words* are the same. Thus, all four of the axioms are versions because they share two key substantive words in common—"division of labor" and "normative consensus." As the illustration suggests, a key substantive word may be thought of as the substantive term without the temporal quantifier (including words that refer to change). So versions of an intrinsic statement differ only in their constituent types of temporal quantifiers *and/or* unit terms.

A "division" of a theory comprises three or more intrinsic statements, interrelated in that: (1) they contain the same unit term, (2) one of the substantive terms in each statement is also present in at least one of the other statements, and (3) all of the statements enter into the derivation of at least one particular theorem. So if additional versions of the axioms, postulates, and transformational statements should be formulated, the illustrative theory would become only a *division* of a more extensive theory.

It must be emphasized that the unit term is the same in all intrinsic statements of a division. Further, two sets of intrinsic statements with the same unit term would not be separate divisions of a theory if one particular complete substantive term (including the temporal quantifier and reference to change) is present in both sets. If one or more complete substantive terms is present in both sets and the unit term is the same for the two sets, then the two are components of the same division of the theory. Conversely, two sets of statements with the same unit term are distinct divisions of a theory if: (1) none of the complete substantive terms (including references to change) in one set are present in the other, (2) at least one theorem can be derived from each set, and (3) the theorem or theorems of one set assert a type of space-time relation that is different from that asserted by the theorem or theorems of the other set. Note especially that two divisions of a theory may or may not have the same unit terms, but no set of statements can be a division unless the unit term is the same in each intrinsic statement. So even though the substantive terms may be identical for two sets of statements, the two could be divisions if the unit term is not the same in both sets.

The foregoing does not imply that all of the intrinsic statements in a division assert the same type of space-time relation or contain the same temporal quantifiers. To illustrate, suppose that after formulating axiom A1b, the theorist formulates a version of axiom A2 as axiom A2b: Among countries, the greater the "absolute increase in the division of labor during T_{0-20}," the greater the "differences among individuals in their sustenance activities at T_{20}." Now whereas axiom A1b asserts a type C-VI-1 space-time relation, the type for axiom A2b is C-III-1. Nonetheless, axioms A1b and A2b could be members of the same division because (1) the same unit term appears in the two statements and (2) one particular substantive term ("absolute increase in the division of labor during T_{0-20}") appears in both statements.

What appear to be different divisions of intrinsic statements may actually be one division. For example, suppose the theorist had worded axiom A2b as follows: Among countries, the greater the "absolute increase in the division of labor during T_{0-20}," the greater the "differences among individuals in their sustenance activities at T_0." That version of A2b and A1b would still be components of a division, but the division would also include all of the original theory, because two of the direct intrinsic statements in the original theory (axiom A2 and postulate P2) also include the substantive term "differences among individuals in their sustenance activities at T_0." Further, without additional axioms and/or postulates, A1b and the second version of A2b would be superfluous, for they would not generate additional theorems.

For each direct intrinsic statement in one division there is a corresponding statement or version in all other divisions. Viewed that way, each division of a theory is essentially a duplication of any other division, the only difference between any two divisions being the unit term and/or the temporal quantifiers of corresponding key substantive words. However, if the unit term is the same in two divisions, then the theorems of one division must assert types of space-time relations not asserted by the theorems of the other division.

As suggested previously, key substantive words (for instance, "normative consensus") are identical for any two divisions of a theory. If the same unit term is present in both divisions, then the type of temporal quantifier of any key substantive word in one division must be different from the type of temporal quantifier of the same key substantive word in the other division. But if the key substantive words are not the same in two sets of intrinsic statements, the two sets are not divisions of the same theory.

The formulation of divisions of intrinsic statements has both advantages and disadvantages. On the positive side, in addition to furthering range and/or intensity, divisions of a theory facilitate tests, interpretation of tests, and the modification of a theory in light of test results. When a theorist promulgates a theory, he asserts that it is testable; but the resources of an investigator may be so limited that he can undertake only partial tests of the theory. Since each division of the theory generates at least one theorem, the arrangement facilitates partial tests (that is, a test of one division). Indeed, regardless of resources, each division is to be tested separately, as though it is a theory in itself, which in a sense it is.

As we shall see, with just a few axioms and postulates, the interpretation of test results can be very complicated. The grouping of direct intrinsic statements into divisions simplifies the interpretation of test results. No less important, divisions make it much less difficult to modify the theory after tests. The test results may be such that some divisions are rejected entirely, others are modified, and still others are retained without alteration. Further, if test results vary markedly from one division of a theory to the next, the probable reason is obvious. Such variation indicates that the theory should be reformulated so as to exclude some unit terms and/or some types of space-time relations.

The one major disadvantage of divisions is that they reduce

the unity of the theory, even though the relation between any two divisions is not purely nominal. For each intrinsic statement in one division there is a counterpart in all other divisions, and the two statements include the same "key substantive words," if not the same temporal quantifiers. The only difference between two intrinsic divisions may be the unit term. Nonetheless, divisions reduce the testability of a theory. For example, if there are two divisions and each generates only one theorem, then the integration of the two divisions would result in four additional theorems.

The integration of two divisions is a simple matter. An integrating statement may take the form of an axiom that links a construct in one division with a construct in another, a postulate that links a construct in one division with a concept in another, *or* a proposition that links a concept in one division with a concept in another. Given such an integrating statement, the two divisions become one and the number of theorems increases exponentially. Although the advantages of integration are obvious, whether or not an integrating statement should be made is strictly a matter of the theorist's judgment. The only formal limit on his judgment is that divisions with different unit terms cannot be integrated.

Three Roman numerals are used to signify divisions of a theory and the difference among those divisions. The first Roman numeral represents the number of divisions, the second the number of unit terms, and the third the number of different *sets* of theorems derived from the divisions (a set of theorems is one or more theorems derived from a division, and two sets are different if they do not assert the same type of space-time relation). Thus, II-II-II would signify two divisions, two unit terms, and two unique sets of theorems. The identification implies that the two divisions do not have the same unit term and that the two sets of theorems are unique. By contrast, IV-I-IV would imply that the unit term is the same in all four divisions but each of the four sets of theorems is unique. As a final example, IV-II-III implies at least two of the divisions have the same unit term and that two of the four sets of theorems assert the same types of space-time relations.

Note that when there are no divisions of a theory, all of the intrinsic statements are taken as one division. Thus, the illustrative theory would be designated as I-I-I, signifying that there is only one group of intrinsic statements.

An inclusive typology.   All of the foregoing distinctions can be combined to form an inclusive typology. In the identification of a theory as to type, the first four Roman numerals denote in sequence: the order of the theory, the number of divisions, the number of different unit terms, and the number of sets of theorems that assert different types of space-time relations. The Arabic numerals complete the identification of a theory, denoting in sequence the number of axioms, postulates, propositions, and transformational statements. So the complete type identification of the illustrative theory is II-I-I-I-2-2-0-2, signifying a theory of the second order, with no divisions, only one unit term, only one set of theorems, two axioms, two postulates, no propositions, and two transformational statements.

It is appreciated that the identification is cumbersome, but it facilitates denotation of certain structural features of a theory. Moreover, abbreviated designations will ordinarily suffice. If the theory is second-order, the first Roman numeral is omitted; and if there is only one set of intrinsic statements (that is, no divisions), the second, third, and fourth Roman numeral is omitted. Finally, as explained previously, unless designated otherwise, there are no propositions and the number of transformational statements can be inferred from the number of postulates or from the number of propositions if there are no postulates. So the abbreviated designation of the illustrative theory is simply 2-2.

Certain variations in the way of abbreviations should be noted. If only one Roman numeral appears in an abbreviated type identification, it designates the order of the theory and signifies no divisions of the theory. Thus, III-0-0-2-3 would signify a third-order theory with only one group of intrinsic statements, no axioms, no postulates, two propositions, and three transformational statements. As another example, IV-II-II-0-0-3-3 would signify a second-order theory with four divisions, two unit terms, and two unique sets of theorems, no axioms, no postulates, three propositions, and three transformational statements. Observe that since only three Roman numerals are shown, the theory is second-order. Further, since the number of transformational statements is shown, the identification reveals that each concept is not linked to one and only one referential, and hence the number of transformational statements is not one greater than the number of propositions.

Use of the typology. The inclusive typology is a convenient nonpictorial way to represent some structural features of a theory, and as such it facilitates the explication of formal theory construction. But the typology also can be used in the assessment of a theory, especially certain aspects of its predictive power.

The identification of the theory as to type does not reveal its testability, predictive accuracy, or discrimination, none of which can be assessed without actual tests. True, one may speak of "potential" testability, but as far as the structure of a theory is concerned, potential testability is strictly a matter of the number of theorems.[31] That number can be inferred from the identification of a theory as to type. The formula is

$$Th = D[T(T-1)/2$$

where *Th* is the number of theorems, *D* the number of divisions, and *T* the number of transformational statements.

The predictive range of a theory is revealed directly by its type. It is simply the number of different unit terms that appear in the intrinsic statements, and that number appears as the third symbol in the type identification.

The intensity of a theory is the number of different types of space-time relations asserted in the theorems. The exact number can be determined only by inspection of the theory, for each division may generate several theorems, and it may be that not all of the theorems in the same division assert the same type of space-time relation. For that matter, even if there are no divisions of a theory, the theorems may not assert the same type of space-time relation. In any case, the intensity of a theory may be furthered by auxiliary statements, and those statements are not revealed by the type identification. However, *minimal* intensity is revealed by the fourth Roman numeral in the type identification (that is, the number of unique sets of theorems).

The scope of a theory is determined by the combined number of constructs, concepts, and referentials, excluding duplications created by variation in unit terms and temporal quantifiers. That number can be inferred from the type identification of a theory by the formula:

$$(A + P + Pr + T) + 1$$

where A is the number of axioms, P the number of postulates, Pr the number of propositions, and T the number of transformational statements.

Finally, parsimony as an aspect of predictive power is the ratio of the number of theorems to the toal number of axioms, postulates, and propositions. Since parsimony is the same for all divisions of a theory (presuming that there are divisions), it can be inferrred from the type identification as follows:

$$[T(T-1)/2]/(A + P + Pr)$$

Limitations of the Illustrative Theory

Since the theory presented in this chapter was formulated only to serve as a simple illustration, it is not to be taken seriously. No tests of it have been undertaken, and its predictive accuracy is probably unimpressive. Nonetheless, it is worthwhile to assess the theory, first as though it were "genuine" rather than illustrative, and second with regard to aspects of formal theory construction not revealed by the illustration.

Predictive power.   Although one may assess the testability of a theory in various ways, until tests have actually been conducted any assessment is questionable. So the testability of a theory is to be judged by only one consideration—the number of tests reported. As such, the testability of the illustrative theory, like so many others in sociology, is at present nil. However, unlike most sociological theories that deal with quantitative properties, formulas are specified and the theorem is explicit as to the relation that should hold among the referents. Moreover, the extrinsic part of the theory would stipulate that the referentials are to be applied to published data, and such data are available. As for "potential" testability, there is no doubt that the theory could be tested. But if "potential" testability is judged solely by the structure of the theory (that is, formally), it is minimal, as there is only one theorem. Further, since there is only one theorem, the notion of discrimination is not applicable.

Because no tests have been undertaken, nothing can be said for the predictive accuracy of the theory. Moreover, if tests should indicate that the predictive accuracy of the theory is only moderate, there is at least one contending theory on the suicide rate (Gibbs and Martin, 1964) with more than moderate predictive accuracy, so it is unlikely that the theory would be retained if tests were undertaken.

One most conspicuous shortcoming of the theory is its limited range. Indeed, since only one unit term appears in the intrinsic statements, range is minimal if judged absolutely and far less than that of contending theories.

What has been said of the range of the theory applies also to its intensity. It is minimal because only one type of space-time relation (C-I) is asserted. However, the intensity of contending theories is also minimal, largely because they were formulated without any stipulation akin to temporal quantifiers.

The scope of the illustrative theory is 7, which is three greater than the minimum (a theory comprising only one proposition and two transformational statements). It is very difficult to judge the scope of contending theories that are stated discursively, but it appears that any of them, especially Durkheim's theory on the suicide rate, far exceed the illustrative theory.

The parsimony of the illustrative theory is .25. That value is unimpressive by any standard, but it should be noted that the parsimony of discursively formulated theories cannot even be judged systematically. Further, the only formally stated theory on the suicide rate (see Gibbs and Martin, 1964) is even less parsimonious than the illustrative theory.

Illustrative limitations. The most conspicuous limitation of the theory as an illustration is its simplicity. For reasons indicated in a later chapter, a type 0-2 or N-2 theory (only two transformational statements) should be avoided if at all possible; so the illustration is decidedly not a desirable standard for theory construction.

Still another limitation is not obvious. By taking an interpretation of Durkheim's observations as a point of departure, the illustrative effort may have created the impression that the construction of a theory always commences at the "top," that is, with axioms or postulates. On the contrary, the construction of a theory may commence at any level, meaning any type of statement; as such, it does not preclude what Glaser and Strauss (1967) have designated as "grounded" theory, meaning a theory that incorporates putative empirical uniformities. As already suggested, the formulation of a theory may commence with a proposition, and it may even take what later is identified as a theorem as the point of departure. To illustrate, suppose that an investigator happens to discover a direct relation between two sets of values corresponding to ARCOD and RFYSR for countries. Now in a sense the investigator would already have the referential formulas, transformational statements, unit terms, and the concepts, for it is most unlikely that anyone applies a formula to data without some concept in mind. So, conceivably, the investigator could take four additional steps—formulate the equivalent to postulate P1, postulate P2, axiom A1, and axiom A2 in the illustrative theory. If so, he has formulated the same theory but by a different route.

There is nothing wrong with taking a putative empirical uniformity as the point of departure in formulating a theory. It at least may lead to propositions, postulates, and/or axioms that integrate the uniformity with other theories. However, it should be recognized that in a strict sense such a theory does not create more order than already recognized in the uniformity taken as the point of departure. Accordingly, in constructing a theory from the "bottom" (that is, partially by induction), the goal should be a set of intrinsic statements from which one can derive theorems that extend beyond the putative uniformity which initiated the formulation of the theory.

Corollaries

As a final consideration, the sign rule implies a relation between all substantive terms in the theory (or a division thereof). For example, in the illustrative theory an inverse relation between "normative consensus at T_0" and 'occupational differentiation at T_0' is implied, and such a relation could be expressed in a formal derivation. The derivation would take the form of a postulate, since it would link a construct and a concept. However, such a statement is identified as an *implied* postulate to distinguish it from a postulate in the form of a direct intrinsic statement. Similarly, one may speak of implied axioms and implied propositions. Finally, if a theory contains any axioms, then relations between constructs and referentials are implied, and such a statement is identified as *circumventive* to signify that intervening postulates and transformational statements have been by-passed.

Generically speaking, implied axioms, implied postulates, implied propositions, and circumventive statements are all corollaries. They receive scant attention for one simple reason—no corollary enters into tests of a theory. In other words, the primary goal in theory construction is the derivation of theorems, and corollaries are of equal significance only if one accepts Zetterberg's version of the axiomatic method, according to which all direct statements are postulates and all derivations are theorems, regardless of their constituent substantive terms.[32] Corollaries may be useful for some purposes, but they do not facilitate tests of a theory, and hence in the present scheme they are considered to be of minor importance.

Notes for Chapter 6

[1] As put by Popper: "science is not a system of concepts but rather a system of *statements*" (1965:35).

[2] The meaning of some terms in a deductively formulated theory is designated by "postulation" (see Northrop, 1947:141), and in some theories axioms are "implicit definitions" of their constituent terms (see Popper, 1965:72). However, no such interpretation alters a fundamental principle in the proposed mode of theory construction—the definition of a construct is not regarded by the theorist as either complete or empirically applicable. Above all, axioms and other types of direct intrinsic statements are not to be interpreted as conventions or taken as true by definition; rather, they are empirical assertions that cannot be tested directly.

[3] The presumption is that derivations are made through a logic of quantitative relations and not a logic of classes or qualitative attributes.

[4] The definition differs markedly from Dubin's conception (1969:chap. 7), that propositions are derived from "laws," and each proposition takes the

form (evidently) of a prediction about a specific value. Thus, evidently, if a law should assert a relation among urban areas between population size and density, then for each size value (such as 245,000) a proposition predicts a particular density value (such as 6,735 per square mile). Given grave doubts about the quality of sociological data, predictions of the form advocated by Dubin are pointless. In any case, Dubin does not stipulate rules by which a proposition is derived from a law.

[5] For comments on this view, see Hempel (1952:31). The character of axioms and postulates is contrary not only to the tenets of radical operationalism but also to what Popper (1965:40) designates as the "positivistic dogma of meaning," according to which a statement is meaningless if it is not possible to verify it. Popper cites Schlick and Waismann's arguments as illustrations, but it does not follow that the dogma is "positivistic." Certainly the dogma is alien to the proposed mode of theory construction, which some critics will regard as positivistic.

[6] In other words, the one and only "justification" for a deductive theory is that it may identify or create order. Of course, some critics demand more justification, and the debate is one of the most durable in the philosophy of science (see Feigl and Maxwell, 1961:esp.1-42,57-89,140-195; and Feigl and Scriven, 1956:3-154).

[7] The problem of bridging the gap between theory and research is not "measurement error" (Blalock, 1968a:12). Whatever the "measurement error" of a formula (or procedure) may be, the formula has no connection with a theory unless reference to it is made in a transformational statement.

[8] The following passages are indicative of what passes for rules of correspondence. "Actual correspondence between the nominal meaning of concepts in the model and the operational meaning of measures in the operational system must exist. Therefore, identity of meaning between these levels will be considered to have been established if and only if both isomorphism of relational structure *and* adequate correspondence of nominal and operational meaning obtain" (Willer, 1967:83). "In the case of the theory model, correspondence between nominal and operational definitions can be inferred if thinking with the model gives essentially identical results to those found by the application of the operational system" (Willer, 1967:91). Let us hope that those who promulgate such "rules" of correspondence never try their hand at devising a criminal code.

[9] The myth is nourished by the idea that there can be a "logic of discovery" (see Hanson, 1961, and Feyerabend, 1961, for a debate of the question).

[10] Since the theorist thinks (or should think) of a transformational statement as an empirical assertion, a concept can be linked to more than one referential. Accordingly, even if critics persist in thinking of a referential formula as an "operation," contrary to the tenets of operationalism a concept is not to be equated with a referential formula, let alone one particular referential

formula. See Blalock (1968a:11) for a discussion of the issue in connection with operationalism.

[11] Consider statements by Dubin, who makes extensive use of the term "indicator" in his mode of theory construction. "An empirical indicator is an operation employed by a researcher to secure measurements of values on a unit" (1969:184). "Race is an absolute indicator; sex is an absolute indicator; age is an absolute indicator. The resultant values of employing these absolute indicators, such as white, Negro, Oriental, or male, female, or 21 years, 33 years, 47 years, can have no other referent than race, sex, and age, respectively" (1969:197-198). "Worker absenteeism in an industrial plant may be taken as an empirical indicator of morale" (1969:200). Are race, sex, age, and worker absenteeism operations? Dubin's terminology is all the more confusing since he earlier (1969:53) identified age as an enumerative unit (property of a thing). So one and the same term, "age," is used to designate a property, an operation, and a particular value.

[12] Consider the question as it applies to "indicators." Stinchcombe (1968) ignores it altogether. Zetterberg (1965:115) writes as though the relation between an indicator and the theory (or some component of it) is purely logical. However, he recognizes four possibilities in the way of a logical relation, but stipulates no rules by which three of them are established in any particular instance. For example, with reference to one of Zetterberg's illustrations (1965:117), by what rule of syntax or semantics does the statement "I like it here in X-town" imply that the individual making the statement is satisfied with his work? Blalock (1969:151) attempts to avoid the question by suggesting that indicators are components of an "auxiliary" theory. Dubin states: "The empirical indicators employed in a study should be homologous with the class of unit built into the theoretical model. This is a logical test that the empirical indicator must meet" (1969:197). But he specifies no rules of syntax or semantics by which one moves from a class of unit to the designation of an indicator.

[13] Blalock (1969), Dubin (1969), Stinchcombe (1968), Willer (1967), and Zetterberg (1965) do not speak directly to the question, but it appears that they would have investigators decide what is an appropriate operational definition or indicator.

[14] Schrag (1967) refers to this notion as the "transitive" rule.

[15] Even if the commonly recognized requirements of an "axiomatized system" (see Popper, 1965:71) are accepted, it does not follow that there can be only one logic or mode of deriving theorems. However, a theory constructed in accordance with the proposed mode of formalization can and should satisfy those commonly recognized requirements; specifically, the intrinsic statements should be (1) consistent, (2) independent, (3) sufficient, and (4) necessary. If a theory is constructed in accordance with the proposed mode of formalization, a failure to satisfy any of the four requirements can be detected readily, which is not the case for the discursive mode of theory construction.

[16] See Costner and Leik, 1964, and Blalock, 1969, especially 12-21.

[17] As an instance see Maris, 1969:51.

[18] Northrop goes so far as to state that "the mathematical physicist's method of verification . . . commits a formal logical fallacy of the hypothetical syllogism" (1947:108). See also Northrop, 1947:146-147. Having recognized that a theory cannot be taken as verified by tests of the theorems without the fallacy of affirming the consequent, Northrop invokes another criterion: "One must go further and show as far as is possible that the theory in question is the only one which is capable, through its deductive consequence, of taking care of the natural history data" (1947:149). In other words, a theory must be "unique in its capacity to give rise deductively to the confirmed consequences" (1947:148). Restating Northrop, a set of direct intrinsic statements may be taken as true if the derived theorems are confirmed and derivable only from those direct intrinsic statements. The "uniqueness" criterion is widely accepted, but it cannot be used to verify a theory. One of the problems is identical to that entailed in the notion of a "unique" explanation. Even if a set of theorems could be "confirmed," one cannot possibly know, let alone demonstrate, that the theorems are derivable from one and only one set of direct intrinsic statements. Moreover, Northrop ignores the dilemma of induction in speaking of theorems as "confirmed." No statement in the form of a universal generalization can be confirmed (see Popper, 1965), and for that reason alone the truth of direct intrinsic statements cannot be established by evidence in support of the derived theorems, as the theorems themselves cannot be confirmed conclusively.

[19] Northrop (1947:146) joins others (for example, Popper, 1965) in holding that the premises of a theory can be falsified by evidence contrary to the consequences (theorems). However, he speaks of the falsification argument as "formally valid," and that qualification is crucial. Insofar as there is any basis for doubt about evidence contrary to the theorems, then the direct intrinsic statements are not conclusively falsified. As will be shown in Chapters 8 and 9, evidence contrary to theorems is inherently questionable, especially in sociology. So the argument that a theory can be falsified is accepted only insofar as one regards evidence contrary to theorems as conclusive (that is, beyond question). The falsification argument is complicated further by recognition that even conclusive evidence contrary to the theorems does not falsify any particular intrinsic statement. Given such evidence, the only justifiable conclusion is that *one or more* of the direct intrinsic statements is false. As shown in Chapter 9, tests can be interpreted in such a way that at least in some cases certain direct intrinsic statements can be identified as *probably* false, but the interpretation is never free of uncertainties.

[20] As shown in Chapters 8 and 9, the idea of a directly testable universal generalization is questionable. In attempting to test a universal generalization, additional assertions must be made, and those additional assertions make the interpretation of a test difficult and inherently debatable.

[21] Nothing is gained by presuming that assertions about the relations between sociological variables should be in "probabilistic" form. Indeed, given

the debates over the meaning of probability and the argument that probabilistic or stochastic statements are not falsifiable even in principle (see Popper, 1965:chap.VIII), there is good reason to avoid probabilistic assertions in theory construction, especially when the intrinsic statements assert relations between quantitative properties. To be sure, the asserted relation is "problematical" in one sense or another, but an assertion in probabilistic form does not make the relation any less problematical. Sociologists are preoccupied with probablistic statements in the belief that the relations among sociological variables are not sufficiently close to warrant assertions in any other form. However, the implied argument is that a nonprobabilistic assertion about a relation should be rejected when evidence indicates that the relation is not close. On the contrary, if the predictive accuracy of a theory is judged relative to condenders, then nonprobabilistic assertions may be retained despite evidence that the relations are not close in an absolute sense. Further, a relativistic or competitive criterion for evaluating theories encourages the modification or qualification of nonprobabilistic assertions with a view to maximizing predictive accuracy. It is specious, then, to argue that any assertion about aggregate or collective properties (see Lazarsfeld and Menzel, 1961), such as rates, is probabilistic by the very nature of the properties. The distinction between probabilistic and nonprobabilistic assertions is a matter of form rather than an empirical consideration. In that connection, examine the simplest form of a probability statement: "the probability for a case of P to be a case of Q is r" (Hempel, 1958:39). When such a statement is made about sociological phenomena, it can be misleading. It may mean that *up to this point* the proportion of cases of P found to be cases of Q is r. As such, the statement is not even an assertion; rather, it is a report of findings, and those findings give rise to an assertion only through crude induction. Even if a sociological theory could be constructed so as to derive such a statement *as an assertion*, the assertion would be dubious for a reason other than the question of falsifiability. When it comes to sociological phenomena, there is no basis whatever to suppose that any proportion akin to r is even approximately invariant. On the contrary, what research findings suggest is variance, and the challenge in theory construction is to identify the correlates of that variance, which is what theorists do when they assert a relation between structural properties and a rate or even a relation between two rates. Such assertions are not and need not be stated in probabilistic form.

[22] There is only one way to exclude the word "causes" from theorems when it is a relational term in the direct intrinsic statements. If rules of inference are formulated, then the causal assertions in the direct intrinsic statements can be translated into the language of space-time relations and that language used in the statement of theorems. However, the translation entails all of the problems and issues examined in Chapter 2. Those problems and issues are not circumvented by using nomic terms other than "causes," for instance, depends on, determines, and produces.

[23] If one of a set of direct intrinsic statements is derivable from the others, the statements are not "independent," which is one requirement of an "axiomatized system" (Popper, 1965:71). In discussing such requirements,

Popper speaks of axioms, but an axiom may be construed as only one type of direct intrinsic statement. Nonetheless, the requirements stipulated by Popper apply to the direct intrinsic statements of a theory, even though Popper might not consider a transformational statement to be an axiom. Popper and other philosophers of science evidently presume that the language of theory construction can be the same for all fields, a presumption that is denied here. Even in the same field it may not be feasible to employ the same language of theory construction, as the types of statements that a theorist makes about qualitative properties may not be appropriate for quantitative properties.

[24] "Explanation in science is indeed not syllogistic, at least in the more interesting cases, namely, those involving quantification, relations, or both. But I think it may be said safely that however the half-educated plain man may use the term, no scientist since Galileo has so narrowly conceived 'deduction' and certainly no logician for almost a century" (Brodbeck, 1962:240).

[25] In using the logic of classes sociologists presume (erroneously) that the putative conventions of a natural language (for example, English) are sufficient for derivations. Such derivations can be made, but they are idiosyncratic rather than systematic, which is to say that the rules of derivation are not explicit and hence there is no assurance that two theorists, working independently, would derive the same statements from a "general" statement. For example, Zetterberg derives the statement "Persons tend to issue prescriptions that maintain the rank they enjoy in their social structure" from the statement "Persons tend to engage in actions that maintain the evaluations they receive from their associates" (1965:80-81). Without explicit and formal rules (which are not stipulated by Zetterberg), it is difficult to see how the first statement "logically" follows from the second. Certainly the syntax of the English language is not an adequate substitute for formal and explicit rules of derivation, nor for that matter is the syntax of sociology's technical vocabulary. Yet sociologists persist in making derivations without explicit and formal rules. What has been said of Zetterberg's illustrative derivation also applies to his "axiomatic format with definitional reduction" (1965:94-96), whereby a list of "propositions" is reduced by combining propositions with definitions. There are various myths associated with the mode of derivation described by Zetterberg. For example, Stinchcombe (1968:16) argues that Durkheim derived various "factual" statements from his theory of egoistic suicide. Quite the reverse is true. The "factual" statements led him to the theory; they were not derived from it.

[26] The point is all the more important in connection with the derivation of theorems. Sociologists who write on the logic of derivations consider only qualitative properties. It is difficult to imagine any such rule or form that is applicable to quantitative properties, and the importance of the difference cannot be exaggerated in theory construction. Consider the following assertion: Among metropolitan areas within the same legal system, the greater the total crime rate, the greater X. Now even if armed robbery is by definition a subclass of crime and if the armed robbery rate is by definition a component of the total crime rate, the subsequent assertion does not "follow" from the

first? Among metropolitan areas within the same legal system, the greater the armed robbery rate, the greater X. The second assertion would follow in part from the first only if a third assertion is made as follows: Among metropolitan areas within the same legal system, the greater the total crime rate, the greater the armed robbery rate. The third assertion is not true by definition; rather, it is an empirical assertion. So at least in some instances an assertion about the relation between quantitative variables cannot be derived from another assertion of the same kind without still a third assertion. Such a mode of derivation is all the more important when the quantitative properties in the premises are not taken as "observable" or, stated another way, are not construed as empirically applicable.

27 Of the writers on theory construction in sociology, Zetterberg (1965: 70-74) has made the most systematic analysis of types of relations between variables. However, he recognized only one temporal distinction (sequential versus a coextensive relation), which is a gross oversimplification; and it is difficult to apply some of his distinctions to the relation between quantitative variables. A less conspicuous deficiency is Zetterberg's failure to integrate his typology of relations with rules for deriving assertions. Certainly such rules do not somehow follow from the typology. Nonetheless, his treatment of the subject is more extensive and systematic than that provided by Dubin (1969), Stinchcombe (1968), Blalock (1969), or Willer (1967).

28 The variety of space-time relations is not fully recognized even in those few studies that have considered types other than C-I. See Pelz and Andrews, 1964.

29 If each concept is linked to a referential but some concepts are linked to more than one referential, then the number of transformational statements exceeds the number of postulates. So the number of transformational statements can be excluded from a type designation only when each concept is linked to one and only one referential. If an abbreviated designation of type excludes the number of transformational statements, the number of propositions must also be excluded, and the reverse is also true. Hence an abbreviated designation never comprises three Arabic numbers, and an abbreviated designation is not feasible if there are any propositions and/or if the number of transformational statements does not equal the number of postulates. In any case, the number of axioms and the number of postulates always are shown, since some constructs may be linked to more than one concept (that is, the number of postulates may exceed $N + 1$ axioms).

30 Zetterberg (1965:65) rightly rejects such "monovariate" assertions as components of a theory.

31 There is a parallel between the present notion of testability and Popper's idea of degree of falsifiability; he even speaks of degree of testability as though synonymous with degree of falsifiability. Observe, however, that Popper considers *potential* testability and not actual tests. The distinction is unimportant only if one chooses to ignore research resources. Further,

Popper's explication (1965:chap.VI) of degree of testability or falsifiability is couched in terms of the conventional logic of classes, and it is not at all clear how his analysis could be applied to theories which assert relations between quantitative properties. Like so many philosophers of science, Popper writes persistently as though there is or can be only one mode of theory construction for all fields and all phenomena. Consider his dictum that a theory "should allow us to deduce, roughly speaking, more *empirical* singular statements than we can deduce from the initial conditions alone" (1965:85). The statement of the criterion is cryptic and, whatever the interpretation, it may be that the applicability of the criterion is contingent on the mode of theory construction.

[32] Zetterberg (1965) does not conceive of the "axiomatic method" as a means of using terms in theory construction that are not defined so as to be complete or empirically applicable. Rather, his version of the method is designed to accomplish two purposes: (1) to reduce the number of "propositions" (premises), and (2) to maximize the number of derivations. To that end, given a list of initial statements, some are taken as postulates and others as theorems, with the distinction having nothing to do with the constituent terms of the statements. Indeed, Zetterberg provides no explicit rules for "selecting" postulates from among the initial statements, being content to say: "One generally strives to use as few postulates as possible" (1965:97). Further, contrary to the principle of "sufficiency" (see Popper, 1965:71), Zetterberg derives some theorems from postulates *and* other theorems, not from postulates alone (1965:98).

CHAPTER 7 ADDITIONAL ILLUSTRATIONS AND HIGH-ORDER THEORIES

The exposition of a theory should be preceded by a brief review of related work, especially contending theories; but in sociology such reviews are usually undertaken to document the importance or significance of the theory, and pronouncements along that line reflect little more than opinion. When there are contending theories, commentaries on them should focus primarily on their predictive power.

This chapter presents another and more complete illustration of a second-order theory. It relates in part to the division of labor, the subject of two classic studies, one by Adam Smith (1805) and the other by Emile Durkheim (1949). Accordingly, should the reader crave pronouncements about the importance of the division of labor, they are found in the works of Durkheim and Smith. In any case, taking those works as theories, they are contenders and a commentary on their predictive power is in order.

Whatever the scope, range, and intensity of the two theories may be, testability is minimal in both cases, and hence an assessment of their predictive accuracy is not feasible. In Adam Smith's theory even the unit term is debatable; it could be countries, industries, organizations, or types of products. Certainly Smith did not explicitly designate the appropriate unit, nor did he distinguish between longitudinal and cross-sectional comparisons, much less specify temporal dimensions. Above all, to test the theory, one must do what Smith did not do—link the two major terms— division of labor and extent of the market—to referentials and stipulate the requisite kind of data.

The unit term in Durkheim's theory is society, but it must be equated with "country" to realize empirical applicability. Further, given his evolutionary perspective, Durkheim may have intended his assertions to be longitudinal rather than cross-sectional; in any case, he used a bewildering variety of relational terms and specified no temporal dimensions. Nor did he link the major substantive terms (division of labor, material density, and ratio of restitutive to repressive law, and so forth) to referentials, even though they designate quantitative properties. One of the terms—material density—can be equated with population density; but Durkheim vacillated between reference to material density and reference to moral density, thereby making the theory ambiguous from the outset. Certainly the two are not synonymous, but the choice between them is debatable. If moral density is chosen, it is difficult to imagine anyone doing what Durkheim did not do—linking the term to an empirically applicable referential formula. On the other hand, if population density is chosen, then a direct relation among countries between population density and the divison of labor appears doubtful.

In light of the foregoing, it is understandable that critics have not conducted tests in assessing either Durkheim's theory or Smith's theory; they evidently accept or reject them as a matter of personal conviction. In any case, an attempt to test either theory would be a thankless task. The investigator would have to complete the theory before conducting tests and, given negative results, apologists for Durkheim or Smith could brand the completion as incorrect.

Of course, some critics will regard the foregoing commentary as a blatant insensitivity to "great works," but that reaction reflects a studied indifference to predictive power in assessing theories. No particular inconsistency is entailed; despite the lip service given to the idea, there is little evidence that sociologists truly want testable theories. On the contrary, they evidently prefer to accept or reject a theory for reasons of "intellectual satisfaction" alone, as that criterion leaves personal opinion sacrosanct.

Other Preliminary Considerations

The illustrative second-order theory in this chapter promises very little predictive power; its range and intensity are minimal, restricted as it is to one type of unit (countries) and one type of space-time relation (C-I). Even its scope is unimpressive. Nonetheless, the limitations are obvious and such that critics can agree on them in assessing the theory.

Although the theory's predictive power clearly is limited, it is testable. But some critics will ignore testability and dismiss the theory after making cryptic noises about explanation, causation, significance, and so on. They are free to do so, of course, but invoking those notions perpetuates anarchy in the assessment of sociological theories.

Sociologists who construct theories formally are handicapped by convention, since formal construction excludes the rhetoric that characterizes sociological theory. The compromise is to present alternative versions of a theory, the first being discursive and the second a formal restatement. That strategy is adopted here, but even in the discursive version the rhetoric is substantially less than in the typical sociological theory.

Finally, some critics will insist that the theory merely refers to the "obvious." The component assertions are certainly not contrary to common beliefs about the division of labor; nonetheless, there is an issue. To believe is one thing, but to bring evidence to bear on a belief is quite another. A formally constructed theory is not merely a collection of beliefs; rather, it is a way of expressing assertions so that evidence can be brought to bear on them systematically.

Discursive Version

Even though the term "division of labor" has been used extensively in the social sciences for nearly 200 years, there are still divergent conceptions of it. Those who use the term commonly fail to distinguish the *degree* and the *bases* of the division of labor. Whereas degree refers to differences among members of a population in their sustenance activities *and* the related exchange of goods and services, a basis of the division of labor is any correlate of those differences. Thus, when only women engage in a particular type of sustenance activity, to that extent sex is one basis (correlate) of the division of labor. Although there can be no basis of the divison of labor without individual differences in sustenance activities, the bases and the degree are nonetheless distinct. To illustrate, even if men and women do not engage in the same types of sustenance activities, that condition does not insure a high degree of division of labor, for it could be that all men engage in the same sustenance activities and so do all women. For that matter, the degree of division of labor could be at a maximum even though there are no bases (that is, no *discernible* correlates of differences in sustenance activities).

As suggested in Chapter 6, the division of labor is multidimensional. One dimension is the differences among individuals in sustenance activities; the other is the exchange of goods and services. However, there is reason to believe that the two dimensions are related directly, so the division of labor can be inferred (in theory at least) from differentiation in sustenance activities alone, that is, without considering the exchange of goods and services. Accordingly, the initial assertion in the present theory is that of a direct relation between the division of labor and differences among individuals in their sustenance activities. Although the assertion may appear to be tautological, it expresses an assumption—that the two dimensions of the division of labor vary directly.

Sustenance activities and occupational differentiation.   It is not feasible to gather data on sustenance activities, especially at the macroscopic level. For one thing, there is no typology of sustenance activities, let alone one that promises substantial agreement among investigators. For another, even if such a typology could be constructed, enormous resources would be required to apply it internationally. Thus, at the macroscopic level differences among individuals in their sustenance activities can be analyzed only indirectly, that is, inferentially.

If only some members of a population engage in a particular type of sustenance activity, they come to be a socially recognized category. To illustrate, when only some members "cut human hair" as a sustenance activity and do so as virtually an exclusive means of livelihood, then the natural language of that population will contain a term that identifies those members as an aggregate. Such a term identifies an occupation, meaning a socially recognized category distinguished solely by reference to type of sustenance activity.

There is no assurance that differences in sustenance activities inevitably give rise to socially recognized categories. As an illustration, even when only some members of a population "cut human hair" as a sustenance activity, the difference may never give rise to a distinctive social label, such as "barber" in English-speaking populations. However, the assumption is that differences in sustenance activities are socially recognized and that the recognition is manifested in occupational titles.

Occupational differentiation.   By definition, occupational differentiation is a function of the number of occupational categories and the "evenness" of the distribution of individuals among them. Given only one category, then obviously there are no occupational differences among individuals; with 500 categories, there are differences but the amount depends on the distribution of individuals. To the extent that individuals are concentrated in any one category, occupational differentiation is less than maximum.

Given the foregoing conceptualization, the degree of occupational differentiation is expressed by the following formula:

$$(1 - [\Sigma X^2/(\Sigma X)^2])/[1 - (1/N)]$$

where X is the number of individuals in a given occupational category and N is the number of categories. The formula may be considered as an operational definition or indicator of occupational differentiation.

Sustenance activities and industries. When individuals differ in their sustenance activities, there is at least some variety in the sustenance activities of the population, but a variety of sustenance activities could be present in a population without individual differences. During one day each individual could engage in a series of distinct sustenance activities (for example, herding animals, felling trees, transporting objects); hence there would be a variety of sustenance activities in the population but no differences among individuals. However, it is assumed that a great variety of sustenance activities is realized only through differences among individuals.

One manifestation of variety in sustenance activities is diversity in the goods and services produced. Now the relation may appear as true by definition, but that is not so. Since some goods may be produced by alternative types of sustenance activities (consider the distinction between "producing" meat by hunting or by the domestication of animals), the variety of sustenance activities could be greater in one population than another even though the variety of goods produced is much the same in the two. Nonetheless, the variety of sustenance activities is assumed to be reflected in the variety of goods and services produced.

As individuals come to differ in respect to the kinds of goods and services they produce, those differences are recognized socially by terms that designate industries rather than occupations. Thus, even though two individuals are identified (occupationally speaking) as secretaries, one may be an employee of the Ford Motor Company and the other of Harvard University; as such, they are in the same occupation but in different industries (manufacturing and education). However, the line of reasoning is much the same for industries as for occupations. Specifically, the variety of goods and services is assumed to be reflected in industry differentiation.

Industry differentiation is a function of the number of industry categories and the evenness of the distribution of individuals among the categories. Given that conceptualization, the degree is expressed by the following formula:

$$(1 - [\Sigma X^2/(\Sigma X)^2])/[1 - (1/N)]$$

where X is the number of individuals in a given industry category and N is the number of categories. Stated another way, the formula is an operational definition or an indicator of industry differentiation.

The division of labor and technology. Consider the implausibility of a commercial pilot not only operating a jet airliner but also building and maintaining it. Similarly, no one physician could invent, build, use, and maintain an instance of each item in the medical technology of the United States. Then consider the ultimate implausibility: one individual capable of inventing, constructing, using, and maintaining all technological items in the United States.

The technology of some populations may be so simple that one individual could master all of its components, but in most countries technological complexity is such as to require a substantial division of labor. Without a division of labor, technological complexity is limited by the ability of exceptional individuals; but with it there is virtually no limit to technological complexity, for each individual can master minute technological components.

Although the foregoing may suggest that the division of labor "causes" technological complexity, the relation is much more complicated. A high degree of division of labor is a necessary condition for a complex technology, but the reverse is also true. If the technology is simple, specialization is not needed, for the typical individual can master most if not all of the technological items. Further, given a simple technology, productivity per worker in agriculture will not yield a surplus, which precludes the development of other industries and occupations. So the causal relation between the division of labor and technological complexity is reciprocal.

Technological complexity is defined as the *variety* of man-made or man-modified objects that reduce the expenditure of energy in the pursuit of physical goals. For various reasons it is not feasible to contemplate gathering data on technological complexity even in one country; so attention is directed to correlates of technological complexity. Specifically, the technology of a country cannot be truly *efficient* unless it is complex. So the two properties of a technology—complexity and efficiency—are presumed to vary directly.

By definition, a technology is efficient to the extent that it reduces the expenditure of human energy in the pursuit of physical goals. The general idea is fairly clear, but it is inconceivable that one could "measure" the efficiency of a nation's technology. However, it is hardly more conceivable that a technology could be very efficient without extensive use of inanimate energy, that is, energy from any source other than living humans, other animals, or plants.

Various national and international statistical agencies report data on the use of inanimate energy. The formula employed to express use per capita is $\Sigma [k (X_1 - X_2)] / P$, where X_1 is the number of units (such as kilograms, cubic meters, or kilowatt hours) of some type of inanimate energy source produced or imported in a year; X_2 is the number exported, placed in reserve stocks, or taken aboard vessels with a foreign destination; k is a constant value that equates one unit of the type of inanimate energy with a kilogram of coal; and P is the population of the country at some point during the year. The formula, like all previous ones, may be considered as an operational definition or indicator.

Three relations. Although various terms have been used in the foregoing observations, each of them are linked, directly or indirectly, to an operational definition or indicator. As such, relations among the operational definitions or indicators can be derived.

One line of reasoning implies that the operational definition or indicator of occupational differentiation is related directly to that of industry differentiation. It also follows, again in terms of operational definitions or indicators, that both industry differentiation and occupational differentiation are directly related to the per capita use of inanimate energy.

Formal Version: Intrinsic Part

Axioms

A1: Among countries, the greater the "variety of goods and services produced during $_0 T_0$," the greater the "variety of sustenance activities at T_0."

A2: Among countries, the greater the "variety of sustenance activities at T_0," the greater the "differences among individuals in their sustenance activities at T_0."

A3: Among countries, the greater the "differences among individuals in their sustenance activities at T_0," the greater the "division of labor at T_0."

A4: Among countries, the greater the "division of labor at T_0," the greater the "technological complexity at T_0."

A5: Among countries, the greater the "technological complexity at T_0," the greater the "technological efficiency at T_0."

Postulates

P1: Among countries, the greater the "variety of goods and services produced during $_0 T_0$," the greater the 'industry differentiation at T_0.'

P2: Among countries, the greater the "differences among individuals in their sustenance activities at T_0," the greater the 'occupational differentiation at T_0.'

P3: Among countries, the greater the "technological efficiency at T_0," the greater the 'per capita use of inanimate energy during $_0 T_0$.'

Transformational Statements

T1: Among countries, the greater the 'industry differentiation at T_0,' the greater the ARCID at T_0.

T2: Among countries, the greater the 'occupational differentiation at T_0,' the greater the ARCOD at T_0.

T3: Among countries, the greater the 'per capita use of inanimate energy during $_0 T_0$,' the greater the RUIE for $_0 T_0$.

Theorems[1]

Th1 (from T1, P1, A1, A2, A3, A4, A5, P3, T3): Among countries, the greater the ARCID at T_0, the greater the RUIE for $_0 T_0$.

Th2 (from T1, P1, A1, A2, P2, T2): Among countries, the greater the ARCID at T_0, the greater the ARCOD at T_0.

Th3 (from T2, P2, A3, A4, A5, P3, and T3): Among countries, the greater the ARCOD at T_0, the greater the RUIE for $_0 T_0$.

Formal Version: Extrinsic Part

Unit term. A country is a territorial unit occupied (in a residential sense) by a politically autonomous population. It may or may not comprise a continuous surface space, that is, some of its territorial components may be separated by water (for instance, the Hawaiian Islands are territorial parts of the United States) or land (East Pakistan and West Pakistan). Finally, a population is politically autonomous if only members of that population make two types of decisions: (1) whether the population as a collectivity shall engage in warfare with other populations and (2) whether the residential movement of people across the territorial boundary of the population is prohibited or limited.

The foregoing definition recognizes conspicuous features of those territorial units commonly recognized as countries, and the definition is to be applied in that sense. Note, however, that an instance of a society, tribe, or culture may or may not qualify as a country.

Substantive terms: constructs. Six of the terms in the intrinsic statements are constructs, each of which is defined below.

Variety of goods and services produced: the number of different types of goods and services produced by members of a population.

Variety of sustenance activities: the number of different types of sustenance activities in which members of a population are engaged.

Division of labor: differences among members of a population in their sustenance activities and the related exchange of goods and services.

Differences among individuals in their sustenance activites: the amount of variation among members of a population as to the way they expend energy in the pursuit of food or in the production of goods or services.

Technological complexity: the variety of man-made or man-modified objects that reduce the expenditure of human energy in the pursuit of physical goals.

Technological efficiency: the amount that human energy is reduced by the use of man-made or man-modified objects in the pursuit of physical goals.

One of the temporal quantifiers, T_0, designates some particular point in time; so for one country it could be April 1, 1970, but for another, June 2, 1971. The other temporal quantifier, $_0T_0$, designates a year that includes T_0. Accordingly, if T_0 for a country is April 1, 1970, then $_0T_0$ would be any one-year period which includes that date. The same temporal quantifiers apply to all substantive terms (constructs or otherwise), and they are so important for referentials (ARCOD, ARCID, and RUIE) that further commentary on the subject is made later.

Substantive terms: concepts. Three concepts enter into the intrinsic statements, each of which is defined below.

Occupation differentiation: a function of the number of occupational categories and the evenness of the distribution of individuals among the categories.

Industry differentiation: a function of the number of industry categories and the evenness of the distribution of individuals among those categories.

Per capita use of inanimate energy: the number of units of energy from sources other than living human beings, animals, or plants used per member of a population.

Substantive terms: referentials. ARCOD, ARCID, and RUIE designate formulas that apply to published data, meaning that only published data are to be used in tests of the theory. Once it becomes feasible to gather international data on occupational composition, industry composition, and use of inanimate energy, such data can be used to test a theory with the same axioms and postulates. However, it would be necessary to change the referential formulas and the stipulations pertaining to data, and those changes would modify the theory.

ARCOD designates the following formula:

$$(1 - [\Sigma X^2 /(\Sigma X)^2]) / [1 - (1/N)]$$

where X is the number of individuals in a given occupational category and N is the number of occupational categories, as tabulated in census reports or in publications of an international statistical agency (such as the United Nations Statistical Office). The presumptions are: (1) occupational data will continue to be published for some countries, (2) investigators can equate certain non-English words with "occupation" and further agree substantially in the translation, (3) they can achieve substantial agreement as to those countries and years for which census reports tabulate occupational data, (4) referents computed for a particular country and year by independent investigators will be sufficiently concordant, and (5) occupational data are sufficiently reliable and comparable.[2]

Most national census agencies tabulate occupational data only for "economically active" individuals, the international equivalent of "in the labor force" or "gainfully occupied." Hence, investigators should not regard the following categories as occupations: housewives, pensioners, retired persons, students, those with no occupation, and any category of individuals who are classified by their relation to someone else (conceivably, a child could be tabulated as a farmer because he depends on his father, a farmer). If included in a census list of occupations, such categories should be excluded before applying the ARCOD formula.

Still other distinctions are introduced in Table 7-1. The table represents thirty-two types of census data on occupations to which the formula can be applied, with each type representing a *version* of ARCOD. Any version can be used but, with a view to furthering the comparability of data, only one version is to be used in each test. However, the suggestion is not that one kind of census data is preferred over others; on the contrary, some of the

Table 7-1
VERSIONS OF ARCOD

			The inclusive set with the least number of occupations or occupational categories		The inclusive set with the greatest number of occupations or occupational categories	
			Including categories designated as unknown, inadequately described, unclassified, other, remaining, or residual***	Excluding categories designated as unknown, inadequately described, unclassified, other, remaining, or residual***	Including categories designated as unknown, inadequately described, unclassified, other, remaining, or residual***	Excluding categories designated as unknown, inadequately described, unclassified, other, remaining, or residual***
Including the unemployed*	Census age minimum**	Both sexes	$ARCOD_1$	$ARCOD_9$	$ARCOD_{17}$	$ARCOD_{25}$
		Males	$ARCOD_2$	$ARCOD_{10}$	$ARCOD_{18}$	$ARCOD_{26}$
	Minimum age 14 or 15	Both sexes	$ARCOD_3$	$ARCOD_{11}$	$ARCOD_{19}$	$ARCOD_{27}$
		Males	$ARCOD_4$	$ARCOD_{12}$	$ARCOD_{20}$	$ARCOD_{28}$
Excluding the unemployed in whole or in part*	Census age minimum**	Both sexes	$ARCOD_5$	$ARCOD_{13}$	$ARCOD_{21}$	$ARCOD_{29}$
		Males	$ARCOD_6$	$ARCOD_{14}$	$ARCOD_{22}$	$ARCOD_{30}$
	Minimum age 14 or 15	Both sexes	$ARCOD_7$	$ARCOD_{15}$	$ARCOD_{23}$	$ARCOD_{31}$
		Males	$ARCOD_8$	$ARCOD_{16}$	$ARCOD_{24}$	$ARCOD_{32}$

*Unless indicated otherwise, the unemployed are presumed to be included.

**Whatever minimum age is stipulated in the census report, if any.

***If any such designation is combined with an occupational title (e.g., former and other), then the category is to be included.

distinctions may not be relevant, which is to say that test results could be approximately the same for different versions of ARCOD.

It would simplify matters to stipulate only one version, but the restriction would reduce testability. To illustrate, if it were stipulated that the formula is to be applied to only one kind of census data (for example, only employed males over 15 years of age in 600 particular occupational categories), then it could be that the requisite data are not tabulated by any national census agency.

Given thirty-two versions of ARCOD, some countries can be included in more than one test. Again, however, it is necessary that each test be based on only one version of ARCOD. To illustrate, if it is possible to compute only $ARCOD_5$ for a particular country and only $ARCOD_7$ for another, those two countries should never be included together in one test. Of course, in some cases the investigator will be able to manipulate census data on occupations so as to compute several versions of ARCOD for a country.

The ARCOD version that will provide the greatest predictive accuracy is not known. After initial tests it may be feasible to exclude all but one version of ARCOD. However, alternative versions of ARCOD do not make the theory untestable; regardless of the ARCOD versions considered, all tests could indicate minimal predictive accuracy. In any case, it is one thing to stipulate alternatives but quite another to specify no formula whatever.

Two distinctions in Table 7-1 require special comment. Some census data on occupations include categories designated as unknown, inadequately described, unclassified, other, remaining, or residual. Their inclusion or exclusion is probably not important, so the distinction is recognized in Table 7-1 primarily to eliminate doubts of investigators.

The other distinction is more important. In some national census reports occupational data are tabulated in several tables and, without special instructions to guide their choice, independent investigators may not use the same one. First, a table must be inclusive. Thus, if a table lists only the occupations of "government workers," it is not suitable, nor is a table that per-

tains to some territorial division of a country. But the occupational categories may not be the same in two tables even though
both are inclusive. A census report commonly includes two occupational tables—one for gross occupational categories and one
for a more detailed list. Ideally, all tests of a theory would be
based on detailed occupations, with no international variation in
the titles; but such a limitation would reduce testability drastically. Nonetheless, the number of categories must be comparable
in some sense from country to country. Accordingly, in each test
all ARCOD referents are to be based on either the greatest number
or the least number of occupational categories tabulated. Should a
national census report contain only one inclusive occupational
table, then the country is to be excluded from tests.

National census reports are the primary data sources, but
occupational statistics published by an international agency can be
used. However, no test is to be based on data obtained *only in part*
from a publication by an international agency. Further, the notion
of the "least" or "greatest" number of occupational categories
takes on a special meaning for data published by an international
agency. As one instance, the United Nations *Demographic Yearbook, 1964* tabulates occupational data for numerous countries,
but the data are tabulated only for gross categories. Nevertheless,
for that source, the gross categories represent the "greatest number," and hence an ARCOD referent computed from that data
would be one of versions 17-32 as shown in Table 7-1.

The temporal quantifiers are especially crucial for ARCOD
and other referentials. For each country T_0 designates a particular
census year (such as 1970, 1971). Thus, if 1970 census data on the
United States are used to compute some version of ARCOD, then
T_0 for that country is 1970, and T_0 for other countries could be
the last census year. However, strictly speaking, T_0 need not be
the same for all countries; it could be 1970 for the United States,
1971 for New Zealand, and 1967 for some other country.

ARCID designates the following formula:

$$(1 - [\Sigma X^2 / (\Sigma X)^2]) / [1 - (1/N)]$$

where X is the number of individuals in a given industry category
and N is the number of categories. As for data, all statements

concerning ARCOD, occupations, and occupational statistics (including sources) apply also to ARCID, industries, and industry statistics. The versions of ARCID parallel the ARCOD versions; thus, the equivalent of ARCOD$_{18}$ is ARCID$_{18}$, the only difference being that in the one case the data pertain to industries and in the other to occupations.

RUIE designates the following formula:

$$\Sigma [k(X_1 - X_2)] / P$$

where X_1 for a country is the number of units (kilograms, cubic meters, or kilowatt hours) of a particular type of inanimate energy source produced or imported during a year; X_2 is the number exported, placed in reserve stocks, or taken aboard vessels with a foreign destination; k is a constant value for the type of inanimate energy source which equates one unit of that energy (whatever the unit may be) with one kilogram of coal; and P is the total population of the country at some point during the year (as reported by the national census agency or an international statistical agency).

The value of each k is to be taken from any publication by an international agency, for substantial agreement is realized on those values. As for the other data, inanimate energy includes all energy from sources other than living human beings, animals, or plants. As such, the relevant data pertain to energy derived from inflammable objects or substances, moving water, wind, geothermal phenomena, solar radiation, nuclear reactions, or gravitational forces. Fortunately, numerous national agencies publish data on the production, exportation, and importation of inanimate energy sources; and at least one international statistical agency, the United Nations Statistical Office, publishes such data. There are doubts about the reliability and comparability of energy data, but RUIE referents of countries differ so much that "inaccuracies," even if known, could be ignored.

The RUIE formula can be applied to any data that purport to represent production, importation, and exportation of inanimate energy sources, provided the data have been compiled by a national or international statistical agency. If per capita figures are reported for countries in one publication, they can be used as RUIE referents without further computations. As a final consideration, the temporal quantifier for RUIE is $_0T_0$, which signifies a one-year period that includes T_0. So, if T_0 for ARCOD or ARCID is some date during 1970, then $_0T_0$ for that country would be the calendar year 1970.

Commentary on the Two Versions

The discursive version of the theory is not a caricature, but its exposition does differ from what is conventional for sociology, and those differences should be noted. First off, it is far more brief than the typical sociological theory, especially those constructed in the grand tradition. Extensive elaboration of a theory may appear desirable, but in sociology it tends to obscure the central assertions. In any case, the brevity of the discursive version is due primarily to the virtual exclusion of rhetoric, especially gratuitous observations on human nature, social life, and human societies. Such observations are a trademark of sociological theory, but the only purpose they serve is catharsis. Still another difference is that unlike the typical sociological theory, the discursive version does incorporate formulas. Since the theory deals with quantitative properties of social units, formulas are necessary for tests, tradition in sociology notwithstanding.

Despite the differences, the discursive version shares much in common with a typical sociological theory, and those similarities are best described by comparing the discursive and formal versions. Unlike the formal version, no particular unit term is employed consistently throughout the discursive version, and the central assertions are ambiguous for that reason alone. For that matter, the central assertions are not identified so as to reveal their position in the theory or the character of their constituent terms. The lack of concern with types of terms or statements is mitigated by the specification of formulas; otherwise, systematic tests of the theory would be precluded altogether. However, the empirical applicability of the formulas is questionable, primarily because requisite data are not described in detail and nothing is said about the procedure for obtaining them. Further, even if investigators could compute three sets of values by application of the stipulated formulas, the theory is not stated so that predictions about the relations among the referents can be derived *systematically*. It is suggested that the relations will be direct, but it is not clear how the relations were derived. The uncertainty stems from the variety of relational terms in the major assertions. Consistency in relational terms is essential for sytematic derivations, but it is only a necessary condition; additionally, there must be explicit rules of derivation, and such rules are alien to a discursive exposition of a theory.

Again, the discursive version of the theory is not a caricature. What has been said of it would apply to virtually all sociological theories; indeed, the discursive version is stated more systematically than the typical sociological theory, and certainly the central assertions are more explicit.

Some considerations excluded from the formal version.
The range and intensity of the illustrative theory are minimal; there is only one unit term (countries) and the temporal quantifiers exclude assertions about change. A theorist may or may not indicate the reasons for restricting the range and intensity of a theory, but his judgment should not become a critical issue in either case. After all, a theory is evaluated in terms of the assertions made, not those omitted.

One reason for the minimal range of the illustrative theory is that data on inanimate energy are not published for intranational territorial divisions (such as metropolitan areas). Another reason is more nearly theoretical. There is evidence that along with economic development the territorial divisions of a country specialize in the production of particular types of goods and services. Such specialization precludes a high degree of industry and/or occupational differentiation within the territorial divisions, but neither is presumed to be a correlate of technological efficiency at the *intranational* level; rather, it may only reflect the position of a unit in the territorial division of labor. So the conclusion is that some of the relations asserted in the theory will be very close only when each unit is ecologically autonomous, that is, when there is no movement of goods, services, or people across the boundary. No territorial unit is ecologically autonomous, but countries approximate that condition more than do other units. Accordingly, the theory is restricted to countries in an attempt to maximize predictive accuracy.

Although a theorist need not stipulate reasons for restricting the range of a theory, eventually he may make the reasons explicit and perhaps extend the theory in a particular way. For example, the theorist could make the following assertion as an *auxiliary statement*: The relation between ARCID at T_0 and RUIE during $_0T_0$ is more direct within a universe of countries than it is among the metropolitan areas of those countries. The auxiliary statement asserts a *differential unit relation*, and it is only one of several that could be made. In additional statements ARCOD could be substituted for ARCID and/or urban areas for metropolitan areas. Still another possibility is that the auxiliary statement could end with the following designation: "among the metropolitan areas of any country."

All such auxiliary statements extend the range of a theory, but they are not somehow implied by the theory. The unit terms and substantive terms that appear in auxiliary statements are strictly a matter of the theorist's judgment, and that judgment should not ignore data considerations. Whatever the reason, a theorist may refrain from making any auxiliary statements. If so, the range of the theory remains limited, but no theory should be rejected merely because its range is limited. It is difficult enough to formulate a testable theory, and to demand more is grossly unrealistic.

The theory's intensity is minimal, for only type C-I relations are asserted. In part, the omission reflects uncertainty about other types of relations, but a practical consideration is paramount. Although the theory tacitly asserts that the designated kind of census and energy data are sufficiently reliable and comparable, that is so only for cross-sectional comparisons, where differences among countries are commonly so great as to justify the assumption that their rank-order would not be appreciably different if absolutely reliable and comparable data could be used. However, short-run changes (for example, ten years) in the ARCID, ARCOD, and RUIE referents are relatively small; and in that context the data are not considered to be sufficiently reliable or comparable.

Needless to say, when the predictive accuracy of two theories is approximately the same, then the theory with the greatest intensity is preferable. But it would be pointless to reject a theory without regard to contenders, for surely a theory with minimal intensity is better than no theory at all.

The extrinsic part of the illustrative theory contains no instructions as to the selection of countries for tests, and that consideration introduces the question of sampling procedures. The mode of theory construction does not stipulate any particular sampling procedure or alternative procedures to the exclusion of others. Instruction as to the selection of units is a substantive consideration best left to the judgment of the theorist. The only requirement is that a theorist not stipulate a procedure that involves the selection of units by considering the relation between referents, nor should an investigator use such a procedure at his discretion.

When a theorist stipulates a sampling procedure (or, more broadly, a selection procedure), that stipulation is part of the theory. Presuming that the theorist is not indifferent to testability, he will consider practical problems (such as research resources) in contemplating alternative procedures, and for that reason alone he is the only one who can or should make the choice. The feasibility of a particular sampling procedure (conventional or otherwise) depends on the theory in question (for example, whether tests are to be based on published data). In that connection, a theorist may not stipulate a sampling or selection procedure, leaving the choice to the discretion of investigators. If so, investigators must select units in some systematic manner and report the procedure in detail. In particular, they must report how the universe of potential units was identified (for instance, all countries in which a census agency conducted a national census between 1960 and 1971, inclusive) and how particular units within that potential universe were selected if the tests are not based on the entire universe. Thus, since it is not stipulated otherwise, investigators would be free to select countries for tests of the illustrative theory as they see fit, provided, of course, that the data stipulated in the explication of the referential formulas are available in published form for each country selected. However, the investigator would be required to stipulate how the universe of potential countries was identified (all countries for which the stipulated data are available for the period 1955-1964) and how selections (if any) were made from that universe (a 20 percent random sample).

Several objections to the prescribed strategy are anticipated. For one thing, if the selection of units is left to the discretion of investigators, then there is nothing to prevent a deliberate selection in such a way as to insure either positive or negative test results. However, science is a human enterprise, and no stipulation of a sampling procedure can curb a dishonest investigator. An investigator can always attempt to select units in such a way as to produce either negative or positive tests; he can then in reporting test results flatly lie about the selection procedure. Nonetheless, critics will persist by arguing that some sampling procedures are "better" than others. It is doubtful if one sampling procedure is superior to all others regardless of the theory in question. Certainly the merits of various alternatives are debatable, especially when practical considerations are entertained.

But there is a larger issue. No sampling procedure can resolve the dilemma of induction. More specifically, if a test is based on a sample, no "test of significance" or related notion can justify a generalization beyond the universe from which the sample was drawn, whatever the procedure may have been. It could do so only if it were possible to draw a random or representative sample from an infinite universe, and that possibility is denied. So sampling and the related probability theory are nothing more than labor-saving devices, not substitutes for substantive theory. Indeed, criteria for judging substantive theories determine the relevance of sampling procedures. If predictive accuracy is judged by an absolute criterion that requires invariant relations, then apart from practical considerations sampling procedures are not needed in tests. Even when a relative criterion is used (that is, invariant relations are not required to retain a theory), the predictive accuracy of a theory should be judged by a series of independent tests rather than by the "statistical significance" of any particular test. The assessment of predictive accuracy by a relative criterion (relative to contenders) does not resolve the dilemma of induction, but neither does a test of significance. Further, with a view to maximizing predictive accuracy, a relative criterion is much more realistic. Any significance level is bound to be arbitrary, and conventional levels in sociology (for example, .01 or .05) are such that tests may be accepted as positive even though absolute predictive accuracy is negligible, and in that sense reliance on a level of statistical significance does not encourage attempts to maximize predictive accuracy.

Theories of a Higher Order

Since there are only two substantive terms in each intrinsic statement of the illustrative theory, the mode of formalization may appear to presuppose that relations between sociological properties are invariant. To the contrary, even an approximate invariant relation is most unlikely. Any relation probably varies from condition to condition, and it is equally probable that maximum predictive accuracy never will be realized. However, a theorist may stipulate what would be an ideal condition, that is, one in which the predictive accuracy of the theory would be maximized. The difficulty is, of course, that the ideal condition may never be realized, so assertions in that connection are largely conjectural. As such, the notion of ideal conditions may be used to circumvent the demand for testable theories. The curious line of reasoning may be something like this: since a theory cannot be tested under ideal conditions, then it is pointless to demand tests at all. The reasoning is all the more dubious in the case of sociology, where theorists habitually fail to specify an ideal condition for tests.

The notion of ideal conditions is accepted, but it should not be construed as justification for the promulgation of untestable theories. Granted that an ideal condition cannot be realized, it can be approximated to varying degrees, and a truly sophisticated theory incorporates assertions about variation in a relation from one type of condition to the next. Such assertions are testable, and the results are suggestive of the relation that would hold under ideal conditions.

The notion of ideal conditions may be incorporated in the initial version of a theory, as when a theorist designates countries as the unit on the presumption that they are more ecologically autonomous than other types of units. So restriction of unit terms is one way to maximize predictive accuracy, but a mode of theory construction should permit alternatives. Specifically, the theorist should be able to word intrinsic statements so as to assert that a relation between two properties is contingent on another property or on another relation between properties. Such statements are components of "higher-order" theories.

A third-order theory. Some intrinsic statements of a third-order theory assume one of four forms: (1) among universes of U's, the greater X for a universe, the greater the direct relation between Y and Z within the universe; (2) among universes of U's, the greater X for a universe, the less direct the relation between Y and Z within the universe; (3) among universes of U's, the greater the intrauniverse variance of X, the greater the direct relation between Y and Z within the universe; or (4) among universes of U's, the greater the intrauniverse variance of X, the less direct the relation between Y and Z within the universe. U is a unit term (for instance, countries), while X, Y, and Z are substantive terms (constructs, concepts, or referentials).

Each form asserts that a relation between two properties is contingent on one or more other properties. Note especially that in tests of a third-order theory *universes of units* are compared rather than units. Thus, if countries are taken as units, one step in tests of a third-order theory is the grouping of countries into universes (as stipulated in the extrinsic part of the theory), with the universes differing as to the magnitude of X or intrauniverse variance in X. In all tests a measure of association (for example, a coefficient of correlation) between two sets of referents (corresponding to the Y and Z properties) is computed for each universe, and a prediction is derived from the theory as to variation in that measure from one universe to the next.

The idea underlying a third-order theory is that the relation between two properties (Y and Z) varies appreciably among universes, with the relation in a given universe contingent on some property, X, of that universe. Now X may be a property of units as well as universes. Thus, if the highest population density of any country in a universe is 250 per square mile, then that value is an attribute of the universe as well as an attribute of a particular country, and the same would be true of the lowest population density. But universes have attributes that do not characterize the component units, such as average population density and intra-universe variance in population density.

Whatever X is, the theorist stipulates that universes are created in such a way as to maximize differences among them with regard to the magnitude of X or intrauniverse variance in X. In either case, the presumption is that some of the universes will approximate an ideal condition more than do the other universes. Accordingly, if the Y-Z relation is contingent on X, then that relation should vary in a predictable manner from one universe to the next.

Universes of units.   In testing a third-order theory, investigators must create universes of units in accordance with the theorist's instructions. Those instructions are distinct from the intrinsic statements, that is, two theorists might make the same third-order assertions but stipulate different procedures for creating universes. Since the assertions and instructions are distinct, an extended commentary on the latter has been delayed to this point.

The predictive accuracy of a third-order theory may depend on the procedure following in creating universes of units, and for that reason alone the choice between alternatives is strictly a matter of the theorist's judgment. His decisions should not be governed by the opinions of "methodologists," nor by the idea that there can be a specific procedure for all theories. Accordingly, the procedures described here are only suggestions.

As a first step, the theorist *stipulates* the number of universes and the number of units in each universe, and that stipulation reflects his assessment of the field's research resources. Obviously, if investigators are to create 10 universes with 20 units in each, then a test of the theory cannot be conducted without securing data and computing referents for 200 units, a requirement that could be unrealistic in light of research resources. Further, when a particular number of universes and units in each are prescribed, investigators cannot modify the procedure so as to adjust for limited research resources. With the foregoing considerations in mind, a theorist should recognize that the testability of a theory can be enhanced by stipulating a minimum number of universes and units (for instance, at least three universes with at least five units in each).

When a theorist asserts that the relation between two properties, Y and Z, is contingent on the magnitude of a third property, X, his instructions can be simple. Suppose that the theorist stipulates that at least four universes with five or more units in each are to be created so as to maximize differences among them regarding the magnitude of X referents. Given that stipulation, an investigator could select as few as twenty units for a test. Once an X referent has been computed for each unit, the units would then be ordered by the magnitude of the referents and each quartile taken as a universe.

When universes are created as just described, the differences among them as regards the magnitude of X are relative, not absolute. However, the theorist can stipulate absolute criteria. For example, when the theorist stipulates that only urban areas with a population exceeding one million are to be included in one universe, only those with a population of between 500,000 and a million in another, and those of less than 500,000 in still another universe, then his instructions call for the creation of universes by absolute criteria. Preliminary to the specification of absolute criteria, the theorist may have to engage in exploratory research to determine the appropriate "cutting points" for differentiating universes. To illustrate, although the theorist may suspect that a direct relation between population size and density appears only when urban areas reach some particular areal size, before framing a third-order assertion he may have to estimate the minimum areal size through exploratory research.

All previous illustrations have suggested that a third-order assertion introduces a third property, that is, the relation between Y and Z is asserted to be contingent on a third property, X. However, the magnitude of Y or Z could be introduced as a universe property. Suppose the theorist asserts a direct relation between Y and Z but suspects that the relation is close only among units with a Y referent above some particular value. If so, his assertion would take the following form: Among universes of U's, the greater Y for a universe, the greater the direct relation between Y and Z within the universe. In such a case the theorist is likely to stipulate that universes be created by reference to absolute Y values. Of course, the same strategy would be followed if the theorist believes that a direct relation between Y and Z appears only among units with Y values below some particular value, in which case his assertion would be: Among universes of U's, the greater the Y for a universe, the less direct the relation between Y and Z within the universe.

The strategy is especially important in contemplating a non-monotonic relation between two properties, such that below a certain Y value the relation between Y and Z changes from direct to inverse or from inverse to direct. Consider the case where the theorist suspects that the Y-Z relation changes from inverse to direct above a certain value of Y. The form of the assertion would be: Among universes of U's, the greater Y for a universe, the more direct the relation between Y and Z within the universe. In such a case the theorist probably will stipulate that universes be created by reference to certain absolute values of Y, and it is equally probable that he will have to engage in exploratory research to determine those values.

When a theorist's instructions call for the creation of a universe so as to maximize differences among them regarding *intra-universe variance* in some property, the explication of procedure is much more complicated. Reconsider one of the forms of a third-order assertion: Among universes of U's, the greater the intra-universe variance in X, the greater the direct relation between Y and Z within the universe. Suppose a theorist makes such an assertion and stipulates the creation of at least three universes with five or more units in each. His instructions should be worded so that they apply regardless of the number of universes or units, but to simplify matters consider how the instructions would apply when an investigator decides to create the minimum number. The theorist's instructions should be worded so that the investigator would: (1) compute an X referent for each of the fifteen units; (2) arrange the units in all possible combinations taken five at a time and designate each combination as a tentative universe; (3) compute the minimum difference (MD) between any two X referents within each tentative universe; (4) arrange all of the tentative universes by the order of the MD values; (5) select the tentative universe with the greatest MD value and designate it as universe I; (6) eliminate all tentative universes that contain a unit included in universe I; (7) from the remaining tentative universe select the one with the greatest MD value and designate it as universe II; (8) eliminate all remaining tentative universes that contain a unit in universe II; and (9) continue the process until the last universe selected represents the lowest MD value. Of course, when only three universes are created, the process ends with step 8, that is, the remaining units would constitute a universe with a low MD value.

The reader will have noted that the first universe selected is characterized by the greatest MD value, but the selection process could have commenced by selecting the universe with the lowest MD value. It does make a difference, but the procedure is not arbitrary, for it can and should be modified in light of the assertion under consideration. In the illustrative case, the assertion is: Among universes of U's, the greater the intrauniverse variance in X, the greater the direct relation between Y and Z within the universe. What the theorist suspects is that a close direct relation between Y and Z holds only in an ideal condition, and he attempts to approximate that condition as nearly as possible by selecting the universes with the greatest variance in X. The assertion is not that an inverse relation between Y and Z will appear in the universe characterized by the least variance in X (the Y-Z correlation in that universe may be positive but approach zero). However, if the theorist thinks of the Y-Z relation as generally inverse but contingent on intrauniverse variance in X, he would put his assertion in a different form: Among universes of U's, the greater the intrauniverse variance in X, the less direct the relation between Y and Z within the universe. In such a case, he would word his instructions so that the first universe selected would be the tentative universe with the lowest MD value, and the last universe selected would be one with the highest MD value.

There may be instances where the theorist suspects that the Y-Z relation varies from inverse to direct, depending on intrauniverse variance in X. The assertion could take either of the two forms, but the instructions for creating universes would be different, for the goal is both to maximize and to minimize intrauniverse variance in X. In such a case, the investigator would be instructed to consider the tentative universe with the greatest MD value and the one with the lowest MD value. If the two universes do not include any unit in common, then both are selected. If they contain one or more units in common, both would be excluded and the investigator would then consider the tentative universe with the second greatest MD value and the one with the second lowest MD value. In brief, the investigator considers successive pairs of universes, with one unit representing a high MD value and the other representing a low MD value, and no pair is selected if they contain any units in common.

Some comments should be made on the suggested procedures for creating universes so as to maximize and/or minimize intrauniverse variance in a property. The identification of tentative universes may appear laborious to the point of being infeasible, even with as few as ten units and taking all possible combinations five at a time. However, the presumption is that investigators have access to a computer, and any of the recommended selection procedures can be programmed readily when the total number of units does not exceed twenty. Still another consideration is the interpretation of variance. Note again that intrauniverse variance is reckoned by a MD value, that is, the *least* difference between *any two* referents in a universe. A more conventional approach would be to compute the standard deviation of the referents in each universe; but variance can be reckoned several ways, and there is a rationale for preferring MD, though not one that a theorist need accept. Consider the case where an investigator creates four universes with five units in each, commencing with an initial list of twenty units. Once all combinations taken five at a time have been identified, the standard deviation of the X referents could be computed for each combination, and the combination with the greatest standard deviation would be selected as representing the greatest intrauniverse variance in X. However, of the five units in the universe, two might be the highest X values in the initial list $(N = 20)$ and two of them might be the lowest X referents. As such, the distribution in that universe would be extremely kurtotic. Moreover, the difference between the units with the two highest X referents could be minute, and the same could be true of the two lowest X referents. Given doubts about the reliability and comparability of sociological data or "measurement error" in general, the ordinal difference between some of the units in the universe would be most questionable. By contrast, if the recommended procedure were used, the universe could comprise units with X referents that approximate the following ranks in the initial list: 1, 5, 10, 15, and 20. Given such a universe, the difference between the X referents of *any two* units would be substantial, and hence there would be less doubt about the "true" ordinal differences among the units as regards their X referents.

Reference has been made to doubts about the reliability and comparability of sociological data, and a third-order theory takes on special significance in that connection. Rather than maximize variance in X, a theorist may maximize variance in Y or Z. To that end he would make a third-order assertion in the following form: Among universes of U's, the greater the intrauniverse variance in Y, the greater the direct relation between Y and Z within the universe. That form is especially important when the theorist has doubts about the reliability and comparability of the referents. By maximizing variance in the Y referents, the theorist hopes to increase the relation between the "true" Y referents (those referents that would be computed if the referential formula were applied correctly to absolutely reliable and comparable data) and the actual Y referents. For illustrative purposes, consider five units (U's) with two Y referents for each, the first being the "true" and the second the actual referent: U_1, 85, 62; U_2, 80, 63; U_3, 50, 55; U_4, 45, 58; and U_5, 40, 50. *Rho* between the "true" and actual referents ($N = 5$) is $+.80$, but when a universe is selected so as to maximize variance in the actual referents (taking U_1, U_3, and U_5), *rho* between the "true" and actual referents in that universe is $+1.00$. Of course, the relation between the "true" and actual referents is unknown and unknowable, but the theorist assumes that maximizing variance increases the ordinal relation between the "true" and actual referents. He further assumes that maximizing the variance in Y also maximizes the variance in Z, and the expectation is a closer correlation between Y and Z referents than would be the case without maximization of variance. Given doubts about the reliability and comparability of sociological data, the strategy should be a common practice.

As an alternative to the suggested procedures, a theorist may stipulate that universes are to be created so as to realize some absolute amount of intrauniverse variance, for example, in one universe MD is to be 75 or more, in another between 10 and 74, and in still another less than 10. Investigators could identify such universes by examining all possible combinations of units, but the testability of the theory is likely to be reduced substantially. Investigators may not have the resources to compile an initial list of units such that universes can be created in accordance with the stipulated criteria. In any case, it is most unlikely that a theorist can stipulate absolute criteria of intrauniverse variance without first engaging in exploratory research.

When universes of units are created by any one of the foregoing procedures, the presumption is that one of them approximates some ideal condition more than do the others. However, the theorist is free to stipulate any method for creating or identifying universes, and he may do so without thinking of an ideal condition. For example, he may specify that each universe is to be a "natural" aggregate of units, but he must clearly identify those aggregates, such as "all urban areas in a country" or "all voluntary associations in an urban area." In specifying that "natural" aggregates of units are to be treated as universes, the theorist presumes that the Y-Z relation varies appreciably among such universes and that the variation is contingent on some property.

As a final consideration, note that universes can be created within a natural aggregate of units. For example, the theorist may stipulate that the metropolitan areas of a country are to be divided into universes in such a way as to maximize differences among the universes. If so, the initial phrase of his third-order assertion would be "Among universes of metropolitan areas in a country."

An issue.   The idea that a relation between two properties may be contingent on other properties is relevant in contemplating Blalock's objections (1969) and those of Costner and Leik (1964) to symmetric or covariance statements in theory construction.

One of their objections is that such statements make derivations ambiguous or debatable, and as an illustration Blalock analyzes three statements numbered as follows: "(7) The greater the X (division of labor), the greater the Y (solidarity). (8′) The greater the Z (consensus), the greater the Y (solidarity). (1) Therefore, the greater the X (division of labor), the greater the Z (consensus)" (1969: 15). He then states: "Most readers will undoubtedly object that proposition 1 does not obviously follow from 7 and 8′, and in so doing will in effect admit that the symmetry-asymmetry question is crucial" (1969: 15).

Blalock distorts the issue at the outset by failing to make the rules of derivation and assumptions about the asserted relations explicit. Given the rules prescribed in Chapter 6, statement 1 does follow from 7 and 8′. As for the empirical validity of statement 1, the theorist should not make statements 7 and 8′ unless he assumes that the relations are close. If he believes that neither relation is close except under certain conditions, then he should extend both statements to a third-order form, with two possibilities being: (1) among universes of countries, the greater the intrauniverse variance of X, the less direct the relation between Z and Y within the universe; (2) among universes of countries, the greater the intrauniverse variance of Z, the less direct the relation between X and Y within the universe. Of course, the way a theorist extends a second-order theory depends on his conception of the relations among the properties. His conception may include causal notions, but the issue is not whether one should think in causal terms. The issue is whether one should use a causal language in the exposition of a theory.

Now consider a "causal" version of statements 7 and 8′ as suggested by one of Blalock's cryptic diagrams (1969: 15). (7a) X causes Y. (8′a) Z causes Y. Blalock does not stipulate rules that would permit any derivations from the two versions and, given commonly held conceptions of causation, the statements are inconsistent. Of course, he may argue that "exclusive" causation is not asserted in either statement, but his conception of causation is so incomplete that the implications of his causal statements and diagrams are conjectural. Suppose that Blalock conceives of causation as "partial" rather than exclusive; consider, then, two other possible assertions: (1) X causes Y and (2) X causes Z. What relation between Y and Z could be derived? Since Blalock stipulates no rules of derivation, the question cannot be answered. Moreover, if either statement asserts "partial" causation, it is difficult to see how any relation between Y and Z could be derived. The question of derivations is all the more important because it is only through derivations that evidence can be brought to bear on intrinsic statements. But Blalock must do more than formulate rules of derivation; he must also make his conception of causation complete and translate each of the various types of causal relations into the language of space-time relations, and in so doing come to grips with the issues and problems discussed in Chapter 2.

A fourth-order theory. Intrinsic statements of a fourth--order theory take one of two forms: (1) among universes of U's, the greater the direct relation between W and X in a universe, the greater the direct relation between Y and Z in the universe; or (2) among universes of U's, the greater the direct relation between W and X in a universe, the less the direct relation between Y and Z in the universe. Either form asserts that the relation between two properties is contingent on the relation between two other properties.

In tests of fourth-order theories there must be at least two universes of units. For each universe there is a correlation coefficient that expresses a relation between W and X referents and another that expresses a relation between Y and Z referents. The prediction (hypothesis) derived from the theory takes one of two forms: (1) that the two sets of coefficients vary directly or (2) that the two sets vary inversely.

A fourth-order theory differs from a third-order theory in that it asserts *a relation between relations*, but there is a less conspicuous contrast. In both cases, the universes of units may be taken as "natural" sets or they may be created with reference to some property. Thus, suppose that the fourth-order intrinsic statements take this form: Among universes of metropolitan areas, the greater the direct relation between W and X in a universe, the greater the direct relation between Y and Z in the universe. Given that assertion, the theorist may stipulate that each universe of metropolitan areas is to be a "natural" aggregate, but he must clearly identify those aggregates, for instance, "all metropolitan areas of any country." Such a designation does not entail reference to a fifth property, so the theorist evidently presumes that both correlations, W-X and Y-Z, will vary from one "natural" universe to the next. Alternatively, the theorist may be able to stipulate a procedure for grouping units so as to maximize differences among universes as to the intrauniverse relation between W and X.

As still another alternative, the theorist may elect to create universes of units by reference to still other properties. Thus, he could specify that the metropolitan areas of a country are to be grouped into universes so as to maximize differences among the universes as to the absolute or relative magnitude of V. He is especially likely to stipulate such groupings if the units are countries, for it could be argued that there are no "natural" universes of countries. In any event, the instructions for gouping units are exactly like those for a third-order theory. However, in the case of a fourth-order theory, if the units are grouped into universes by reference to some property, that property is not stipulated in the intrinsic statements. Nonetheless, it is part of the theory, for the theorist implies that the "relation between relations" is contingent on that property. If the units are grouped by reference to the magnitude of V, then the theorist's assertion can be expressed as follows: $V - [(W - X) - (Y - Z)]$, that is, the relation between two relations is somehow contingent on the magnitude of V. The expression becomes even more complicated if the units are grouped by reference to a fifth and sixth property, for example, variance in V and the magnitude of T, variance in both V and T, the magnitude of V and the variance in T, or the magnitude of both V and T. Such assertions are simply too cumbersome to express in an intrinsic statement; nevertheless, by the stipulation of universes, more than four properties may be included in a fourth-order theory. Of course, when units are grouped by reference to properties, the implied assertion is that the "relation between relations" will hold at least among such universes, that is, it would not hold if the units are grouped randomly (without regard to any property); and that implied assertion can be examined in special tests of the theory.

It is appreciated that a fourth-order theory is complicated and alien to contemporary sociology. However, once a theorist attempts to extend a second-order theory, he may conclude that some of the second-order relations are contingent on the relation between other properties, and a mode of theory construction should not only facilitate the articulation of such assertions but also provide a procedure for tests. The latter consideration is all the more important because a fourth-order assertion cannot be tested by partial or multiple correlation.

Illustrative Instances of a Third-Order Theory

Although a third-order theory can be formulated *de novo*, the following illustration is an extension of a second-order theory. In the illustrative second-order theory pertaining to the division of labor, the only "condition" of a relation is the unit term; but regardless of the unit term it is improbable that any relation between two sociological properties is invariant. So in formulating a second-order theory, it is unrealistic to expect maximum predictive accuracy. Of course, one may attempt to maximize the predictive accuracy of a second-order theory by further qualification of the unit terms, modification of the referential formulas, and/or alteration of the specifications pertaining to data. However, none of those changes allows for the possibility that the relation varies in some regular way from one condition to the next, and that possibility can be considered only by extension of the theory. But no theory should be rejected simply because it is second-order. After all, that type of theory may create or identify some order, and it sets the scene for subsequent extensions. However, no extension of a second-order theory should be formulated until attempts have been made to maximize its predictive accuracy. Higher-order theories take the unit terms and referential formulas of a second-order theory as given; the presumption is that those components need not be modified. So the extension implies some confidence in the initial theory, even though its predictive accuracy is not satisfactory, and for that reason alone work on second-order theories is fully justified.

First illustration. One possible extension of the illustrative second-order theory has already been suggested. In limiting the unit term to countries, such units are presumed to be more ecologically autonomous than other types of units. However, no country is ecologically autonomous; therefore, maximum predictive accuracy of the theory would not be expected. The line of reasoning suggests that some of the second-order relations will vary from one universe of countries to the next if those universes differ as to ecological autonomy, and that idea is the starting point for an illustrative third-order theory.

"Intrinsic part"

Postulate I: Among universes of countries, the greater the "maximum ecological autonomy during $_0T_0$," the greater the 'direct relation among countries between industry differentiation at T_0 and the per capita use of inanimate energy during $_0T_0$.'

Postulate II: Among universes of countries, the greater the "maximum ecological autonomy during $_0T_0$," the less the 'maximum international exchange of goods for any country in the universe during $_0T_0$.'

Transformational statement I: Among universes of countries, the greater the 'direct relation among countries between industry differentiation at T_0 and the per capita use of inanimate energy during $_0T_0$,' the greater rho (ARCID, T_0 — RUIE, $_0T_0$) for the universe.

Transformational statement II: Among universes of countries, the greater the 'maximum international exchange of goods for any country in a universe during $_0T_0$,' the greater the RMITPC during $_0T_0$ for the universe.

Theorem I: Among universes of countries, the greater the RMITPC during $_0T_0$ for a universe, the less rho (ARCID, T_0 — RUIE, $_0T_0$) for the universe.

"Extrinsic part"

Even when a third-order theory is an extension of a second-order theory (that is, not formulated *de novo*), the extrinsic part should define the unit terms, the substantive terms, and stipulate a procedure for grouping units into universes. However, since the second-order theory has been examined previously, the present extrinsic part is limited to definitions of substantive terms not heretofore considered and to a procedure for grouping countries.

The construct in postulates I and II are defined through two statements. Ecological autonomy refers to the isolation of a territorial unit from other units as far as the movement of population or any form of human contact is concerned. Accordingly, the construct "maximum ecological autonomy during $_0T_0$ " refers to the maximum amount of isolation of any country in a universe during a calendar year.

The notion of international exchange of goods (postulate II) considers the origin of objects or substances made or modified by human beings. Taking any territorial unit as the point of reference, three classes of objects or substances are distinguished: (1) those wholly made or modified and wholly consumed or used in the unit, (2) those wholly made or modified in the unit but wholly consumed or used elsewhere, and (3) those wholly made or modified outside the unit but consumed or used in the unit. Accordingly, the international exchange of goods for a country is the ratio of instances of classes 2 and 3 to instances of class 1 during some period. The concept 'maximum international exchange of goods for any country during $_0T_0$,' thus refers to the maximum $(2 + 3) / 1$ ratio in a universe of countries during a calendar year.

The referential ITPC, signifying international trade per capita, is a component of RMITPC; and the formula is:

$$ITPC = [(I + E)S] /P$$

where I is the monetary value of all imports of a country during a calendar year as expressed in terms of the national currency, E is the monetary value of all exports during the same calendar year, S is the ratio of the value of one unit of the national currency to one unit of the currency of some other country taken as the standard for comparison, and P is the resident population of the country at some point during the calendar year. The S value would not be the same for all countries (for example, the ratio of the Swiss franc to the United States dollar is not the same as the ratio of the Mexican peso to the United States dollar), but the same national currency must be taken as the standard for all countries; that is, the value of imports and exports must be expressed in the same monetary unit for all countries. Any currency can be taken as the international standard, but monetary equivalents must be stipulated in an official publication of an international agency. Figures on the monetary value of imports and exports are to be obtained only from official national publications or official publication of an international agency (such as the United Nations Statistical Office). If an international agency reports the per capita value of imports and exports (combined), those figures can be used in tests of the theory without application of the formula. However, in such a case, each test should be restricted to the figures reported in one particular official publication.

Tests of the theory must consider at least ten countries for which the following referents have been computed: ARCID, T_0; RUIE, $_0T_0$; and ITPC, $_0T_0$. The countries are to be ordered by the magnitude of the ITPC referents, with the country having the greatest magnitude heading the list. The initial list of countries then is divided into at least two universes, so that the range of ITPC referents in any universe is different from the range in any other universe. For example, if the initial list comprises twenty countries, then all groups below the median ITPC value could be taken as one universe and the remaining ten countries (those with ITPC referent above the median) would be a second universe. The countries must be grouped in such a way that each universe comprises *at least five countries* and the number of countries is the same in each universe. When the number of countries in the initial list is a prime, the country representing the median ITPC referent should be excluded from all universes.

Once the countries are grouped in accordance with the foregoing instructions, then RMITPC (referential of maximum international trade per capita) for a universe is simply the greatest ITPC referent in the universe.

The referential *rho* (ARCID, T_0 — RUIE, $_0T_0$) designates the formula for the rank-order coefficient of correlation as applied to the ARCID and RUIE referents *in a universe of countries*. As such, each *rho* coefficient is a referent for a particular *universe* of countries.

Second illustration of a third-order theory.   By applying the rules of derivation to the illustrative second-order theory, the following *implied* proposition is derived: Among countries, the greater the 'industry differentiation at T_0,' the greater the 'per capita use of inanimate energy during $_0 T_0$.' The preceding illustration of a third-order theory took that implied proposition as the point of departure, primarily for the sake of simplicity. Note in particular that any expressed or implied axiom, postulate, proposition, or transformational statement of a second-order theory can be taken as the point of departure in formulating a third-order theory, and the latter need not be restricted to the extension of only one second-order statement.

When the relation asserted in a theorem or in an implied proposition of a second-order theory is judged as contingent, the judgment suggests that some relations asserted in the direct intrinsic statements are also contingent. Presuming that the theorist can identify such statements, his formulation of a third-order theory should commence with them rather than an implied proposition or theorem. Reconsider axiom A3 of the illustrative second-order theory: Among countries, the greater the "division of labor at T_0," the greater the "differences among individuals in their sustenance activities at T_0." Now suppose the theorist suspects that the relation asserted in axiom A3 is contingent on "ecological autonomy." To the extent that a country specializes in the production of particular types of goods and services, residents engage in similar sustenance activities. However, given extensive specialization, differences among individuals in their sustenance activities will not be proportionate to the exchange of goods and services (the other component of the division of labor), for exchange is both external (international) and internal (intranational). Accordingly, the theorist could commence a third-order theory with axiom I: Among universes of countries, the greater the "maximum ecological autonomy during $_0 T_0$," the greater the "direct relation among countries between the division of labor at T_0 and differences among individuals in their sustenance activities at T_0." Postulate Ia would be: Among universes of countries, the greater the "direct relation among countries between the division of labor at T_0 and differences among individuals in their sustenance activities at T_0," the greater the 'direct relation among countries between industry differentiation at T_0 and per capita use of inanimate energy during $_0 T_0$.'

Postulate Ia reflects two considerations: (1) the direct rela-

tion between industry differentiation and per capita use of inanimate energy is an implied proposition in the second-order
theory and (2) axiom A3 enters into the derivation of that implied
proposition. So the theorist reasons that if the relation between
the "division of labor" and "differences among individuals in their
sustenance activities" varies among universes, so does the relation
between 'industry differentiation' and 'per capita use of inanimate
energy,' for the former relation enters into the derivation of the
latter. Thus, when one second-order relation is derived in part
from another, the magnitude of the two relations are presumed to
vary directly or inversely among universes. Four principles are
involved. First, if a direct relation between Y and Z has been
derived in part from a direct relation between W and X, then
among universes, the greater the direct relation between W and X,
the greater the direct relation between Y and Z. Second, if an
inverse relation between Y and Z has been derived in part from a
direct relation between W and X, then among universes, the
greater the direct relation between W and X, the less direct the
relation between Y and Z. Third, if a direct relation between Y
and Z has been derived in part from an inverse relation between W
and X, then among universes, the greater the direct relation between W and X, the less direct the relation between Y and Z.
Fourth, if an inverse relation between Y and Z has been derived in
part from an inverse relation between W and X, then among universes, the greater the direct relation between W and X, the greater
the direct relation between Y and Z.

Up to this point, only an axiom (I) and a postulate (Ia) have
been formulated as components of another illustrative third-order
theory. When those two statements are taken with postulate II,
transformational statement I, and transformation statement II of
the first illustration, the same theorem can be derived (theorem I
in the first illustration). It might appear that taking a second-order
axiom as the point of departure has accomplished nothing, for the
same theorem is derived. However, since the second-order axiom
(that is, axiom A3) enters into the derivation of more than one
implied proposition, the present illustration is incomplete. To
complete it, two additional statements would be made.

Postulate Ib: Among universes of countries, the greater the "direct relation among countries between the division of labor at T_0 and differences among individuals in their sustenance activities at T_0," the greater the 'direct relation among countries between occupational differentiation at T_0 and the per capita use of inanimate energy during $_0T_0$.'

Transformational statement III: Among universes of countries, the greater the 'direct relation among countries between occupational differentiation at T_0 and the per capita use of inanimate energy during $_0T_0$,' the greater *rho* (ARCOD, T_0 — RUIE, $_0T_0$) for the universe.

If those two statements are combined with axiom I, postulate II (first illustration), and transformational statement II (first illustration), the derivation is theorem II: Among universes of countries, the greater the RMITPC during $_0T_0$ for a universe, the less *rho* (ARCOD, T_0 — RUIE, $_0T_0$) for the universe. Then from postulate Ia, transformational statement I, postulate Ib, and transformational statement III, we have theorem III: Among universes of countries, the greater *rho* (ARCID, T_0 — RUIE, $_0T_0$) for a universe, the greater *rho* (ARCOD, T_0 — RUIE, $_0T_0$) for the universe. So we see the advantage of commencing with the extension of a second-order axiom rather than a second-order implied proposition—it furthers the scope of the third-order theory.

Observe that a theorem pertaining to the relation between ARCID and ARCOD is not derived. It is not derived because axiom A3 of the second-order theory does not enter into the derivation of an implied proposition that links industry differentiation and occupational differentiation. Accordingly, the theorist has tacitly assumed that the relation is invariant. But suppose he suspects that the relation between industry differentiation and occupational differentiation or between ARCID and ARCOD is not invariant. If so, he would attempt to extend one or more of the following relations in the original second-order theory: A1, A2, P1, P2, T1, T2.

Third illustration. One complexity of a third-order theory is terminological or, more precisely, notational. In each third-order statement there is a "relational" substantive term, that is, an expression which relates what were originally (in the second-order theory) two constructs, a construct and a concept, two concepts, a concept and a referential, or two referentials. There is no problem if the component terms are of the same type (two constructs, two concepts, or two referentials), but suppose a theorist extends a postulate or a transformational statement. Consider the form of a third-order statement as an extension of a second-order postulate: Among universes of U's, the greater X for a universe, the greater the "direct relation between Y and Z" within the universe. In the second-order theory Y was a construct and Z a concept, and when the two are combined, the substantive term is a relational construct. Accordingly, if X is a construct then the foregoing illustrative statement would be an axiom, for the relational substantive term ("direct relation between Y and Z") is a construct. But if X is a concept, then the statement would be a postulate.

Now consider a third-order statement as an extension of a second-order transformational statement: among universes of U's, the greater K, the greater the direct relation between 'L and M.' Originally (that is in the second-order theory) L was a concept and M was a referential; hence the relational substantive term should be identified as a concept. Accordingly, if K is identified as a construct, then the statement is a postulate.

Reconsidering the illustrative second-order theory, suppose the theorist suspects that the relation in postulate P3 is not invariant and further suspects that the relation is contingent on some property designated by the construct X. As such, he would commence the third-order theory with the following axiom (II): Among universes of countries, the greater the "maximum X," the greater the "direct relation among countries between technological efficiency at T_0 and per capita use of inanimate energy during $_0 T_0$." Since "maximum X" is taken to be a construct, the next two steps would be to link "maximum X" and Y (a concept) in a postulate (Ic) and link Y and Z (a referential) in a transformational statement (IV).[3] Now postulate P3 in the second-order theory enters into the derivation of two implied propositions, each of which may be said to enter into the derivation of two theorems. So two additional statements are needed.

Postulate Id: Among universes of countries, the greater the "direct relation among countries between technological efficiency at T_0 and per capita use of inanimate energy during $_0T_0$," the greater the 'direct relation among countries between industry differentiation at T_0 and per capita use of inanimate energy during $_0T_0$.'

Postulate Ie: Among universes of countries, the greater the "direct relation among countries between technological efficiency at T_0 and per capita use of inanimate energy during $_0T_0$," the greater the 'direct relation among countries between occupational differentiation at T_0 and the per capita use of inanimate energy during $_0T_0$.'

Given axiom II, postulate Ic, transformation statement IV, postulate Id, and transformational statement I, the theorem (IV) is: Among universes of countries, the greater Z for a universe, the greater rho (ARCID, T_0 — RUIE, $_0T_0$) for the universe. Then given axiom II, postulate Ic, transformational statement IV, postulate Ie, and transformational statement III, the theorem (V) is: Among universes of countries, the greater Z for a universe, the greater rho (ARCOD, T_0 — RUIE, $_0T_0$) for the universe. Finally, given postulate Id, transformational statement I, postulate Ie, and transformational statement III, theorem VI is: Among universes of countries, the greater rho (ARCID, T_0 — RUIE, $_0T_0$) for a universe, the greater rho (ARCOD, T_0 — RUIE, $_0T_0$) for the universe.

Theorems IV and V may suggest a direct relation among countries in a universe between ARCID, T_0 and ARCOD, T_0. However, those two referentials should not be confused with the relational referentials in theorems III and IV. To be sure, a direct relation between ARCID, T_0 and ARCOD, T_0 would be predicted, but that prediciton was derived as theorem Th2 in the second-order theory. In other words, an extension of postulate P3 in the second-order theory has no implications for theorem Th2, as the latter is not derived in part from the former.

Fourth illustration. Suppose the theorist suspects that the relation in transformational statement T3 of the second-order theory is contingent on some property designated by the construct W. The extension of the theory would include the following postulate ("If" in the illustrative series): Among universes of countries, the greater the "maximum W," the greater the 'direct relation among countries between per capita use of inanimate energy during $_0T_0$ and RUIE during $_0T_0$.' The next two steps would then be to link "maximum W" to P (a concept) in a postulate (Ig) and to link P to Q (a referential) in a transformational statement (V).[4] Then two additional transformational statements would be needed.

VI: Among universes of countries, the greater the 'direct relation among countries between per capita use of inanimate energy during $_0T_0$ and RUIE during $_0T_0$,' the greater rho (ARCID, T_0 — RUIE, $_0T_0$) for the universe.

VII: Among universes of countries, the greater the 'direct relation among countries between per capita use of inanimate energy during $_0T_0$ and RUIE during $_0T_0$,' the greater rho (ARCOD, T_0 — RUIE, $_0T_0$) for the universe.

Given postulate If, postulate Ig, transformational statement V, and transformational statement VI, the theorem (VII) is: Among universes of countries, the greater Q for a universe, the greater rho (ARCID, T_0 — RUIE, $_0T_0$) for the universe. Then given postulate If, postulate Ig, transformational statement V, and transformational statement VII, the theorem (VIII) is: Among universes of countries, the greater Q for a universe, the greater rho (ARCOD, T_0 — RUIE, $_0T_0$) for the universe. Finally, given transformational statements VI and VII, theorem (IX) is: Among universes of countries, the greater rho (ARCID, T_0 — RUIE, $_0T_0$) for a universe, the greater rho (ARCOD, T_0 — RUIE, $_0T_0$)for a universe.

An alternative to partial correlation. The procedure for stating and testing a third-order theory is different from conventional techniques in sociology. In particular, no use is made of partial correlation.

Suppose the intrinsic statements of a third-order theory assert that the relation between properties Y and Z is somehow contingent on property X. One interpretation is that where all units have the same X value the Y-Z relation will be different from the case where X varies markedly among the units. Given that interpretation, it might appear that partial correlation should be used in tests, with the prediction being that $r_{yz.x}$ will differ from r_{yz}. But such a prediction is at one and the same time ambiguous and vacuous. It is ambiguous in that the direction of the difference is not stipulated, and it is virtually certain to be true since by chance alone the two correlations are likely to differ. Moreover, it is extremely difficult to word intrinsic statements so that predictions can be derived from them as to both the direction and the magnitude of the difference between $r_{yz.x}$ and r_{yz}. The question of magnitude is especially difficult, for some difference is a virtual certainty, but the quality of sociological data precludes a prediction as to the exact amount. Of course, one may derive a prediction about a partial correlation without reference to the difference between r_{yz} and $r_{yz.x}$, but some difference is suggested.

As an alternative to partial correlation, the intrinsic statements of a third-order theory may assert that the relation between properties Y and Z varies from one *universe* of units to the next, with the relation *in a universe* contingent on the *intrauniverse variance* in property X. Such an assertion takes one of two forms: (1) among universes of U's, the greater the intrauniverse variance in X, the greater the direct relation between Y and Z in the universe; or (2) among universes of U's, the greater the intrauniverse variance in X, the less direct the relation between Y and Z in the universe. Either form is less cumbersome than wording a theory so that predictions about partial correlations can be derived from it, and either form avoids the problems entailed in deriving a prediction about the difference between a partial and a zero-order coefficient of correlation.

The import of the two forms of a third-order assertion can be clarified by translating each into a prediction. Given an assertion of the first form, the prediction is: in a universe of units with no variance in X among the units, the magnitude of rho_{yz} or tau_{yz} will be *less* than in a universe with variance in X among the units. For an assertion in the second form the translation is: in a universe of units with no variance in X, the magnitude of rho_{yz} or tau_{yz} will be *greater* than in a universe with variance in X among the units. To be sure, it may not be feasible to create a universe of units such that all units have the same X value, but units can be grouped into universes so as to maximize differences among the universes as regards intrauniverse variance in X. Given such groupings, the assertion is that the Y-Z relation in a universe is contingent on (that is, associated with) intrauniverse variance in X.

The method of "universe comparisons" is not just a substitute for partial correlation, as the former can reveal relations that would not be revealed by the latter. Suppose that for forty countries r_{xy} and r_{xz} is .00; so r_{yz} would equal $r_{yz.x}$. Now suppose that the forty countries are divided into four universes with ten countries in each, and the countries have been grouped so as to maximize differences among the universes as to intrauniverse variance in X. Even though $r_{yz} = r_{yz.x}$ among the forty countries, it would be mathematically possible for r_{yz}, rho_{yz}, and tau_{yz} to vary from one universe to the next and be inversely or directly related to intrauniverse variance in X. In other words, the Y-Z relation may be contingent on intrauniverse variance in X even though neither Y nor Z is contingent on X (that is, r_{xy} = .00 and r_{xz} = .00). If it be objected that such a condition only indicates that the data do not meet the assumptions of partial correlation, then those assumptions are too restrictive, and hence the preference for universe comparisons.

Still another limitation of partial correlation is revealed by reconsidering the other two forms of a third-order assertion: (1) among universes of U's, the greater X for a universe, the greater the direct relation between Y and Z in the universe; or (2) among universes of U's, the greater X for a universe, the less direct the relation between Y and Z in the universe. Both forms preclude the conventional use of partial correlation. To test either assertion it would be necessary to group the units into universes in such a way as to maximize the differences among universes with regard to the magnitude of X. In each universe X would be "held constant" within certain limits, but the assertion is that the Y-Z relation depends on the level at which X is "held constant," a contingency that would not be revealed by partial correlation, which assumes that the Y-Z relation is approximately the same at all magnitudes of X. To illustrate, suppose that an X referent has been computed for each of forty countries and the countries have been grouped into quartiles by reference to the magnitude of the X referents, with each quartile identified as a universe. Now suppose that the correlation between Y and Z is computed for each universe. It is mathematically possible for the Y-Z correlation to vary substantially from one universe to the next even though among all countries ($N = 40$) the correlation between X and Y or between X and Z is .00. In other words, the relation between Y and Z may be contingent on the magnitude of X even though X is not related to either variable. If it be objected that the pattern cannot be revealed by partial correlation because the data would not meet the underlying assumptions, then the reply again is that the assumptions are too restrictive.

So, to summarize, the preference for "universe comparisons" over partial correlation is simply that the former has more general application. The method is not conventional, but it can reveal patterns that would not be revealed by partial correlation.

Fifth illustration. Up to this point each illustrative extension of a second-order intrinsic statement has introduced only one additional property. Now consider the extension of axiom A3 in the illustrative second-order theory as follows in axiom (1): Among universes of countries, the greater "W and X" the greater the "direct relation among countries between the division of labor at T_0 and differences among individuals in their sustenance activities at T_0." The second construct in the statement has been considered before, and it can be linked to the relational referential *rho* (ARCID, T_0 — RUIE, $_0 T_0$) through postulate Ia (second illustration), and transformational statement I (first illustration). It also can be linked to the referential *rho* (ARCOD, T_0 — RUIE, $_0 T_0$) through postulate Ib and transformational statement III of the second illustration. However, the first construct in the statement, "W and X," introduces a new consideration, for it is a "composite" substantive term, combining what would otherwise be two distinct constructs. The mode of theory construction does not prohibit *composite* substantive terms.[5] In this instance the theorist would formulate a postulate in one of two forms: (1) among universes of countries, the greater "W and X," the greater 'S and T,' or (2) among universes of countries, the greater "W and X," the greater 'P.' In the first form the composite construct is linked to a composite concept ('S and T') but, as suggested by the second form, it is not necessary that both substantive terms be composites.

If postulate (2) should be formulated, the character of the subsequent transformational statement would be no different from those previously considered. However, if postulate (1) is formulated, the subsequent transformational statement would link a composite concept and a referential. Now it might appear that in such a transformational statement the referential would also be a composite, but the theorist is to stipulate one and only one referential. So transformational statement (1) would be: Among universes of countries, the greater 'S and T,' the greater Rc for the universe. In such a statement Rc is a referential that designates only one particular formula in the extrinsic part of the thoery. To be sure, the referential formula may have composite structure, such as $Rc = (Sr) + (Tr)$ or $Rc = (Sr)(Tr)$, and the theorist may think of Sr as the referential of concept 'S' and Tr as the referential of concept 'T.' If so, he will structure the explication of the referential formula so that two sets of computational instructions are given—one for Sr and one for Tr. Those instructions may include the stipulation of ancillary formulas (for example, $Sr = X/N$); if so, the referential formula will represent the combination of two or more distinct "variables." One may think of each variable as a referential in itself, but nonetheless in a transformational statement there is only one referential (Rc in this particular illustration), and in the extrinsic part of the theory there is only one referential formula.

The specification of a referential formula is not governed by technical rules that transcend particular theories. On the contrary, each referential formula is a product of the theorist's imagination and judgment. True, the theorist may use a standard technique (such as the conventional procedure for computing composite index numbers), but his judgments are not limited by conventions. He even may weight the component variables, for instance, $Rc = 2.5(Sr) + (Tr)$, having arrived at the weights inductively through exploratory research.

To complete the illustration, when the theorist makes transformational statement (1), which links the composite concept 'S and T' with the referential Rc, he can derive three theorems.

Theorem (1) [from axiom (1), postulate (1), transformation statement (1), postulate Ia, and transformational statement I]: Among universes of countries, the greater Rc for a universe, the greater rho (ARCID, T_0 − RUIE, $_0T_0$) for the universe.

Theorem (2) [from axiom (1), postulate (1), transformational statement (1), postulate Ib, and transformational statement III] : Among universes of countries, the greater Rc for a universe, the greater rho (ARCOD, T_0 — RUIE, $_0T_0$) for the universe.

Theorem (3) [from postulate Ia, transformational statement I, postulate Ib, and transformational statement III]: Among universes of countries, the greater the rho (ARCID, T_0 — RUIE, $_0T_0$) for a universe, the greater rho (ARCOD, T_0 — RUIE, $_0T_0$) for the universe.

Multiple correlation. Given an assertion that three or more properties are interrelated, it might appear that the conventional multiple correlation (R) can and should be used to test it. That technique is rejected for essentially the same reasons given in the case of partial correlation.

Consider an assertion in the following third-order form: Among universes of urban areas, the greater W and X for a universe, the greater the direct relation between Y and Z in the universe. Now suppose that a W, X, Y, and Z value have been computed for each of forty urban areas. No multiple correlation (for example, $R_{y.zwx}$) for those values of the forty urban areas would be relevant in tests of the assertion. The data would become relevant only after the urban areas are divided into universes and the Y-Z correlation computed for each universe. According to the assertion, the Y-Z correlation in a universe will be contingent on the magnitude of the W and X values in the universe, which is to say that the correlation should vary from one universe to the next in a predictable manner. However, the predicted pattern could not be inferred from any multiple correlation based on all forty urban areas considered together. Suppose that for all forty urban areas considered together the coefficient of correlation between any two of the four sets of values (W, X, Y, and Z) is .00. Now consider what the Y-Z correlation could be in two universes of urban areas when each W and X value in one universe is greater than any W or X value in the other universe. It is mathematically possible for the correlation to be of much greater magnitude in one universe than in the other.

When a relation between two properties is asserted to be contingent on more than one other property, it is difficult to group units into universes so as to maximize interuniverse differences with regard to all contingent properties. So, continuing the illustration, the urban areas could be grouped so as to maximize differences among universes with regard to either the W values or the X values, but it would be difficult to maximize differences with regard to both W and X. However, there are five alternatives as follows: (1) designate "natural" universes, such as "all urban areas of a country"; (2) group the urban areas as to the absolute or relative magnitude of W; (3) group them as to the relative or absolute magnitude of X; (4) group them by selecting the urban areas for each universe at random; (5) link W and X as a *composite substantive term* to Rc referents for the urban areas and group them by the magnitude of those referents.

Whatever the theorist's choice, he should give detailed instructions in the extrinsic part of the theory for grouping the units into universes. His choice may reflect practical considerations, but it is nonetheless a part of the theory and not a "technical" question.

Sixth illustration. All previous illustrations are oversimplifications in one particular respect. In each case only one second-order relation has been extended; but, obviously, two or more relations in the same second-order theory may be taken as contingent and perhaps contingent on different properties. If so, the third-order theory is "complex."

In the third illustration the relation in postulate P3 of the second-order theory is asserted to be contingent on property X, and in the fourth illustration the relation in transformational statement T3 is asserted to be contingent on property W. For the sake of simplicity the two illustrations were presented as though independent, but that is not the case, since postulate P3 and transformational statement T3 enter into the derivation of at least one theorem in common.[6] However, even though the third and fourth illustrations are not independent, a more inclusive theory would not be constructed by simply merging the intrinsic statements of the two illustrations. On the contrary, all complex third-order theories are to be formulated in such a way that each relational referential [for example, *rho* (ARCID, T_0 − RUIE, $_0T_0$)] is linked in the theorems to one and only one contingent referential. That would not be the case if the intrinsic statements of the two illustrations are combined to form a more inclusive theory. The merger would not change the four theorems (III, IV, V, and VI in the illustrative series), but in that set of theorems each relational referential is linked to more than one contingent referential.

When the relations asserted in two or more direct intrinsic statements are taken as contingent on different properties, the third-order statements that assert such contingencies must be of a particular form.[7] Before designating that form two principles should be made explicit. First principle: When two or more contingent properties enter into a third-order theory, no relation (direct or inverse) between them is asserted or implied, and the sign rule should not be used to derive such a relation. Thus, returning to the third and fourth illustrations, even though the relation in postulate P3 is asserted to be contingent on property X and the relation in transformational statement T3 is asserted to be contingent on property W, it does not follow that a relation between X and W is asserted or implied. Second principle: If a relational referential is presumed to be contingent (directly or indirectly) on two or more properties, then the association between that referential and any one of those properties, alone is not assumed to be close. Returning again to the third and fourth illustration, consider the relation between *rho* (ARCID, T_0 — RUIE, $_0 T_0$) and property X. A direct relation is implied, for it is asserted that the relation in transformational statement T3 is contingent on X, and transformational statement T3 enters into the derivation of theorem Th1. Now if all of the other intrinsic relations that enter into the derivation of theorem Th1 were presumed to be close, then the relation between X and *rho* (ARCID, T_0 — RUIE, $_0 T_0$) would be presumed to be close. But postulate P3 also enters into the derivation of theorem Th1, and by asserting that the relation in that postulate is contingent on W, the theorist in effect asserts that *rho* (ARCID, T_0 — RUIE, $_0 T_0$) is indirectly contingent on W. So that *universe* referential is contingent on both X and W; hence it would be closely related to both only if X and W are closely related. But the theorist may or may not think of X and W as closely related, and in any case the assertions made in formulating a third-order theory need not imply that the two are closely related.

If the theorist believes that a relational referential is contingent (directly or indirectly) on two or more properties, those properties should appear as one *composite* substantive term in statements linking them to the referential. Thus, the combination of the third and fourth illustrations would require one axiom, three postulates, and three transformational statements as follows.

Axiom III: Among universes of countries, the greater "X and W," the greater the "direct relation among countries between technological efficiency at T_0 and per capita use of inanimate energy during $_0T_0$ and between per capita use of inanimate energy during $_0T_0$ and RUIE for $_0T_0$."

Postulate Ih: Among universes of countries, the greater the "direct relation among countries between technological efficiency at T_0 and per capita use of inanimate energy during $_0T_0$ and between per capita use of inanimate energy during $_0T_0$ and RUIE for $_0T_0$," the greater the 'direct relation among countries between industry differentiation at T_0 and RUIE for $_0T_0$.'

Postulate Ii: Among universes of countries, the greater the "direct relation among countries between technological efficiency at T_0 and per capita use of inanimate energy during $_0T_0$ and between per capita use of inanimate energy during $_0T_0$ and RUIE for $_0T_0$," the greater the 'direct relation among countries between occupational differentiation at T_0 and RUIE for $_0T_0$.'

Postulate Ij: Among universes of countries, the greater "X and W," the greater 'Y and P.'

Transformational statement VII: Among universes of countries, the greater 'Y and P' for a universe, the greater R_{yp} for the universe.

Transformational statement VIII: Among universes of countries, the greater the 'direct relation among countries between industry differentiation at T_0 and RUIE for $_0T_0$,' the greater rho (ARCID, T_0 − RUIE, $_0T_0$) for the universe.[8]

Transformational statement IX: Among universes of countries, the greater the 'direct relation among countries between occupational differentiation at T_0 and RUIE for $_0T_0$,' the greater rho (ARCOD, T_0 − RUIE, $_0T_0$) for the universe.[9]

Given the foregoing seven direct intrinsic statements, three theorems are derived by application of the sign rule.

Theorem X (from AIII, PIh, PIj, TVII, and TVIII): Among universes of countries, the greater Ryp for a universe, the greater rho (ARCID, T_0 — RUIE, $_0T_0$)for the universe.

Theorem XI (from AIII, PIi, PIj, TVII, and TIX): Among universes of countries, the greater Ryp for a universe, the greater rho (ARCOD, T_0 — RUIE, $_0T_0$) for the universe.

Theorem XII (from PIh, PIi, TVIII, and TIX): Among universes of countries, the greater rho (ARCOD, T_0 — RUIE, $_0T_0$) for a universe, the greater rho (ARCOD, T_0 — RUIE, $_0T_0$) for the universe.

In the way of explication, observe that none of the direct intrinsic statements appear in the third or fourth illustration, and that comparison is indicative of the radically different character of a third-order theory when the relations asserted in two or more of the direct intrinsic statements are presumed to be contingent on different properties. Then note that the present illustration makes use of two *implied* transformational statements in the second-order theory, one linking industry differentiation at T_0 and RUIE for $_0T_0$, and the other linking occupational differentiation at T_0 and RUIE for $_0T_0$. The first *implied* transformational statement appears as a relational substantive term in postulate I and again in transformational statement VIII. The second *implied* transformational statement appears as a relational substantive term in postulate Ii and again in transformational statement IX. In formulating postulate Ih and Ii the theorist has reasoned as follows: since both postulate P3 and transformational statement T3 enter into the derivation of the implied transformational statements, then the relations in those implied transformational statements are contingent on the relations in postulate P3 and transformational statement T3, which in turn (as asserted in axiom III) are contingent on properties X and W. Finally, all relations in the direct intrinsic statements are asserted to be close but only because the contingent properties have been combined in a composite substantive term.

Although a third-order theory can be formulated *de novo*, the present illustration underscores the advisability of formulating it as an extension of second-order theory. Investigators can test a third-order theory by considering only the theorems, but the audience is much more likely to grasp the structure of the theory when it is presented as an extension.

Illustrations of a Fourth-Order Theory

Suppose the theorist believes that axiom A3 in the illustrative second-order theory is not invariant, that is, the relation varies markedly from one universe to the next. Suppose he further believes that the relation is contingent not on a property but on the relation between two properties designated as M and N. In other words, he believes that a close direct relation between the "division of labor" and "differences among individuals in their sustenance activities" holds only in a universe of countries where there is close relation (direct or inverse) between M and N.

Given those beliefs, the theorist would commence his formulation of a fourth-order theory with the following axiom (number IV in the entire illustrative series): Among universes of countries, the greater the "direct relation among countries between M and N," the greater the "direct relation among countries between the division of labor at T_0 and differences among individuals in their sustenance activities at T_0." The next two steps would be to link the relation between M and N to the relation between S and T (two concepts) in the form of a postulate (Ik) and then to link the relation between S and T to rho $(U-V)$, a relational referential, in transformational statement XI.[10] The theorist would then formulate what has been identified in the third-order illustrative series as postulate Ia, postulate Ib, transformational statement I, and transformational statement III.

Given axiom III, postulate Ik, transformational statement XI, postulate Ia, transformational statement I, the theorem (XIII) is: Among universes of countries, the greater rho $(U-V)$ for a universe, the greater rho (ARCID, T_0 — RUIE, $_0T_0$) for the universe. Then given axiom IV, postulate Ik, transformational statement XI, postulate Ib, and transformational statement III, theorem XIV is: Among universes of countries, the greater rho $(U-V)$ for a universe, the greater rho (ARCOD, T_0 — RUIE, $_0T_0$) for the universe. Finally, given postulate Ia, transformational statement I, postulate

Ib, and transformational statement III, we have a repetition of theorem XII: Among universes of countries, the greater *rho* (ARCID, T_0—RUIE, $_0T_0$) for a universe, the greater *rho* (ARCOD, T_0—RUIE, $_0T_0$) for the universe.

In a test of the fourth-order theory two correlation coefficients would be computed for each universe of countries, with each coefficient identified as a referent. The prediction derived from the theory would assert a direct relation among the universes between the two sets of correlation coefficients.

The foregoing illustration is an oversimplification in that it extends only axiom A3 of the second-order theory. Now suppose that the theorist also extends postulate P3, presuming the relation asserted in the postulate to be contingent on the relation between two properties identified as *C* and *D*. If so, the fourth-order theory would be "complex" and radically different from the previous illustrations. Since axiom A3 and postulate P3 both enter into the derivation of theorems Th1 and Th3 of the second-order theory, then the two relational referentials, *rho* (ARCID, T_0—RUIE, $_0T_0$) and *rho* (ARCOD, T_0—RUIE, $_0T_0$), are both contingent on *two* relations between properties, one being the relation between *M* and *N* and the other being the relation between *C* and *D*. So the situation is like a third-order theory in which a relational referential is presumed to be contingent (directly or indirectly) on two properties. The only difference is that in a third-order theory the contingent properites are combined in a composite substantive term (for example, "*X* and *W*" in a universe), whereas in a fourth-order theory two relational substative terms are combined as follows: "direct relation among countries in a universe between *M* and *N* and between *C* and *D*." Otherwise, a complex third-order theory and a complex fourth-order theory are formulated in the same way.

Second-Order Theories Reconsidered

For the sake of simplicity, no composite substantive terms were introduced in illustrations of a second-order theory. As we have seen, a composite substantive term may be necessary in a "complex" third-order theory, but that type of term is not precluded from a second-order theory. Nor does the use of a composite substative term in a second-order theory blur distinctions among theories as to order. Briefly summarizing the distinctions, the intrinsic statements of a second-order theory make assertions about the relation between properties; some intrinsic statements of a third-order theory assert that a relation between properties is associated with one or more other properties; and the intrinsic statements of a fourth-order theory assert that a relation between properties is associated with one or more relations between other properties. Another way to explain the differences is to give the alternative forms of the intrinsic statements that enter into each type of theory, including a first-order theory.

First-order: All U's are W.

First-order: All U's are W and X.

Second-order: Among U's, the greater W, the greater X.

Second-order: Among U's, the greater W and X, the greater Y.

Third-order: Among universes of U's, the greater W, the greater the direct relation between X and Y.

Third-order: Among universes of U's, the greater W and X, the greater the direct relation between Y and Z.

Fourth-order: Among universes of U's, the greater the direct relation between W and X, the greater the direct relation between Y and Z.

Fourth-order: Among universes of U's, the greater the direct relation between S and T and between W and X, the greater the direct relation between Y and Z.

Returning to the initial consideration, the use of composite substantive terms in a second-order theory does not require any special rules or principles. Indeed, as shown below in the final illustration, the structure of a second-order theory in which some of the substantive terms are composites is in no way different from previous illustrations of that type of theory.

Axiom 1: Among U's, the greater "A and B," the greater "C."

Axiom 2: Among U's, the greater "A and B," the greater "D and E."

Postulate 1: Among U's, the greater "C," the greater 'F and G.'

Postulate 2: Among U's, the greater "D and E," the greater 'H.'

Transformational statement 1: Among U's, the greater 'F and G,' the greater Rfg.

Transformational statement 2: Among U's, the greater 'H,' the greater Rh.

Theorem: Among U's, the greater Rfg, the greater Rh.

Observe that in some statements both substantive terms are composites, and in the others only one substantive term is a composite. However, each transformational statement links a concept or a composite concept to one and only one referential, and each referential would designate one and only one referential formula in the extrinsic part of the theory.

Notes for Chapter 7

[1] Observe that the derivation of each theorem need not be explained in detail if the audience is familiar with the mode of formalization, the sign rule in particular. However, in each case it is desirable to stipulate the direct intrinsic statements that enter into the derivation of a theorem. That stipulation does more than further understanding of the theory; it also serves as a check on two requirements of any theory—that the direct intrinsic statements be sufficient and necessary. If a direct intrinsic statement does not enter into the derivation of any theorem, it is not necessary. If the theorist cannot stipulate the direct intrinsic statements from which a theorem was derived, the statements taken as a set are not sufficient. Theorems are also relevant in considering still another requirement—consistency. Two direct intrinsic statements are inconsistent if they differ *only* in the relational terms ("greater ... greater" in one but "greater ... less" in the other). Note also that when two theorems differ only in their relational terms, some of the direct intrinsic statements may be inconsistent.

[2] This observation and several others could be excluded from the extrinsic part of the theory. They further the explication of the mode of theory construction, but the extrinsic part could be made even more concise by excluding them.

[3] Postulate Ic: Among universes of countries, the greater the "maximum X," the greater 'Y.' Transformational statement IV: Among universes of countries, the greater 'Y,' the greater Z for the universe.

[4] Postulate Ig: Among universes of countries, the greater the "maximum W," the greater 'P.' Transformational statement V: Among universes of countries, the greater 'P,' the greater Q for the universe.

[5] A composite substantive term is not alien to the notion of a third-order

assertion, that is, a third-order assertion may contain what would be four distinct substantive terms in a second-order assertion. The criterion of a third-order assertion is not strictly numerical; rather, the idea is that a relation between two properties is taken as contingent on *one or more other properties*.

[6] If the second-order theory were such that postulate P3 and transformational statement T3 do not enter into the derivation of any theorem in common, the extensions of the two could be treated separately, as in the third and fourth illustrations. Further, if both postulate P3 and transformational statement T3 were taken as contingent on the *same* property or properties, there would be no particular problem. Thus, if both the relation asserted in postulate P3 and the relation asserted in transformational statement T3 are taken as contingent on property X, then theorems Th1 and Th3 are indirectly contingent on that property, and the extension would end with theorems III and IV of the third illustration.

[7] The presumption is that the direct intrinsic statements enter into the derivation of at least one theorem in common.

[8] The statement is not a repetition of transformational statement I in the first illustration.

[9] The statement is not a repetition of transformational statement III in the second illustration.

[10] Postulate Ik: Among universes of countries, the greater the "direct relation among countries between M and N," the greater the 'direct relation among countries between S and T.' Transformational statement XI: Among universes of countries, the greater the 'direct relation among countries between S and T,' the greater the $rho(U\text{-}V)$ for a universe.

PART THREE: TESTS OF THEORIES

CHAPTER 8 TERMINOLOGY AND PROCEDURE

Sociologists agree on few matters, but all bemoan the gap between research and theory. Yet that consensus appears paradoxical, because for generations sociologists have tolerated the promulgation of untestable theories. But the condition is no paradox if it is recognized that sociologists have quite different ideas as to what constitutes a test. Indeed, a sociologist may bemoan the gap between theory and research without making any reference to tests, and the omission implies that "general observations" are sufficient evidence for assessing theories. Some critics of sociological theory deny that general observations are adequate; they argue that evidence must be generated by *systematic tests*. The proposed mode of formalization represents essential agreement with the critics. In brief, the idea of "bringing evidence to bear on a theory by general observations" is vague, and a defensible assessment of predictive accuracy requires systematic tests.

Critics of sociological theory have failed badly in one respect —they have not articulated a complete conception of a test, much less a standard procedure.[1] It will not do to look to the philosophy of science or any of the physical sciences for a test procedure. There can be no one procedure for testing theories, for a procedure is useful only if it is specific, and a specific procedure cannot recognize the problems and conditions of work in all fields.

The Notion of Correspondence with Facts

Regardless of the field, no conception of a test is complete unless it speaks to issues entailed in the correspondence "theory" of knowledge.[2] According to that notion, there are "facts" which exist independently of any particular theory;[3] as such, a test is nothing more than an observation that the theory is or is not consistent with a particular fact. Three assumptions are entailed: (1) that in conducting tests investigators will agree in their identification of "relevant" facts, (2) that they will agree further as to whether a particular fact is consistent with the theory, and (3) that the "facts" themselves are not questionable. The opposing argument is that facts do not exist independently of theory; extending the argument, some theories may be said to create facts, especially when a theorist specifies a novel formula. If one grants that the formula and the values computed by its application are "facts," then in what sense do they exist independently of the theory? Certainly they did not exist prior to the theory! Further, tests are based not on facts in the sense of unquestioned experience, but rather on assertions. As such, the assertions themselves are subject to question, and thus it is misleading to say a test proves that a theory is consistent with reality, corresponds to the facts, or is valid.

All that one should say concerning a test is that a prediction derived from the theory about a particular body of data is either consistent or inconsistent with an assertion made by an investigator about the same data. Moreover, the investigator's assertion cannot be of the same form as an assertion in the theory proper (an axiom, postulate, proposition, transformational statement, or theorem). Thus, suppose that the following assertion is made as an axiom: Among countries, the greater the "normative consensus at T_0," the greater the "social order during $_0T_0$." Now suppose that an investigator makes the following assertion: Among countries, the greater the "normative consensus at T_0," the *less* the "social order during $_0T_0$." Given the conventions of the English language, the investigator's assertion is inconsistent with the axiom, but no one is likely to regard it as evidence, for nothing is said about particular countries or data. Note also that the investigator's assertion is of the same form as the axiom, but, since neither substantive term is considered to be empirically applicable, how could the investigator possibly know that normative consensus and social order are not directly related? Of course, he may say that he knows by intuition, but that is hardly evidence. If an investigator is to make relevant assertions, he must consider only the theorems, and no assertion by an investigator can be taken as a direct refutation or confirmation of axioms, postulates, propositions, or transformational statements. As stressed repeatedly, the truth, validity, predictive accuracy, and so on, of any intrinsic statement cannot be known directly and never with certainty.

Of the five types of intrinsic statements, only theorems enter directly into tests of a theory, but assertions made by investigators in tests never take the same form as a theorem. To illustrate by reference to the second-order theory in Chapter 7, suppose that an investigator reports a test as follows: Among countries, the greater the ARCID at T_0, the less the RUIE during $_0T_0$. Again taking the conventions of the English language, the investigator's assertion is inconsistent with theorem Th1. But no one is likely to take the assertion as evidence, for again nothing is said about particular countries or data. The investigator's assertion may represent an interpretation of findings, but it does not reveal the findings themselves or the rules of interpretation. He may defend the assertion by pointing to a table in which there are two values for each of several countries, and some such presentation is desirable; but it is not sufficient to present a set of values, as those values are not relevant unless connected with the referentials in the theorem. Specifically, if the values have been computed by application of a referential formula, the investigator must so signify by a standard procedure. Further, regardless of the relation among two sets of values, the investigator would not be justified in making the foregoing assertion, that is: Among countries, the greater the ARCID at T_0, the less the RUIE during $_0T_0$. After all, the assertion suggests that he has considered an infinite universe, but surely the test pertains to particular countries. Moreover, how did the investigator decide that the values for that set are inconsistent with the theorem? He may point to some conventional statistic (such as a rank-order coefficient of correlation) that expresses the relation between the values as negative rather than positive, but on what basis would a negative coefficient be inconsistent with the theorem? Here the conventions of the English language are not adequate, meaning that no test is truly systematic unless reported in accordance with explicit rules of interpretation and not the putative conventions of a natural language. A negative coefficient of correlation may appear inconsistent with theorem Th1, but it is so only if a prediction of a positive coefficient has been derived from the theorem by explicit rules.

Now all of the foregoing suggests that the "correspondence-with-the-facts" conception of a test is a gross oversimplification. In some fields it may be that all an investigator need do is report his observations as consistent or inconsistent with the theory. Alternatively, he may present some data and simply say that the data are consistent or inconsistent with the theory. In either case, the acceptance of the investigator's report as a test means more than trust in his integrity and competence. Certainly it does not mean that the field has no systematic procedures for conducting tests of a theory. It may be that the procedure has become so conventional that an investigator need not make it explicit; rather, he can simply present his "facts" and declare them to be consistent or inconsistent with the theory, the presumption being that the audience will somehow know what facts are relevant and also agree with his conclusion.

Turning to sociology, there is no fully accepted systematic

procedure for tests of theories, nor has there ever been one. On the whole, sociologists uncritically accept the "correspondence-with-the-facts" notion in formulating theories and in reporting tests.[4] That notion is not an adequate substitute for a systematic procedure, as it does not assure agreement among investigators in reporting tests, not even those based on the same universe.

Agreement among investigators is likely only when they employ the same procedure, and several conditions in sociology are conducive to divergent procedures. The most conspicuous ones are the terminological problems of the field, the limited research resources, the discursive mode of theory construction, and the widespread belief that someone other than the theorist should select the appropriate operational definitions or indicators for tests of a theory.

The only way to insure systematic tests is to stipulate a procedure in conjunction with a particular mode of theory construction. That proposal is alien to sociology, past and present. Just as sociological theoreticians traditionally have been indifferent to data problems, so have they been indifferent to test procedures. They clearly accept the correspondence notion of a test, that is, one need only formulate a theory and investigators will somehow know how to test it. To the contrary, without the stipulation of a procedure investigators are not likely to agree in their answers to several questions. What "facts" are relevant? How are they to be generated, found, or otherwise obtained? How are the facts to be reported? How are they to be linked to the intrinsic statements of a theory? By what rules are the facts to be construed as consistent or inconsistent with the theory? As we shall see, no test procedure can provide an answer to those questions without stipulating types of terms, types of statements, rules for deriving predictions from a theory, and rules of interpretation, nor can those considerations be divorced from theory construction.[5] Specifically, there can be no standard procedure for tests that is applicable regardless of the way the theory is formulated. On the contrary, theory construction and test procedure are interrelated, and a mode of formalization is incomplete unless it stipulates procedures for testing theories.

Referents and Types of Statements

Given stipulations of referential formulas and definitions of unit terms in the extrinsic part of a theory, investigators supposedly can identify instances of the designated units, obtain the prescribed data, and compute values by applying the referential formulas. As indicated previously, those values are referents, and no test of a theory can be made without them. However, referents in themselves do not constitute a test of theory, meaning that referents, like "facts," do not speak for themselves.

The notion of a referent. Suppose an investigator undertakes a test of the illustrative second-order theory presented in Chapter 7. Whatever else he might do, the investigator would have to report that he applied the stipulated referential formulas to the prescribed kinds of data; otherwise, he cannot claim to have tested the theory. The point is important because much of what passes for tests of a sociological theory is nothing more than criticism. Of course, the typical sociological theory is stated so vaguely that systematic tests are precluded, and one can do little more than criticize it. However, when the theorist has stipulated referential formulas and the requisite data, investigators cannot ignore those stipulations and still claim to have tested the theory.

In reporting or alluding to referents, an investigator should: (1) designate what he has taken to be instances of the unit term or terms, (2) describe how the units were identified and selected for the test, (3) describe the kinds of data gathered or otherwise obtained, and (4) identify the procedure used to gather data or the source of published data. All statements along those lines signify that the investigator has acted in accordance with instructions in the extrinsic part of the theory. However, regardless of those instructions and the statements made by the investigator, an epistemological question is posed: What are the referents reported or alluded to by the investigator?

Since the referents are presented in the context of a test, they could be identified as "facts." However, scientists and philosophers habitually use the word "fact" uncritically, that is, without a definition of it. Further, any definition of a fact is debatable

and casts doubt on the notion itself. Consider a dictionary definition of a fact: "Something known with certainty." If that definition is accepted (and it is in keeping with commonly held conceptions), then there are no facts in science. One may feel that he
knows something with certainty and he may express that conviction publicly, but by conventional standards that "something"
would hardly become a scientific fact. However, although a scientific fact is not equated with personal convictions, it is difficult to
divorce the two and still include the notion of certainty. One may
attempt a definition that avoids reference to certainty and/or personal convictions, but it appears that any such definition will be
arbitrary. In any case, the word "fact" is especially questionable
when used in tests of theories.

Of course, one may argue that a test is essentially a matter of
experience; that is, the theory is either consistent or inconsistent
with experience. That argument, like the "correspondence-with-
the-facts" conception of a test, is a gross oversimplification and
ignores the public character of science. After all, experience is
private, and it becomes scientifically relevant only when public
statements are made about it. But those statements are assertions
about experience, not experience itself.

It is neither defensible nor necessary to say that science deals
with facts or experience. Rather, it deals with public assertions,
and that characterization is consistent with the notion of a referent. Any referent is a tacit assertion, for it is a value *allegedly*
computed by applying a designated referential formula to a designated body of data. Of course, one may identify referents as facts,
but the identification is superfluous. Should critics object that a
referent is not an assertion about experience, it is inconceivable
that data can be gathered or a value can be computed without
experience. Moreover, the objection implies a naive conception of
the way scientific theories are assessed. They are not assessed by
public testimonials in which individuals relate immediate experience, such as: "I experienced a red color." On the contrary, they
are assessed largely by reference to data, which is to say by symbolization rather than immediate experience. To be sure, a symbol
in one way or another may represent an assertion about experi-

ence, but it need not assert immediate or direct experience.[6] So scientific theories are assessed primarily through symbolization, and a referent is merely one type of symbol.

Once a referent is described as a symbol, critics are certain to ask: What does any given referent represent? There is only one adequate answer—it is a value allegedly computed by applying a designated referential formula to a designated body of data. The answer will not satisfy some critics, and they are likely to rephrase the question as follows: What does a referent "mean" and how do we know it is reliable? There are two answers: first, no referent is taken to have any meaning outside the context of a theory; and, second, there is no way to demonstrate the reliability of a referent. Replying another way, the theory gives meaning to a referent, and the theory asserts that it is reliable. What the question suggests is an uncritical adoption of the "correspondence-with-the-facts" conception of a test, as though facts exist independently of any theory. To the contrary, if referents are taken as "facts," then a theory not only stipulates what facts are relevant but also generates them. But, however paradoxical it may appear, the very referents that a theory "creates" may lead to its rejection.

Epistemic statements.   Whatever their interpretation, referents by themselves do not constitute evidence for or against any theory. Should an investigator present two sets of referents and declare them to be consistent or inconsistent with the theory, he has voiced a conclusion, not a test.

Referents become evidence only when a prediction about them is derived from the theory, and such a prediction cannot be derived formally without epistemic statements, meaning statements that link the referentials in a theorem with sets of referents. Two illustrations follow, both of which pertain to theorem Th1 of the illustrative second-order theory in Chapter 7. Epistemic statement 1: Among the countries in Table 8-1, the greater the ARCID at T_0, the greater the referent in column 1. Epistemic statement 2: Among the countries in Table 8-1, the greater the RUIE during $_0T_0$, the greater the referent in column 3.

Epistemic statements are not part of the theory proper; rather, they are made by an investigator when conducting and

Table 8-1
REFERENTS OF INDUSTRY DIFFERENTIATION (ARCID$_{27}$),
OCCUPATIONAL DIFFERENTIATION (ARCOD$_{27}$),
AND USE OF INANIMATE ENERGY (RUIE),
19 COUNTRIES, *CIRCA* 1960

Countries and Years	ARCID$_{27}$ * Col. 1	ARCOD$_{27}$** Col. 2	RUIE*** Col. 3
Canada, 1961	.923	.925	5,663
Chile, 1960	.913	.899	839
El Salvador, 1961	.695	.691	128
Ghana, 1960	.667	.657	99
Greece, 1961	.741	.727	443
Honduras, 1961	.576	.577	160
Hungary, 1960	.862	.848	2,080
Ireland, 1961	.887	.869	1,844
Japan, 1960	.898	.868	1,166
Netherlands, 1960	.915	.900	2,691
New Zealand, 1961	.939	.903	2,029
Norway, 1960	.938	.901	2,740
Philippines, 1960	.623	.614	143
Portugal, 1960	.839	.801	379
Sweden, 1960	.904	.897	3,491
Thailand, 1960	.347	.359	62
Turkey, 1960	.418	.475	246
United Arab Republic, 1960	.731	.747	290
United States, 1960	.897	.906	8,047

*Adjusted referent of census industry differentiation. See text for a designation of the referential formula used to compute the referents. Industry data from United Nations, *Demographic Yearbook, 1964*, Table 9.

**Adjusted referent of census occupational differentiation. See text for designation of the referential formula used to compute the referents. Occupational data from United Nations, *Demographic Yearbook, 1964*, Table 10.

***Referent of the use of inanimate energy in the form of coal, lignite, petroleum products, natural gas, and hydroelectricity, with each unit expressed in kilograms of coal equivalent per capita. Figures from United Nations, *Statistical Yearbook, 1964*, Table 131.

reporting a test. So each epistemic statement is unique, for it refers to a particular set of referents. Nonetheless, the statement is made as a step in a prescribed test procedure, and it does not represent an independent judgment by an investigator as to what is relevant evidence. Reconsider epistemic statement 1. Given any set of referents identified as ARCID, the epistemic statement follows from that identification regardless of the investigator's opinions. He simply makes the statement in accordance with a procedure prescribed by the mode of theory construction.

Given conventional practices, no one will think of referents as operational definitions, but some sociologists may think of them as "indicators." That identification introduces a needless ambiguity. Although the word "indicators" is used frequently in tests of theories or islolated propositions, the usage is uncritical. So if it were said that the referents in column 1 of Table 8-1 are indicators of ARCID or of 'industry differentiation at T_0,' any interpretation of that statement would be conjectural. Zetterberg has shown (1965:116) that the word "indicator" may be used in three different senses, and even his analysis ignores a major ambiguity. Whereas some sociologists would use the word "indicator" to identify a formula by which values are computed, others evidently reserve the word for the values themselves. Thus, to illustrate, some sociologists would regard the ARCID formula as an indicator, while others would identify each value in column 1 of Table 8-1 as an indicator.

Even if the word "indicator" is used to identify the values in Table 8-1, that identification would not be an adequate substitute for epistemic statements. Consider the following expression: The values in column 1 are indicators of ARCID. What is the relational term? Regardless of the answer, the expression terminates the systematic use of relational terms, and it would preclude the formal derivation of a prediction about the referents from the theory. Further, if the verb in the expression is the relational term, it suggests a purely logical relation between ARCID (or the referential formula) and the values. Such a relation is alien to an epistemic statement, which, even though untestable, is an empirical assertion. What it asserts is a relation between something given or "known"—the referents—and something unknown. In stipulating a referential formula and the requisite kind of data, the theorist asserts that investigators will apply the formula correctly and that the designated kinds of data will be sufficiently reliable and comparable. So any referential (for instance, ARCID) designates referents that would be computed were the referential formula *applied correctly to absolutely reliable data.* For any unit at any point in time, that "ideal" referent is unknown and unknowable; nonetheless, the theorist presumes that the actual referents approximate the ideal or, to put it another way, that there is a close direct relation between the actual and the postulated ideal referents. But that relation is not true by definition; so, to use Northrop's terminology (1947), an epistemic statement asserts an epistemic correlation.[7] Such an assertion is not directly testable, since the relation between actual and ideal referents is unknown and unknowable, but it is an assertion nonetheless.

It is now fashionable in sociology to question the distinction between theoretical and empirical languages (see Doby, 1969). The distinction in sociology is blurred, but only because the field has not adopted formal theory construction. In the proposed mode of formalization, the theoretical language comprises constructs, concepts, and referentials, while the empirical language includes unit terms, relational terms, and referents.

Given a distinction between theoretical and empirical languages, the two must be connected to realize tests of a theory. That is the function of an epistemic statement, and it is necessarily a combination of theoretical and empirical terms. If it be objected that an epistemic statement blurs the distinction between the two languages, the objection is misplaced. An epistemic statement does not negate the distinction; rather, it connects the two languages.[8] However, if critics demand that each epistemic statement be "justified," the reply is that made to the demand that transformational statements and referential formulas be justified. Briefly, any attempt to justify an epistemic statement without reference to a theory and its predictive accuracy is merely opinion.

Hypotheses.   As stressed repeatedly, referents are evidence only when a prediction about them is derived from a theory. In conducting and reporting tests, predictions about referents are statements in the form of derived assertions, and such a statement is identified as an hypothesis. Note that an hypothesis is not just any prediction; rather, it is a formally derived prediction.

An hypothesis is derived by the application of the sign rule to a theorem and two epistemic statements.[9] As an illustration, the application of the sign rule to epistemic statement 1, epistemic statement 2, and theorem Th1 of the illustrative second-order theory in Chapter 7 results in the following hypothesis: Among the countries in Table 8-1, the greater the referent in column 1, the greater the referent in column 3.

Descriptive statements.   In testing an hypothesis it might appear that the *actual* relation between sets of referents is known and hence the hypothesis is either consistent or inconsistent with that "fact." The matter is not that simple.

Once an hypothesis has been derived, the next step for the investigator is to *describe* the "actual" relation between the two sets of referents. Investigators working independently should agree in their descriptions, but this is not likely if the investigators use a natural rather than a technical language. So a technical language is utilized, and that language is part of the test procedure as stipulated by the mode of theory construction. The proposed mode stipulates the use of some ordinal measure of association (such as *rho* or *tau*) to express the association between two sets of referents.[10]

It might appear that the test ends with the computation of a *rho* or *tau* value, the idea being that an investigator can pronounce the hypothesis as either consistent or inconsistent with the value. But two additional steps are necessary before that pronouncement. In the next subsequent step the investigator makes a *descriptive* statement that stipulates a value as representing the association between the two sets of referents.[11] An illustration of a descriptive statement follows: Among the countries in Table 8-1, the rank-order coefficient of correlation between the referents in column 1 and the referents in column 3 is +.837.

Although a descriptive statement is as close to a "fact" in the sense of "that which is unquestioned" as sociologists ever come, it is nonetheless an assertion and one not beyond question.[12] Investigators have been known to make a mistake in computing a value, so there is not absolute assurance that all investigators will report the same values in their descriptive statements. Even if a series of values do agree, there is no assurance that the "N + 1" investigator would compute that value. So absolute certainty never is realized, and for that reason a descriptive statement is an assertion.

Given a descriptive statement and the corresponding hypothesis, the consistency or inconsistency of the two statements may appear obvious to the investigator. However, his conclusion should be governed by two rules in the interpretation of tests, both of which presume that the value in the descriptive statement is a coefficient of correlation. Rule 1: If the relational term of the hypothesis is "greater . . . greater" and the sign of the coefficient in the descriptive statement is positive, then the hypothesis and the descriptive statement are consistent and the hypothesis is true; but if the coefficient sign is not positive the two statements are inconsistent and the hypothesis is false. Rule 2: If the relational term of the hypothesis is "greater . . . less" and the sign of the coefficient in the descriptive statement is negative, then the hypothesis and the descriptive statement are consistent and the hypothesis is true; but if the sign of the coefficient is not negative, the two statements are inconsistent and the hypothesis is false.

A series of true hypotheses is evidence of the theory's predictive accuracy, but that criterion can be considered in two ways. A true hypothesis only indicates *congruence*, that is, a correct prediction of the *direction* of a relation.[13] The *magnitude* of predictive accuracy is another matter; it is judged by the magnitude of correlation coefficients in descriptive statements. Since an assessment of a theory considers both congruence and magnitude, investigators should report the truth or falsity of each hypothesis and also the coefficient of correlation in the corresponding de scriptive statement.

Theorems and tests.   As indicated previously, only theorems enter directly into tests, for they are the only intrinsic statements considered in the derivation of hypotheses. However, other types of intrinsic statements enter *indirectly* into derivations of hypotheses, and hence tests are relevant in assessing the theory as a whole.

The derivation of theorems is largely a matter of convenience. Thus, the foregoing illustrative hypothesis could have been derived by application of the sign rule to the epistemic statements and the following direct intrinsic statements of the second-order theory presented in Chapter 7: axioms A1, A2, A3, A4, and A5; postulates P1 and P3; and transformational statements T1 and T3. But it is far less cumbersome to derive hypotheses from theorems and epistemic statements. As such, the test results are evidence for assessing theorems, but an assessment of theorems implies an assessment of the direct intrinsic statements.

The foregoing commentary does not mean that a test of *one* hypothesis necessarily produces evidence for assessing the entire theory. On the contrary, each test is confined to one particular theorem, and some of the direct intrinsic statements may not enter into the derivation of all theorems. So a test may produce evidence for the assessment of only some direct intrinsic statements. But the matter is simplified for the investigator; all he need do is derive each hypothesis from a particular theorem.

An Illustrative Test of a Second-Order Theory

To further the explication of terminology and procedure, this section reports three tests of the second-order theory presented in Chapter 7. The reports are oversimplified in certain respects. As indicated previously, each report of tests should commence with a brief summary of the theory. When the theory has been reported in the literature, the summary can be limited to a statement of the theorems and the referential formulas. No summary is needed in the present illustration. Also, as indicated previously, the report should designate the units selected for the test, the manner of their selection, the kind of data used, and the method used to gather the data or their source (if obtained in published form). Since most of that information is provided by Table 8-1, the present report is limited to epistemic statements, descriptive statements, hypotheses, and interpretation.

Epistemic statements

E1: Among the countries in Table 8-1, the greater the $ARCID_{27}$ at T_0, the greater the referent in column 1.

E2: Among the countries in Table 8-1, the greater the $ARCOD_{27}$ at T_0, the greater the referent in column 2.

E3: Among the countries in Table 8-1, the greater the RUIE during $_0T_0$, the greater the referent in column 3.

Hypotheses

H1 (from Th1, E1, and E3): Among the countries in Table 8-1, the greater the referent in column 1, the greater the referent in column 3.

H2 (from Th2, E1, and E2): Among the countries in Table 8-1, the greater the referent in column 1, the greater the referent in column 2.

H3 (from Th3, E2, and E3): Among the countries in Table 8-1, the greater the referent in column 2, the greater the referent in column 3.

Descriptive statements

D1 (corresponding to H1): Among the countries in Table 8-1, the rank-order coefficient of correlation between the referents in column 1 and the referents in column 3 is +.837.

D2 (corresponding to H2): Among the countries in Table 8-1, the rank-order coefficient of correlation between the referents in column 1 and the referents in column 2 is +.949.

D3 (corresponding to H3): Among the countries in Table 8-1, the rank-order coefficient of correlation between the referents in column 2 and the referents in column 3 is +.904.

Interpretation

All three hypotheses are true, and the predictive magnitudes are +.837, +.949, and +.904. Each value is a rank-order coefficient of correlation that expresses a type C-I space-time relation among nineteen countries, with the sources of the data shown in Table 8-1.[14]

Discrimination.   Since the illustrative theory generates more than one theorem, the tests just reported provide a basis for assessing discriminatory power. The descriptive statements reveal that the relations among the three sets of referents in Table 8-1 are not equally close. Accordingly, if the theory discriminates, it should be possible to predict *ordinal* differences in the magnitude of the three correlation coefficients. Such predictions are derived in accordance with the following principle: The greater the number of substantive terms that enter directly or indirectly into the derivation of an hypothesis, the less the relative magnitude of the correlation coefficient in the corresponding descriptive statement. The rationale of the principle is that the relation between any two variables is contingent on the number of intervening variables.

A theorist is not required to employ the foregoing principle in deriving predictions as to the relative magnitude of correlation coefficients in the descriptive statements. He may stipulate some other principle (one with a different rationale); but, whatever the principle, it must be such that independent investigators can derive the same predictions in any given set of tests. In other words, the principle must be *potentially* applicable to any theory that has been constructed in accordance with the proposed scheme. However, it may be that a theorist does not anticipate any systematic variance in the correlations and/or simply declines to stipulate any principle; if so, the theory has no discriminatory power.

Ten substantive terms enter into the derivation of hypothesis H1, and the correlation coefficient in descriptive statement D1 is .837. In the case of hypothesis H3 there are eight substantive terms, and the coefficient is .904. Finally, seven substantive terms enter indirectly into the derivation of hypothesis H2, and the coefficient is .949. Observe that hypotheses rather than variables are the units of analysis, and two properties of each hypothesis are considered: (1) the number of substantive terms that entered indirectly into its derivation and (2) the magnitude of the correlation coefficient in the corresponding descriptive statement. To the extent that the two properties vary inversely among hypotheses, the theory discriminates; and the discriminatory power of the illustrative theory in this one set of tests is maximal.

Commentary.   The foregoing report of tests is very brief, but that is a virtue of a formal theory. If a theory is stated discursively, then *attempts* to test it cannot be reported briefly, and it seems inevitable that rhetoric and criticism enter into the report.

The subject touches on an important practical consideration. Although sociologists generally recognize the need for tests and replications, very few accounts of tests can be published unless sociologists adopt a parsimonious procedure for stating results. This they cannot do without abandoning the discursive mode of theory construction.

An Illustrative Test of a Third-Order Theory

Since a third-order theory is complicated, an illustrative test of an instance is needed if only to further understanding of that type of theory. This section reports a test of the first illustrative third-order theory in Chapter 7. No summary of the theory is needed, and the information pertaining to units and data can be gleaned from Tables 8-1 and 8-2. However, a complete report would indicate how the countries in Table 8-2 were grouped into universes. Such groupings are always done in accordance with the instructions in the extrinsic part of the theory, but the instructions may be such that the investigator has some options. In this case, the nineteen countries in Table 8-1 were arranged by magnitude of their ITPC referents and grouped into three universes of six countries (excluding Greece, which represents the median ITPC referent). Thus, as indicated in Table 8-2, no ITPC referent in Universe I is less than 199 nor greater than 747 (which is RMITPC for that universe).

Only one theorem was derived from the illustrative third-order theory, which is here repeated as theorem 1: Among universes of countries, the greater RMITPC during $_0T_0$ for a universe, the less *rho* (ARCID, T_0-RUIE, $_0T_0$) for the universe. A comment has already been made on RMITPC, so only the second referential need be explained. It is a relational referential, signifying the rank-order coefficient of correlation between ARCID, T_0 and RUIE, $_0T_0$ referents in a universe of countries. Three *rho* (ARCID, T_0-RUIE, $_0T_0$) referents are shown in column 4 of Table 8-2.

It may appear obvious that the theorem predicts an inverse relation between the referents in column 3 and the referents in column 4. However, as before, such a prediction must be derived formally.

Table 8-2
REFERENTS OF INDUSTRY DIFFERENTIATION AT T_0 ARCID$_{27}$),
PER CAPITA USE OF INANIMATE ENERGY DURING $_0T_0$ (RUIE),
MAXIMUM INTERNATIONAL TRADE PER CAPITA DURING $_0T_0$
(RMITPC), AND THE RELATION BETWEEN
INDUSTRY DIFFERENTIATION AT T_0 AND PER CAPITA USE
OF INANIMATE ENERGY DURING $_0T_0$ [RHO(ARCID$_{27}$-RUIE)]
AMONG COUNTRIES IN THREE UNIVERSES, *CIRCA* 1960

Universe, Country, and Year	ARCID$_{27}$*	RUIE*	RMITPC**	Rho(ARCID$_{27}$-RUIE)
	Col. 1	Col. 2	Col. 3	Col. 4
Universe I			747	+.14
Canada, 1961	.923	5,663		
Ireland, 1961	.887	1,844		
Netherlands, 1960	.915	2,691		
New Zealand, 1961	.939	2,029		
Norway, 1960	.938	2,740		
Sweden, 1960	.904	3,491		
Universe II			198	+.83
Chile, 1960	.913	839		
El Salvador, 1961	.695	128		
Ghana, 1960	.667	99		
Hungary, 1960	.862	2,080		
Portugal, 1960	.839	379		
United States, 1960	.897	8,047		
Universe III			91	+.77
Honduras, 1961	.576	160		
Japan, 1960	.898	1,166		
Philippines, 1960	.623	143		
Thailand, 1960	.347	62		
Turkey, 1960	.418	146		
United Arab Republic, 1960	.731	290		

**Maximum amount of imports and exports in dollar value per capita for any country in the universe. Computed from data in United Nations, *Statistical Yearbook, 1966*, Table 148, and United Nations, *Statistical Yearbook, 1964*, Table 2.

Epistemic statements

E1: Among universes of countries in Table 8-2, the greater the RMITPC during $_0T_0$, the greater the referent in column 3.

E2: Among the universes of countries in Table 8-2, the greater the *rho* (ARCID, T_0-RUIE, $_0T_0$), the greater the referent in column 4.

Hypothesis

H1 (from Th1, E1, and E2): Among the universes of countries in Table 8-2, the greater the referent in column 3, the less the referent in column 4.

Descriptive statement

D1: Among the universes of countries in Table 8-2, the rank-order coefficient of correlation between the referents in column 3 and the referents in column 4 is $-.50$.

Interpretation.

The hypothesis is true, and predictive magnitude is $+.50$. The value is a rank-order coefficient of correlation that expresses a type C-I relation among three universes of countries, with a *N* of six countires in each universe.[15] Sources of the published data used in the tests are shown in notes to Tables 8-1 and 8-2.

Commentary on the theory.   Since the illustrative third-order theory generates only one theorem, it is simple to the point of being unrealistic. If the theory had been constructed for other than illustrative purposes, an additional theorem would have been generated. The second theorem would be exactly like the first, except ARCOD, T_0 would replace ARCID, T_0 in the relational referential. Had the second theorem been derived, another hypothesis would have been tested by the same procedure.

Although investigators do nothing more than report tests of hypotheses, the assessment of a theory should be more than a compilation of test results. The goal in an assessment is modification of the theory so as to maximize predictive accuracy. In this particular test congruence is realized, but the correlation coefficient in D1 does not reveal the values in column 4 of Table 8-2.

Observe that the correlation between ARCID and RUIE is much less in universe I than in II or III but in universe III the correlation is less than among all nineteen countries. That difference may be due to the magnitude of the ITPC referents in universe III, that is, the correlation between ARCID and RUIE might be greater if RMITPC for universe III were 0 rather than 91. But there is another possibility. Observe that variance in ARCID and RUIE is much less in universe III than among all nineteen countries. Given doubts about the reliability of the data, the lower correlation in universe III could be due to the reduction of variance in ARCID and/or RUIE referents. With that possibility in mind, the theorist could modify the theory so that the theorem asserts a relation between *rho* (ARCID, T_0-RUIE, $_0T_0$) and three conditions in a universe of countries: magnitude of ITPC referents, variance in ARCID referents, and variance in RUIE referents.

Comment on Fourth-Order Theories

No illustrative tests of a fourth-order theory are given. However, illustrations are not necessary, since the procedure would be identical to that employed in the test of a third-order theory.

The only difference has to do with the referentials, not procedure. As suggested by Table 8-2, in tests of a third-order theory only one set of referents takes the form of correlation coefficients, with a coefficient for each universe. But in a test of a fourth-order theory both sets of referents would comprise correlation coefficients, with two coefficients for each universe. Thus, whereas only one set of correlations coefficients appear in Table 8-2 (column 4), a table presented in conjunction with a test of a fourth-order theory would include two sets.

Overview

Sociological theories are often condemned as untestable and justly so, but it will not do to use the word "test" uncritically, that is, as though everyone agrees as to what constitutes a test. That is not the case in sociology.[16] One may describe a test in various ways, but the description should be such that investigators will employ the same procedure, for only then are tests truly systematic.

Systematic tests will be realized in sociology only when the field adopts formal theory construction, as there can be no *specific* procedure for tests unless the procedure is integrated with a particular mode of theory construction. The procedure proposed here requires seven major steps for each test. Stated in the briefest manner possible those steps are: (1) identifying and selecting instances of the unit term; (2) gathering or otherwise obtaining data on each unit as stipulated in the extrinsic part of the theory; (3) computing two referents for each unit by application of referential formulas to the data; (4) making epistemic statements that link referentials in a theorem to referents; (5) deriving a hypothesis from a theorem and two epistemic statements; (6) making a descriptive statement about the relation between the two sets of referents; and (7) judging the consistency of the hypothesis and the descriptive statement.

Notes for Chapter 8

[1] The sociological literature on theory construction (for example, Dubin, 1969; Blalock, 1969; Zetterberg, 1965; Stinchcombe, 1968; and Willer, 1967) provides no *specific* procedures for either conducting or interpreting tests. The omission may reflect an uncritical acceptance of the "correspondence-with-the-facts" conception of a test, which is a tacit denial that any special procedures are needed. Alternatively, the omission may reflect the belief that theory construction and test procedures are distinct. The argument pursued here is completely to the contrary in either case.

[2] See Mitchell, 1964, for an extensive critique.

[3] The perspective is manifested in two of Braithwaite's statements: "Man proposes a system of hypotheses: Nature disposes of its truth or falsity. Man invents a scientific system, and then discovers whether or not it accords with observed fact" (1953:368). Significantly, Braithwaite uncritically accepts the notion of "observed facts" and constantly uses the phrase "testable by experience," two practices that are much in keeping with the correspondence theory of knowledge. Virtually all philosophers of science also speak of "testable by experience," but the use of that phrase to characterize assertions in the context of scientific knowledge gives rise to an issue. Some assertions are testable by experience, but the perspective makes tests private and personal. Taking science as an enterprise, a test is a matter of the consistency or inconsistency of public assertions; it is not, strictly speaking, a matter of the correspondence between an assertion and "experience" or "observed fact."

[4] "The word *test* employed in this phase denotes a scientist in contact with the empirical world. It is through the test that he relates the facts he finds in the empirical world to his theoretical predictions about them" (Dubin, 1969:211).

[5] In writing on theory construction, methodology, or research techniques, no sociologist has stipulated such an inclusive procedure. An argument that it is not needed only reflects the uncritical acceptance of the "correspondence-with-the-facts" conception of a test.

[6] Consider Braithwaite's comment on statements made by quantum physicists. "The 'observations' to which they refer need not even be observations in the ordinary sense, which presuppose a basis of immediate experience; they may be traces on a photographic film or holes punched in a tape by an electron counter. So far as the science is concerned, the 'observing' may all be done by machines, the records of these machines being subsequently inspected and interpreted" (1953:7).

[7] Northrop's terminology cannot be used without resolving an ambiguity. Although an epistemic correlation is an assertion of a relation between a "known" or "observable" phenomenon and one unknown or unobservable, the former could be an infinite class or particular instances of the class. If it is an infinite class, then any transformational statement could be taken as an epistemic statement. Similarly, if it be stated that suicide is a manifestation of the death instinct and suicide is taken as "empirical," then the statement asserts an epistemic correlation. But what of a statement that asserts a relation between a "nonempirical" phenomenon and particular events, things, or symbols thereof? Surely the assertion is no less "epistemic" in Northrop's sense than an assertion about infinite classes. The ambiguity stems from Northrop's failure to distinguish between the idea of an "epistemic correlation" and epistemic assertions or statements. He evidently thought of epistemic correlations in connection with infinite classes, but the word is not used here in that sense.

[8] "On the one hand, there should be a general theory expressed in abstract terms; on the other, there must also be a specific *auxiliary* theory necessary for testing purposes" (Blalock,1968a:24). Following Blalock's terminology, epistemic statements would be identified as auxiliary theories, but such identification would be most questionable. For one thing, though apart from the theory itself, each epistemic statement is made in accordance with the theory. Further, since an epistemic statement is unique (that is, it refers to a particular set of values), it is difficult to see how it could be considered as a theory, auxiliary or otherwise.

[9] Note the difference between an hypothesis derived by the prescribed procedure and the notion of an "operational" hypothesis as used in sociology. Typically, an empirical assertion is made and then translated into an "operational" hypothesis. Consider the following illustration: "Hypothesis 4: For a given set of values and given distance between valuer and desideratum, desir-

ing will vary with the activation of different levels in some prepotency hierarchy. In operational terms, this hypothesis states that the intensity of a person's desire for a given object will vary systematically from time to time, and that this intensity at any particular time will be a function of the continuously perceived similarity of the object to other objects strongly desired at that time" (Catton, 1966:147-148). Like many other "operational" hypotheses in the sociological literature, Catton's translation or interpretation does not take the form of a prediction about any particular finite universe; and one must surely wonder how such a prediction could be derived from the "operational" hypothesis. But there is another issue, which is best put in the way of a question: How does Catton derive his "operational" version of the hypothesis? Certainly he does not do so by explicit rules of derivation, and the logical relation between the two statements is conjectural. The criticism is not to be equated with the ambiguous demand for "justification" as it is ordinarily made by sociologists. Of course, if by justification one simply means that it be done in accordance with explicit rules of derivation, there is no issue. In any case, Catton's derivation (if it can be called such) illustrates why "operationalism" is not a substitute for formal theory construction.

[10] Since only predictions of ordinal differences are considered in the test of a theory, each test could be based on a comparison of only two units. As such, there would be only one question: Given the two units and taking the X referents as known, does the hypothesis correctly predict the difference between the Y referents of the two units? However, it is much less laborious to consider differences among several units simultaneously, so *rho* or *tau* is used to indicate the predictive accuracy of the theory. Of the two, *tau* is preferred, since it indicates the proportion of correct predictions that could be made if pairs of units were compared. But *rho* is justified, for it is easier to compute and approximates a *tau* value.

[11] In the present context, a descriptive statement is an assertion about quantitative variables. Nonetheless, the notion is akin to Carnap and Neurath's conception of a "protocol sentence" and to Reininger's conception of an "elementary statement" (see Popper, 1965:95-100). However, the same may be said of an epistemic statement. The point is that philosophers of science do not recognize that a test of a universal assertion about the relation between quantitative properties requires more than one or two types of statements. Popper is a notable exception, for he speaks of basic statements, singular statements, instantial statements, and singular existential statements. But his explication of those types is so garbled (see Popper, 1965:100-105) that it is questionable to suggest any parallel between his scheme and the one proposed here. The difficulty is that Popper does not present systematic illustrations and, evidently, his typology is based on a logic of classes rather than a logic of quantitative relations.

[12] The "generalizibility" of a descriptive statement is never questioned. Each descriptive statement is an assertion about a *particular finite universe*, and it should not be construed as a generalization beyond that universe. Viewed that way, a descriptive statement is akin to the traditional notion of a "find-

ing." Note, however, that you can have a finding with no hypothesis. There is a temptation to interpret a finding or a series of findings as suggesting a generalization about an infinite universe (see Berelson and Steiner, 1964); but such a generalization is arrived at inductively, that is, unlike a theorem, it is not derived.

[13] Recall that a hypothesis is taken as "true" if consistent with the corresponding descriptive statement; otherwise, it is taken as false. Accordingly, should two independent investigators make inconsistent descriptive statements about particular sets of referents, then one and the same hypothesis may be judged as both true and false. The only way to avoid the paradox is to invoke some criterion of *absolute* truth or falsity, for example, an hypothesis is true if it is consistent with each corresponding descriptive statement made by at least three independent investigators. Such a criterion is arbitrary and ignores the dilemma of induction. Assuming that a hypothesis is consistent with each descriptive statement made by three independent investigators, how could we be sure that it would be consistent with descriptive statements made by still other investigators? We could not. Further, assume that a hypothesis is consistent with only some of the descriptive statements made by independent investigators. By what criterion would the hypothesis be taken as true or false? One possibility is a statistical criterion, for instance, an hypothesis is true if it is consistent with more than 50 percent of the corresponding descriptive statements. However, any such criterion is bound to be arbitrary, and a very difficult question remains unanswered: How many descriptive statements are to be made by independent investigators before judging an hypothesis? Obviously, the outcome depends on the answer to that question, but any answer would be arbitrary. In a sense, arbitrariness could be avoided by invoking an absolute "falsity" criterion, such as, an hypothesis is false if inconsistent with *any* corresponding descriptive statement. However, such a criterion would not establish the truth of the hypothesis nor escape the dilemma of induction, for the outcome in judging an hypothesis may depend on the number of corresponding descriptive statements made by independent investigators. All of the problems stem from the idea that there must be some criterion by which the absolute truth or falsity of an hypothesis is established. Such a criterion is not necessary if the goal is an assessment of the predictive congruence of a theory, which is simply the proportion of tests in which the hypothesis is consistent with the corresponding descriptive statement. If more than one investigator makes a descriptive statement about the same referents, then each descriptive statement enters into a separate test. Of course, the proportion of tests in which the hypothesis and the descriptive statement are consistent does not establish the "absolute" truth or falsity of any hypothesis, nor does it resolve the dilemma of induction; but no criterion can do either without being arbitrary.

[14] The initial universe included all countries listed in Tables 9 and 10 of the United Nations, *Demographic Yearbook, 1964*, and Table 131 of the United Nations, *Statistical Yearbook, 1964*. The nineteen countries were selected for

a variety of reasons, which is to say that the sampling method cannot be articulated. Therefore, the tests are only illustrative.

[15] For the reason set forth in note 14, the test is only illustrative.

[16] Tests of "hypotheses" or theories occasionally are reported in the sociological literature. Typically, however, the investigator presents or alludes to a body of data and then simply declares the data to be consistent or inconsistent with the hypothesis or theory. See, for example, Zetterberg, 1965:105.

CHAPTER 9 INTERPRETATION OF TESTS

Although each test requires seven steps, the outcome is simple—the hypothesis is either consistent or inconsistent with the descriptive statement. However, the interpretation of tests entails so many issues and imponderables that the complexity of the subject defies exaggeration; and if one demands certainty, there are several insoluble problems.

The Notion of a Direct Test

The first complexity is the realization that only hypotheses are testable in a direct sense; that is, only an hypothesis is consistent or inconsistent with a descriptive statement. One may argue that if the two are consistent then the theory is also consistent with the descriptive statement and, as such, the test result supports the theory. Extending the argument to the converse, if the hypothesis and the descriptive statement are inconsistent, then the result is contrary to the theory. Nonetheless, even accepting the argument, it does not follow that theories are directly testable. An hypothesis is not a part of a theory; rather, it is a derived prediction, and the derivation precludes equating the truth or falsity of the prediction with that of the theory. Indeed, it is questionable to characterize a theory as true or false. A theory comprises several intrinsic statements and, if labeled false, the label would imply that no component intrinsic statement is true. But a false hypothesis does not demonstrate that all intrinsic statements are false, nor does a true hypothesis reveal that all are true. For that matter, the truth or falsity of any intrinsic statement is unknowable in any direct or conclusive sense.

By illustration, consider Figure 9-1, which depicts the structure of a type 1-2 theory, meaning a second-order theory with no divisions, one axiom, two postulates, no propositions, two transformational statements, and one theorem. Observe first of all that the epistemic statements, the hypothesis, and the descriptive statement are not parts of the theory proper; they enter only into the test. For present purposes assume that the hypothesis and the descriptive statement are consistent. As such, would it follow that the intrinsic statements are all true? No, not even if the dilemma of induction could be ignored.

In Figure 9-1 all theoretic statements (intrinsic and epistemic) assert a direct relation between properties; hence the hypothesis predicts a direct relation between two sets of referents. But suppose that *any two* theoretic relations are actually inverse rather than direct as asserted;[1] there would be a direct relation between the referents, and the hypothesis would still be consistent with the descriptive statement, even though some of the theoretic assertions are false. To be sure, the actual theoretic relations are unknowable, but various combinations of direct and inverse theoretic relations could generate the same hypothesis.

Consider the situation another way. Suppose that an *even* number of direct intrinsic statements are modified so that they

Figure 9-1
THE STRUCTURE OF A TYPE 1-2 THEORY AND ONE TEST

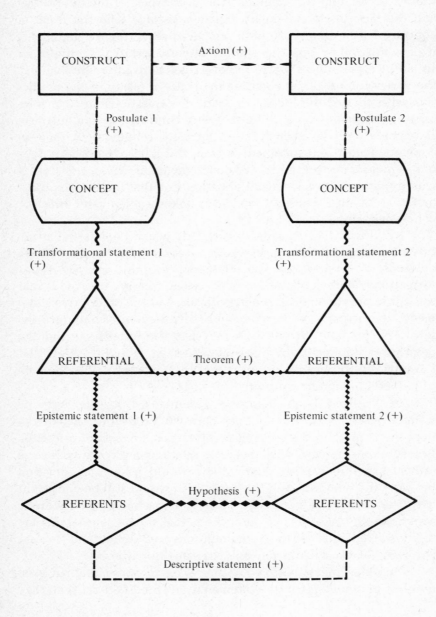

assert an inverse rather than a direct relation. Despite the change the hypothesis would not be altered; it would still predict a direct relation between referents. Accordingly, since the theoretic statements can be altered without altering the hypothesis, the consistency of the hypothesis and the descriptive statement does not prove that the theoretic relations are as asserted.

If the theorem and the hypothesis could be derived from only one set of theoretic statements (that is, a unique derivation) only then would the consistency of the hypothesis and the descriptive statement be compelling evidence. Yet even that consistency would not be conclusive, because the truth of any descriptive statement is a presumption (albeit a very strong one), and the dilemma of induction makes the truth of any intrinsic statement forever unknown.[2] But for present purposes, the crucial consideration is that the same theorem and hypothesis can be derived from different sets of theoretic statements; and for that reason alone the consistency of a hypothesis and a descriptive statement cannot be interpreted with certainty. So to summarize, given a test in which an hypothesis and a descriptive statement are consistent, the only defensible conclusion is that the actual theoretic relations *could be* as asserted in the theoretic statements.

Now consider an instance where the hypothesis and the descriptive statement are inconsistent, as would be the case in Figure 9-1 if the sign of the descriptive statement were negative. Just as positive tests do not incontrovertibly establish the truth of a theory, not even repeated inconsistencies between hypotheses and descriptive statements would prove that all theoretic statements are false. Repeated inconsistencies would only indicate that the relation between properties is not as asserted in one or more of the theoretic statements, and one negative test is especially inconclusive. Although a remote possibility, the descriptive statement could be false, meaning that the direction of the relation between referents is not as asserted in the descriptive statement. Should that be so, then the "facts" are in error, not the theory.

A much more likely possibility is that one or both epistemic statements is false; that is, the relation between the referential and the referents is not in the direction asserted and/or is not close. If so, the negative test would not prove that any of the intrinsic statements are false. On the other hand, a theorist cannot dismiss negative tests by attributing them to false epistemic statements. Given a negative test, the result might have been different had another investigator applied the referential formulas; but, if so, the referential formulas are not empirically applicable, and in that sense the theory is a failure. To say it again, the referential formulas are components of the theory and their use entails the assertion of empirical applicability, which is itself a part of the theory.

Should several independent investigators report the same referents and should each test be negative, it could still be that the epistemic statements are false and the intrinsic statements are true. If the referential formulas have been applied to published data, there is no assurance that the data are sufficiently reliable and comparable. When data have been gathered especially for tests of the theory, then concordance between referents suggests that each investigator gathered the data as stipulated in the extrinsic part of the theory. But even if there is no divergence in the data gathered by independent investigators, there is still no absolute assurance that any of the sets of data are absolutely reliable. However, in designating a kind of published data or in stipulating a procedure for gathering data, the theorist asserts that the data will be sufficiently reliable and comparable, and that assertion is very much a part of the theory. So the general conclusion is that unless one questions the truth of descriptive statements, negative tests cannot be "explained away"; they clearly reveal that something is wrong with the theory.

Finally, suppose it were known that all epistemic and descriptive statements are true; then and only then would negative tests prove that some intrinsic statements are false. But it could be that only one of them is false. To illustrate, suppose that the *actual* relation between the properties in postulate 1, Figure 9-1, is not as asserted (that is, the actual relation is inverse). As such, the descriptive statement would assert an inverse relation and be inconsistent with the hypothesis. However, despite the negative test, with the exception of postulate 1, all the relations could be as asserted in the intrinsic statements.

Some issues. Since tests do not reveal the truth or falsity of a theory or any particular component intrinsic statement, sociologists who demand certainty will reject the proposed scheme. In doing so, they are blissfully ignorant of one point—uncertainty is inevitable regardless of the mode of theory construction—and the proposed scheme only makes recognition of uncertainty inescapable.

At the outset, observe again that the dilemma of induction alone precludes conclusive verification of a theory. If tests are positive (whatever the criterion), there is no assurance that subsequent tests will also be positive. That consideration is reason enough for assessing a theory in terms of its relative predictive power rather than presuming that its "truth" can be established.

Most sociologists are aware that the dilemma of induction precludes verification of a theory.[3] Nonetheless, they cling to the belief that a theory can be proven false, and they are likely to cite Karl Popper's work (1965) in that connection. The burden of Popper's argument is that a theory should be falsifiable, and there is no quarrel with that dictum. However, to say that it is *logically* possible to falsify a particular theory is one thing, but it is quite another to say that tests have falsified it.[4]

Any test of a theory involves descriptive statements which are distinct from the theory itself. So, given a negative test, the finding conclusively falsifies the theory only insofar as one accepts the descriptive statements as "true." But it is difficult to accept the notion that descriptive statements are necessarily true. To be sure, it may be reasonable to take them as true, because, given a particular set of referents, independent investigators can make descriptive statements and the "truth" of any particular one is a matter of agreement with the others. Thus, given two sets of referents, investigators can apply the same conventional measure of association (for example, *rho*) and, if agreement is realized, no one would balk at accepting either descriptive statement as true. Even if only one investigator makes a descriptive statement, the inclination is surely to take it as true; otherwise, there is no basis for judging the corresponding hypothesis. Nonetheless, there always remains the remote possibility that a descriptive statement is false, for investigators have been known to make errors and several investigators could make the same mistake. Consequently, while descriptive statements are accepted as true and with good reason, that judgment falls short of certainty.[5] Correlatively, the truth or falsity of an hypothesis is never more certain than the truth of a descriptive statement, for the latter is employed to judge the former.

Whereas an hypothesis is either true or false, those labels are inappropriate when applied to a theory or any intrinsic statement. Uncertainty is so inevitable that a theory or a component intrinsic statement can be accepted or rejected only tentatively, and those labels are quite different from the conventional notions of true and false.[6] If a theorist tentatively accepts a theory, he merely signifies his willingness to continue tests. His acceptance may mean that the theory appears to have greater predictive power than alternatives, including complete ignorance or no theory at all.

After a series of tests, the theorist may decide that the theory has less predictive accuracy than some contender, but his rejection of it should be tentative. After all, conclusions as to predictive accuracy are always based on a finite universe, and there is no assurance that past and future tests will be congruent. For that reason alone, a theorist is not likely to reject the theory (even tentatively) on the basis of one test; but no particular number of negative tests would justify a categorical rejection.

Even when judging the predictive accuracy of a theory relative to that of contenders, the choice is clouded by uncertainties. The comparative predictive power of two theories is always *ex post facto*, and there is no way to know what the comparison would reveal in light of future tests. So a theorist's decision to reject a theory should be tentative, and subsequent developments (a failure to find an alternative theory or negative tests of an alternative) may lead him to reconsider.

The notion of a tentative assessment of a theory gives rise to an issue. Given the uncertainties, one could argue that predictive power, accuracy in particular, is not an "objective" criterion for assessing a theory or even suitable for invidious comparisons of contenders. In other words, should two sociologists employ the predictive power criterion in assessing a theory, they might disagree in the decision to accept or reject it, even though they considered the same tests of the theory and the same contending theories.[7] The possibility of such a disagreement is admitted readily, which is to say that debates over the predictive power of theories are expected in any field. Even when based on an assessment of predictive accuracy, a decision to accept or reject a theory entails a guess; but the history of science clearly reveals that some theorists "guess" better than others and guessing is simply part of the game.

Granted that uniform agreement in decisions to accept or reject a theory is not assured by the predictive power criterion, the consensus that could be achieved is much greater than that now realized in sociology. The declaration is only an allegation, but it is such only because sociologists do not demand systematic tests, nor do they insist on the evaluation of theories by reference to predictive power. Rather, they invoke such notions as plausibility (a euphemism for successful rhetoric), logical consistency (a criterion that cannot be applied systematically to sociological theories as they are now formulated), scope (the vaguer the theory the better), or such exotic criteria as "bringing men back in" (a plea for reductionism), significance (an open invitation to the opinionated), and consistency with human nature or the nature of social life (impressions are more compelling than systematic tests). Consequently, we seldom know what criterion critics have employed in assessing a sociological theory, and even the critics may be unwilling or unable to make their standards explicit. To state the issue as a question: Who is prepared to argue that if theories are assessed in terms of predictive accuracy, the consensus will not be greater than is now realized? In making such an assessment, critics can at least point to a specific figure, the proportion of tests in which an hypothesis and a descriptive statement are consistent.

Some sociologists are certain to attribute the admitted uncertainty in the interpretation of tests to the proposed mode of theory construction and the emphasis on predictive accuracy; they may even presume that an alternative scheme or criterion would eliminate the uncertainties. That presumption is a delusion. Regardless of the mode of formalization, insofar as the notion of empirical evidence enters into the assessment of a theory, the dilemma of induction alone precludes certainty. Nonetheless, critics are likely to reject such notions as axioms, postulates, theorems, and so on, in preference for isolated, seemingly simple generalizations, the presumption being that tests can be conclusive. That presumption is also a delusion. Given any universal generalization that asserts a relation between quantitative variables, the interpretation of any test is inherently debatable.

Consider the following generalization: Among countries, the greater the population density, the greater the degree of urbanization. The generalization may appear to be clearly false; however, no matter how systematic the test, uncertainty is inescapable. Given the generalization, one may make two statements as follows: (1) in 1961 the population density of India was greater than that of Australia; (2) in 1961 the degree of urbanization in Australia was greater than in India. Those two statements appear contrary to the generalization, but the truth of either is not beyond question. If one is allowed to designate any set of numbers as representing the degree of urbanization and population density in Australia and India, then one is free to accept or reject statements 1 and 2 as he sees fit. True, both substantive terms are used extensively in research, and they evidently have fairly clear meanings. Nonetheless, the "refutability" of the generalization hinges on agreement in designating relevant data, and the only way to assure agreement is to stipulate referentials and referential formulas. But how is the connection between the concepts and referentials to be stipulated? It can be done systematically only through transformational statements. One may choose to forego explicit transformational statements, but nothing is gained by leaving them implicit, nor is complexity alleviated by the use of "measures" rather than "referentials." Then note that statements 1 and 2 by themselves do not contradict the generalization, nor can they without deriving a prediction (whether or not labeled as an hypothesis) about the difference between Australia and India with regard to either population density or degree of urbanization. The consistency or inconsistency of the derived prediction with a descriptive statement would then represent evidence for or against the generalization. However, statements 1 and 2 are not truly descriptive, since neither refers to a particular value. Further, an hypothesis cannot be derived systematically without making statements that connect each referential and a particular value. Such statements may not be made explicit or labeled as epistemic, but they are entailed in the test whether recognized or not. Now suppose that a descriptive statement is made, and it is consistent with the hypothesis. How can one be sure the transformational statements are true? How can one be sure the epistemic statements are true? There is no way, and even the truth or falsity of the descriptive statement is not beyond doubt (after all, numbers can be misread). To summarize, starting with what appears to be a simple generalization, a test of it would involve four additional assertions, a derived prediction, and still another statement, none of which are beyond debate. So, the proposed mode of formal theory construction does not introduce avoidable complexities and uncertainties.

Another issue pertains to the demand that sociologists formulate testable theories. The admitted uncertainty in the interpretation of tests could be taken as a rationale to reject the demand for testable theories. After all, so one might argue, since tests are not conclusive, demanding them is pointless. Extending the argument, the demand could be construed as counterproductive in sociology, the fear being that unreliable data will lead to the premature rejection of a theory.

It is admitted readily that demands for testable theories are often naive, especially those made by sociologists whose panacea is operationalism. They are also naive when they represent an uncritical acceptance of the correspondence "theory" of knowledge and hence never recognize, nor appreciate, the uncertainty entailed in the interpretation of tests. However, the uncertainty in no way justifies the formulation of untestable theories in sociology. A theory must be evaluated by some standard, so what is the alternative to tests? If one rejects predictive accuracy as a criterion, then tests of a theory are superfluous, and the issue is joined. In any case, one cannot justify the fear that tests will lead to the premature rejection of a theory. Tests are not undertaken with a view to a categorical rejection of a theory; an assessment should be tentative, and the predictive power of a theory is to be judged relative to that of contenders. As such, the proposed procedure duly recognizes the danger of prematurely rejecting theories; but that recognition in no way suggests postponement of tests until all doubts about data have been eliminated. An argument for postponement suggests an unrealistic divorcement of theory construction and the data problem; specifically, until sociologists commence demanding testable theories, data problems will not receive the attention they deserve. Untestable sociological theories do not prevail because of a data problem, for they are stated so vaguely that the kind of data needed for a test is not even specified. As a case in point, it is ridiculous to argue that Parsonian theory is not testable because "adequate data" are not available. The theory does not even indicate what would be adequate data.

Interpretation of Tests: A Type 2-3 Theory

If after tests the theorist decides (by one criterion or another) that the predictive accuracy of a theory with only one theorem is inadequate, he has no choice but to reject the theory as a whole. Insufficient predictive accuracy only suggests that one or more of the intrinsic statements is false, but since a type N-2 or 0-2 theory includes only one kind of unit term and one theorem, negative tests cannot be interpreted so as to reveal the intrinsic statements that should be rejected or modified. Accordingly, unless the theorist is to modify intrinsic statements by intuition, he has no choice but to reject the theory as a whole. For that reason, one should avoid formulating theories in which there is only one theorem and no divisions.[8] A much more desirable structure is depicted in Figure 9-2. Of course, that type of theory (2-3) has greater scope than a type 0-2 theory. More important for present purposes, given negative tests it may be possible to identify (other than by intuition) those intrinsic statements that should be modified or rejected.

An assessment of a theory (regardless of the type) should be guided by the *principle of conservation*, according to which the theorist seeks to retain as many intrinsic statements as possible. The principle stipulates a priority: consider the modification of referential formulas before modifying or rejecting axioms or postulates. Obviously, if an axiom is rejected, it may be necessary to abandon some of the postulates and transformational statements; but a referential formula can be modified without any loss of intrinsic statements. Moreover, there is only one way to modify

Figure 9-2
THE STRUCTURE OF A TYPE 2-3 THEORY AND A SET OF TESTS

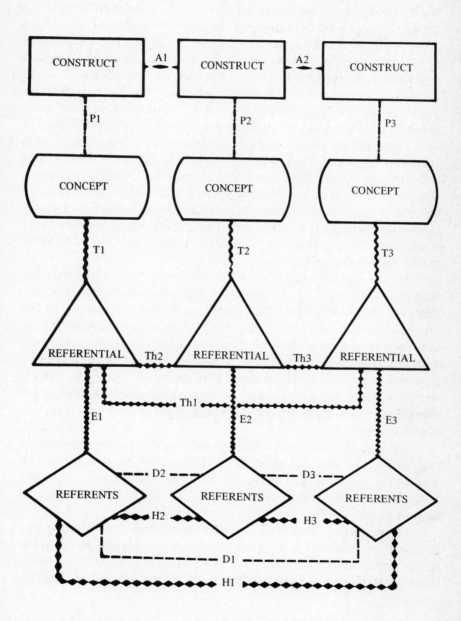

axioms or postulates—reverse the direction of the asserted relation—whereas a referential formula can be modified in several ways. The theorist may alter his explication of the formula, revise the formula itself, designate another kind of data, or stipulate a different procedure for gathering the same kind of data.

Other preliminary considerations. A test always pertains to a particular population and point in time, meaning that each hypothesis is unique. However, hypotheses are classified as to type, depending on the theorems that enter into their derivation. Thus, in Figure 9-2, hypothesis 1 (H1) is derived in part from theorem 1 (Th1), and *in all other tests the hypotheses derived in part from that theorem would be designated as H1.* Similarly, each epistemic statement is unique, but each is also an instance of a type, depending on the constituent referential. Observe that a particular referential is a substantive term in transformational statement 1 (T1), and the same referential is a substantive term in epistemic statement 1 (E1). Accordingly, in any test where that referential enters into an epistemic statement, the statement would be labeled as E1.

If the unit term in a type 2-3 theory is "metropolitan areas within a country," an instance of H1 could be tested for one universe (for example, the United States, 1960), an instance of H2 on another universe (Australia, 1961), and an instance of H3 on still another universe (the United States, 1970). However, when data are available, it is desirable to test an instance of each type of hypothesis on the same universe (for example, Australia, 1961), with the tests taken together designated as a *set*, that is, a set comprises a test of an instance of each type of hypothesis on the same universe.

Ideally, the assessment of a theory would consider a *series* of tests, meaning two or more sets. Thus, again given a theory in which the unit term is "metropolitan areas in a country," a series of tests could comprise a set for Australia, 1961, and a set for the United States, 1960.

Some of the complexities could be avoided if it were realistic to speak of "tests of a theorem" without considering hypotheses or epistemic statements. But even if one speaks in that manner, there must be derived predictions about particular values (that is, referents), and some term should identify those predictions. Certainly the predictions are not identical with the theorem even though the theorem enters into their derivation, and hence they are designated as hypotheses. Further, unlike hypotheses, theorems are not testable in any direct sense. A test should lead to a definite conclusion—the prediction is either true or false—but that conclusion is defensible only for an hypothesis. By contrast, one may conclude that a theorem "could be true" or conclude that it is "probably false," but either conclusion is quite different from that pertaining to an hypothesis. The point is that the truth or falsity of a theorem, like any theoretic statement, is unknowable, and for reasons unrelated to the dilemma for induction. Reconsider Figure 9-2 and suppose that all tests of the type 1 hypotheses are negative, meaning that in each instance H1 and D1 are inconsistent. The result would suggest that theorem 1 is false, but that cannot be known with certainty. Conceivably, in each test one or both of the epistemic statements (instances of E1 or E3) and/or the descriptive statement (an instance of D1) could have been false.

If there is a low degree of concordance in referents computed by independent investigators, then negative test results are expected, which is to say that positive findings are expected only to the extent that the referential formulas are empirically applicable. Accordingly, tests of a theory should not be undertaken until the theorist is satisfied that the referential formulas are empirically applicable to a sufficient degree; otherwise, in assessing the theory he will be haunted by doubts.

When tests of a type of hypothesis are negative despite confidence in the descriptive statements and evidence that the referential formulas are empirically applicable, then it may appear obvious that the corresponding theorem is false. However, the situation is not that simple. If the data are not sufficiently reliable and comparable, then negative tests may ensue even if the theorem is true. After all, each transformational statement and each theorem refer to "ideal" data, that is, absolutely reliable and comparable data. If the data are not reliable and comparable, the epistemic statements are false regardless of the concordance between independent sets of referents (that is, regardless of empirical applicability). However, as indicated previously, uncertainty about the reliability and comparability of data is inevitable; and there is no way that predictive accuracy can be divorced from questions about data. So given negative tests, the theorist may have such grave doubts about the data that he will retain the referential formulas but specify a different kind of data and/or a different procedure for gathering the same kind of data. But there is an alternative. If he comes to doubt the "truth" of a transformational statement, he may modify the referential formula itself, a change that may also entail specification of different kinds of data and/or procedures for gathering data. Subsequent tests may suggest that the theorist made the "correct" choice between alternatives, but he has little basis for making the choice other than intuition.

Since theorems enter directly into the derivation of hypotheses, it might appear that the first step is to reach a conclusion as to their probable truth or falsity. Such a procedure can be followed, but it is not essential. The truth or falsity of a theorem may depend on the direct intrinsic statements from which it was derived, and hence a judgement of a theorem implies a judgment of "cognate" transformational statements, propositions, postulates, and axioms.[9] For that reason alone, it is desirable to interpret tests so as to reach a conclusion about the probable truth of the *direct* intrinsic statements.

As suggested previously, an hypothesis is either true or false, but the appropriate terminology in assessing an intrinsic statement is *probably* true or *probably* false.[10] Each intrinsic statement asserts a close relation between properties; as such, there are four possibilities: (1) the actual relation is close and in the direction asserted by the intrinsic statement, (2) the actual relation is in the direction asserted but not close, (3) the actual relation is neither close nor in the direction asserted, or (4) the actual relation is close but not in the direction asserted. Since test results commonly do not permit a distinction between possibilities 2 and 3, the two are considered as one condition, that is, the actual relation may or may not be in the direction asserted but in either case it is not close. Accordingly, there are three alternative judgments of any theoretic statement: (1) probably true in the "A" sense (the actual relation is close and in the direction asserted), (2) probably false in the "B" sense (the actual relation may or may not be in the direction asserted but it is not close in either case), (3) or probably false in the "C" sense (the actual relation is close but not in the direction asserted). Needless to say, those alternative labels do not solve the basic problem—that the correspondence between the actual and the asserted relation cannot be known directly or with certainty and hence judgment of the probable truth of a theoretic statement must be based on inference.

As a final preliminary consideration, there are two alternative strategies in assessing the predictive accuracy of a theory. In one the initial emphasis is on *predictive congruence*, that is, the consistency of descriptive statements and hypotheses. If predictive congruence is realized in each test, all of the intrinsic statements would be judged true in the "A" sense, but *predictive magnitude* could be unimpressive by any standard. Specifically, even though all of the coefficients of correlation are as anticipated by the hypothesis, it could be that each coefficient barely exceeds zero. So predictive congruence in no way insures predictive magnitude and, viewed that way, predictive congruence is the "weakest" criterion for judging a theory. However, that is precisely the reason for making it the initial criterion. In sociology, where grave doubts about the reliability and comparability of data are endemic, it is grossly unrealistic to expect a predictive magnitude even remotely approaching that achieved in the physical sciences. Indeed, if substantial predictive magnitude is demanded, there is danger of premature rejection of theories.

The designation of predictive congruence as the initial criterion does not preclude subsequent consideration of magnitude. Given two contending theories with approximately the same predictive congruence, further invidious comparisons can be made by reference to predictive magnitude. Of course, the ultimate goal is maximum predictive magnitude, and theories always are subject to modification in an attempt to realize that goal. But modification presupposes a prior judgment—that the theory should be retained at all—and the initial consideration in making that judgment should be predictive congruence. In other words, some predictive congruence is a *necessary* condition for retaining a theory or particular intrinsic statements with a view to modification.

As an alternative to the prescribed strategy, the predictive accuracy of a theory may be assessed exclusively in terms of predictive magnitude. Given a series of tests, an average predictive value can be computed for each type of hypothesis. To illustrate by reference to Figure 9-2, suppose that there are three sets of tests with the following coefficients of correlation: descriptive statements of type D1, +.95, +.85, and +.90; descriptive statements of type D2, +.15, -.10, and +.01; and descriptive statements of type D3, +.20, -.20, and +.15. Now suppose that the relational terms in the three corresponding hypotheses are such that a positive coefficient is predicted for all D1 statements and a negative coefficient for all D2 and D3 statements. As such, the coefficients of correlation are transformed into the following *predictive values*: for D1, +.95, +.85, +.90; for D2, -.15, +.10, -.01; and for D3, -.20, +.20, -.15. The *average* predictive values for the three corresponding types of hypotheses are thus: H1, +.90; H2, -.02; and H3, -.05. Given an average predictive value for a type of hypothesis, it can be assigned to each of the direct intrinsic statements that enter into the derivation of the theorem corresponding to the hypothesis (for instance, in Figure 9-2, Th1 corresponds to H1). Continuing the illustration, the assignment of values in the case of Figure 9-2 would be as follows: T1, +.90, -.02; P1, +.90, -.02; A1, +.90, -.02; P2, -.02, -.05; T2, -.02, -.05; A2, +.90, -.05; P3, +.90, -.05; and T3, +.90, -.05.

The use of predictive values has two advantages. First, judgments of direct intrinsic statements could be made more directly than by any alternative strategy. And, second, the predictive values reflect both congruence and predictive magnitude, which would simplify the assessment even more. But there are several disadvantages. One conspicuous problem is that the predictive values of a statement may be markedly divergent, as they are in the foregoing illustration for all direct intrinsic statements except P2 and T2. Further, regardless of how such divergence is resolved (for example, considering only the highest average predictive value) any absolute criterion (for example, a predictive value of at least +.50) for judging the truth of a statement would be arbitrary. Moreover, the predictive values for a direct intrinsic statement cannot be construed as even an approximation of the actual relation, which could be much closer than indicated by the highest predictive value assigned to the statement. There are techniques for estimating an unknown relation between two variables from known relations (such as estimating the correlation between variables X and Y in a particular case from knowledge of their mutual relation with Z), but they are not defensible when several relations are unknown.[11] Finally, still another objection is that statements would be judged without reference to context or *position* in the theory. For example, in the foregoing illustration, it could be argued that P2 and/or T2 are probably false in the "B" sense, while all other direct intrinsic statements are probably true in the "A" sense. The argument recognizes that only P2 and T2 enter into the derivation of H2 and H3, both of which have low average predictive values (-.02 and -.05). Viewed that way, the probable truth of an intrinsic statement should be judged in context rather than by the uncritical assignment of predictive values to it.

A series of tests.   An assessment of a theory should not be undertaken until after at least one *series* of tests. Even so, a question remains: How many *sets* of tests should be conducted? There can be no inflexible rules in that regard, as the number of tests may be limited by research resources and, obviously, the more tests the better. But there is a rationale for stipulating a minimum number of sets. Given three hypotheses (as in a 2-3 theory), there are eight possible outcomes in any set, one of which is, for example: H1 is consistent with the descriptive statement, H2 is inconsistent, and H3 is inconsistent. The number of possible outcomes (P) is always: $P = 2^h$, where h is the number of types of hypotheses or, viewed another way, the number of theorems. Thus, with three types of hypotheses P is 8, and with six types P is 64.

If predictive congruence is a matter of chance, then each type of possible outcome will approximate Ns/P, where Ns is the number of sets in the series of tests. Observe that if Ns is less than P, then some possible outcomes cannot be realized; hence in a series of tests Ns should equal or exceed P. The prescribed procedure does not require that $Ns > P$, but it is desirable.

The theorist hopes that in every set of tests each hypotheses is consistent with the descriptive statement, and that type of outcome is designated as x. Thus, Nx is the number of sets in which all hypotheses are consistent with the descriptive statement. Conversely, the least desirable outcome is a set in which each hypothesis is inconsistent with the descriptive statement, a type of outcome designated as y.

Nx and Ny are crucial in contemplating a question: Given a series of tests, by what rules of interpretation should a theory be tentatively accepted for further consideration?[12] The suggested rule of interpretation (interpretive rule 3) is: A theory should be considered further if $[Nx - (Ns/P)] > Ny$, or $[Ny - (Ns/P)] > Nx$, or $(Ns/P) > (Nx + Ny)$. If none of the three possibilities are realized, then the consistency of hypotheses and descriptive statements is ostensibly a matter of chance.[13] The theorist may consider the theory further in violation of rule 3, but he has no basis for doing so other than intuition.

If Nx equals Ns, the theory certainly deserves further consideration; but the Nx/Ns ratio only reveals predictive congruence, not predictive magnitude, and the latter should be considered before taking additional steps. Even when Nx equals Ns, predictive magnitude could be so unimpressive that the theorist elects to reject the theory totally; but, again, it is not feasible to stipulate a minimum predictive magnitude. It is simply a matter of what the theorist is willing to settle for, and that in turn should depend on the predictive magnitude of contending theories.

The foregoing does not mean that the theorist should reject a theory when $Ns > Nx$. On the contrary, that criterion would require predictive congruence in each test, a most unrealistic requirement in light of the conditions of work in sociology, especially the quality of data. For that matter, the criterion would be unrealistic in any field, since it ignores the possibility of an isolated false descriptive statement.

If after initial tests the theorist is satisfied with the predictive accuracy of the theory, additional tests are in order. However, even though a theorist is unwilling to reject the theory, he may be so dissatisfied with its predictive accuracy that he wants to modify it before conducting other tests. With that end in mind, he must make a decision concerning each intrinsic statement. There are three alternatives: retain it without modification, modify it, or reject it altogether. His decision should be based on tests of each type of *hypothesis*, and their interpretation should end with a judgment as to whether the type of hypothesis is true in the "A" sense, false in the "B" sense, or false in the "C" sense. Only then should the theorist commence an assessment of intrinsic statements.

The judgment of a type of hypothesis can be made by one of two criteria: (1) average predictive value or (2) predictive congruence. In the case of predictive values, the theorist will specify a range of values for judging a type of hypothesis as true in the "A" sense (for example, $+.75$ to $+1.00$), false in the "B" sense ($-.74$ to $+.74$), and false in the "C" sense ($-.75$ to -1.00). The specification of ranges is not arbitrary if the theorist can estimate the predictive accuracy of contending theories and is unwilling to settle for less. However, if that perspective is adopted, the range of predictive values for "A," "B," and "C" will vary from one theory to the next, depending on the estimated predictive accuracy of contenders; hence it is not feasible to suggest any particular absolute values for making judgments.

Alternatively, the theorist can consider predictive congruence and, if so, additional interpretive rules are needed. As suggested by rules 1 and 2 (Chapter 8), there are only two exclusive outcomes in a test—the hypothesis is judged as either true or false. Hence after a series of tests two numbers are computed *for each type of hypothesis*: Nt, the number of tests in which an instance is judged as true and Nf, the number in which an instance is judged as false. Accordingly, $Nt + Nf = Nh$, where Nh is the total number of tests of that type of hypothesis.

Given Nt and Nf for each type of hypothesis, interpretive rule 4 is applied as follows: A type of hypothesis is judged true in the "A" sense if $[Nt - (Nh/2)] > Nf$. Note the rationale—in approximately one-half of the tests instances of a type of hypothesis would be expected by chance alone to be true. Similarly, a type of hypothesis is judged false in the "C" sense if $[Nf - (Nh/2)] > Nt$, which is interpretive rule 5. Finally, interpretive rule 6: if $Nf > [Nt - (Nh/2)]$ and $Nt > [Nf - (Nh/2)]$, the type of hypothesis is judged false in the "B" sense. In other words, if an hypothesis is not judged true in the "A" sense or false in the "C" sense, then it is taken as false in the "B" sense.

At various steps in the assessment of a theory the focus is on a particular type of hypothesis or a particular intrinsic statement, but at some point it should be possible to reject the theory as a whole. Rule 3 may lead to such a decision, and so may rule 7, as follows: If all types of hypotheses are judged false in the "B" sense by rule 6, then the theory should be rejected. Rule 7 is especially important when rule 3 cannot be applied. If an instance of each type of hypothesis cannot be tested on each universe, the tests cannot be grouped into sets, which precludes the use of Rule 3.

The number of possible outcomes in the judgment of hypotheses is 3^h. Thus, with as few as three types of hypotheses, as in the case of a 2-3 theory, there are twenty-seven possible outcomes in a set of tests. An example of one is: type H1 is true in the "A" sense, type H2 is false in the "C" sense, and type H3 is also false in the "C" sense. Needless to say, with twenty-seven possibilities it is very difficult to formulate a procedure by which one can move from a judgment of hypotheses to a decision concerning each intrinsic statement, that is, to retain, modify, or reject it. There may be instances where the appropriate decision appears obvious. To illustrate, if type H1 is judged as true in the "A" sense but both H2 and H3 are judged as false in the "C" sense, then the theorist need only reverse the direction of the relation asserted in P2 (postulate 2). However, the appropriate revision in other instances may not be obvious; indeed, the complexities defy exaggeration. In any case, an assessment of intrinsic statements should be guided by an explicit system.

The system proposed here specifies a procedure for each type of theory.[14] Given a type 2-3 theory, there are two axioms and three postulates, each of which could be true in the "A" sense, false in the "B" sense, or false in the "C" sense. So there are 243 possible combinations (Pc) as regards the truth or falsity of the axioms and postulates, that is, $Pc = 3^i$, where i is the number of postulates and axioms. Consider one such combination in a type 2-3 theory: A1 is true in the "A" sense, A2 is false in the "C" sense, P1 is true in the "A" sense, P2 is false in the "B" sense, and P3 is true in the "A" sense.

If the truth or falsity of axioms and postulates could be known and if it were assured that all transformational statements, epistemic statements, and descriptive statements are true in the "A" sense, it would be possible to anticipate the judgment of each type of hypothesis in accordance with rules 4-6. For example, given the foregoing illustrative combination of axioms and postulates, both H2 and H3 would be judged false in the "B" sense, while H1 would be judged false in the "C" sense. Now the basic idea is that judgments of types of hypotheses can be interpreted so as to reach a tentative conclusion concerning the direct intrinsic statements. To illustrate, if after a series of tests H1 is judged as false in the "C" sense but H2 and H3 are judged false in the "B" sense, then one conclusion is that P2 and/or T2 is probably false in the "B" sense.

Reasoning from judgments of hypotheses to an assessment of intrinsic statements is fraught with difficulties and uncertainties. Further, some critics will reject the proposed procedure on the ground that it entails the logical fallacy of "affirming the consequent." In a sense it does, but the critics are unaware that inferring the *probable* truth or falsity of the premises from the truth of the asserted consequences is a common practice in even the most advanced sciences, and there is no alternative when the truth of the premises cannot be known directly or given by definition. Furthermore, intrinsic statements are always assessed in terms of their probable truth or falsity, and thus uncertainty in the assessment of intrinsic statements is duly recognized. As an instance, reconsider the previous illustration, where H1 has been judged false in the "C" sense and both types H2 and H3 have been judged false in the "B" sense. The reasoning is as follows: H1 could not be false in the "C" sense if P1, A1, A2, or P3 is false in the "B" sense, so each of those four intrinsic statements are presumed to be either true in the "A" sense or false in the "C" sense. Accordingly, P2 is the only postulate that could account for H2 and H3 being false in the "B" sense, and so it is identified as probably false in the "B" sense. Now consider uncertainties in the assessment. For one thing, P2 could be true in the "A" sense or false in the "C" sense and, if so, transformational statement T2 or numerous instances of E2 would be false in the "B" sense. That possibility would be considered in revising the theory but, more importantly for present purposes, it is the reason why P2 is identified as *probably* false in the "B" sense. Observe also that P1, A1, A2, or P3 would not be identified as probably true in the "A" sense, nor as probably false in the "C" sense. The inference is that one or three of the four statements could be false in the "C" sense, but there is no way to identify that one statement or set of three statements. So the conclusion for each of the four statements (P1, A1, A2, and P3) is that the statement is probably true in the "A" sense *or* probably false in the "C" sense. Needless to say, the conclusion is different from declaring each statement to be true in the "A" sense and also different from declaring each to be false in the "C" sense.

If the theory survives rule 7, the next step is to analyze the consequences of each combination of axioms and postulates, that is, for any combination to identify the types of hypotheses that *would be* judged true in the "A" sense, false in the "B" sense, and false in the "C" sense. As indicated previously for a type 2-3 theory, there are 243 combinations. The consequence of each combination is determined by four principles as follows. Principle I: If any postulate or axiom that enters into the derivation of a theorem corresponding to a type of hypothesis is false in the "B" sense, then after tests that type would be judged false in the "B" sense. Principle II: If no axiom or postulate that enters into the derivation is false in the "B" sense but one or an odd number of them is false in the "C" sense, then the type of hypothesis would be judged false in the "C" sense. Principle III: If no axiom or postulate that enters into the derivation is false in the "B" sense but an even number of them is false in the "C" sense, then the type of hypothesis would be judged true in the "A" sense. Principle IV: If each axiom and postulate that enters into the derivation is true in the "A" sense, then the type of hypothesis would be judged true in the "A" sense.

Application of the principles. Application of the four principles (I-IV) to each of the 243 combinations of axioms and postulates in a type 2-3 theory results in the identification of eleven outcomes in judgments of the three types of hypotheses. Those outcomes are shown in columns 1-3 of Table 9-1. To illustrate, taking one combination in which P1 and A2 are false in the "B" sense and all of the other axioms and postulates (A1, P2, and

P3) are true in the "A" sense, then the outcome is the judgment shown in the first row (columns 1-2) of the table, that is, each type of hypothesis is judged false in the "B" sense. Hence in the first row the letter "B" in column 6 (corresponding to statement A2), and letter "B" in column 7 (corresponding to P1), and the letter "A" in columns 5, 8, and 9 (corresponding to A1, P2, and P3) signify that such an outcome would be realized *if* the truth and falsity of the axioms and postulates are as stipulated. Accordingly, in an actual situation, if all three hypotheses are judged false in the "B" sense, then *one* interpretation is that P1 and A2 *could be* false in the "B" sense and the other direct intrinsic statements *could be* true in the "A" sense. However, that interpretation is not the only possibility, for various combinations of axioms and postulates would result in all hypotheses being judged as false in the "B" sense. To illustrate, if A1 and P3 are false in the "B" sense and the remainder false in the "C" sense, then tests would lead to the judgment that all types of hypotheses are false in the "B" sense. So in the first row the letter "B" is entered in columns 5 and 9, and the letter "C" is entered in columns 6-8. As can be seen, once all combinations that could result in the judgment of each hypothesis as false in the "B" sense are considered and appropriate column entries made in the first row of the table, it becomes obvious that when all types of the hypotheses have been judged as "B" *in an actual situation*, there is only one defensible interpretation—each axiom or postulate could be true in the "A" sense, false in the "B" sense, or false in the "C" sense. Since the theorist has no basis for retaining, modifying, or rejecting one statement rather than another, he should seriously consider rejecting the theory totally.

As one more illustration, consider the following combination of axioms and postulates: P1, A1, and P3 are taken as true in the "A" sense; A2 false in the "C" sense; and P2 false in the "B" sense. Given that combination, tests would lead to the judgment of hypotheses as shown in columns 1-3, second row of Table 9-1. So the letter "B" is in column 8 of that row, "C" is in column 6, and the letter "A" is in columns 5, 7, and 9. But other entries have been made in all except column 8, and those entries indicate that additional combinations of axioms or postulates could result in

Table 9-1
ASSESSMENT OF AXIOMS AND POSTULATES IN A TYPE 2-3 THEORY WITH A STRUCTURE AS SHOWN IN FIGURE 9-2, BASED ON JUDGMENTS OF HYPOTHESES FOLLOWING A SERIES OF TESTS

Judgment of Hypotheses Types 1, 2, and 3				Assessment of Axioms and Postulates*					
False in "B" Sense*	False in "C" Sense*	True in "A" Sense*	Congruence Value (Cv)	A1	A2	P1	P2	P3	Minimal Changes Needed in Modification of the Theory
Col. 1	Col. 2	Col. 3	Col. 4	Col. 5	Col. 6	Col. 7	Col. 8	Col. 9	Col. 10
1,2,3			6	ABC	ABC	ABC	ABC	ABC	Revise two referential formulas or reject the theory as a whole
2,3		1	7	A C	A C	A C	B	A C	Revise T2 referential formula and reverse A1 or A2
	2,3	1	8a	A C	A C	A C	B	A C	Revise T2 referential formula
1,3		2	8b	A C	ABC	A C	A C	ABC	Revise T3 referential formula and reverse A1
1,2		3	9	ABC	A C	ABC	A C	A C	Revise T1 referential formula and reverse A2
1,3		2	10	A C	ABC	A C	A C	ABC	Revise T3 referential formula
1,2		3	12	ABC	A C	ABC	A C	A C	Revise T1 referential formula
	2,3	1	13	A C	A C	A C	A C	A C	Reverse P2
	1,3	2	14	A C	A C	A C	A C	A C	Reverse A2
	1,2	3	15	A C	A C	A C	A C	A C	Reverse A1
		1,2,3	18	A C	A C	A C	A C	A C	None

*A = actual relation may be in direction asserted and close; B = actual relation may not be close; C = actual relation may be close but not in asserted direction.

the same judgment of hypotheses, with an example of one other combination being: A1 is false in the "C" sense, P2 is false in the "B" sense, and all other statements are true in the "A" sense. So in an actual situation, if the judgment of hypotheses is as shown in the second row of the table, then certain interpretations would be justified. For one thing, only P2 is identified as possibly false in the "B" sense. However, each of the other four statements (P1, A1, A2, and P3) could be either false in the "C" sense or true in the "A" sense, and there is no way to determine which is the case for any of those statements. Nonetheless, the theorist is in a much better position to modify the theory than he would if all types of hypotheses have been judged as false in the "B" sense.

Commentary on usage. Table 9-1 is designed not merely as an illustration, but for use in the assessment of theories with a structure as shown in Figure 9-2. Should the theorist decide to retain the theory and revise it, the next step is to equate his judgment of the hypotheses with one of the outcomes in columns 1-3. For example, suppose that the types of hypotheses have been judged in accordance with rules 4-6 with the following outcome: H1 and H2 are judged as false in the "C" sense and H3 as true in the "A" sense. The theorist would find the corresponding outcome in the next-to-last row of the table, and in column 10 of that row a particular revision of the theory is recommended. However, before considering the revisions, several comments should be made on the table itself.

Given judgments of several types of hypotheses, it is laborious to search the table for the corresponding outcome. To facilitate the search, a congruence value (Cv) should be computed for the outcome as follows: $Cv = Bx + 2Cx + 3Ax$, where Bx is the sum of the hypotheses numbers that have been judged as false in the "B" sense, Cx is the sum of the hypotheses numbers that have been judged false in the "C" sense, and Ax is the sum of the hypotheses numbers that have been judged as true in the "A" sense. The presumption is that the theorems and the hypotheses have been identified by numbers exactly as shown in Figure 9-2, that is, all hypotheses derived in part from Th1 (theorem 1) are identified as instances of H1 (hypotheses of type 1), and so forth as regards Th2, H2 and Th3, H3. Now suppose that H1 and H3 have been judged false in the "B" sense, and H2 has been judged true in the "A" sense. As such, Bx is 4 (that is, H3 + H1), Cx is 0, and Ax is 2. Therefore, Cv, the congruence value, is 10. Since the value reflects only an arbitrary numerical designation of types of hypotheses, it is nominal. But a Cv value serves two purposes. First, it summarizes tests; generally, the greater the Cv value, the more positive the tests and the more simple the recommended revisions. Second, it facilitates the use of Table 9-1. Given a Cv value of 9, for example, the appropriate row can be located readily by inspection of column 4, where the corresponding Cv value is in the fifth row. The only complication is that two or more different judgments of types of hypotheses may result in the same Cv value, as in the third and fourth rows of Table 9-1. Hence a letter as a subscript to a Cv value signifies that the same value results from two or more outcomes, and the outcome sought in the table can be determined by inspection of columns 1-3. Thus, if H2 and H3 have been judged false in the "B" sense and H1 true in the "A" sense, then the appropriate row is the third rather than the fourth, even though the Cv value is the same for both rows.

Still another consideration is the omission in Table 9-1 of sixteen possible outcomes in the judgment of hypotheses. Again, with three types of hypotheses there are twenty-seven possible outcomes, but only eleven are shown in the table. The sixteen excluded outcomes share one thing in common—none can be derived from any combination of two axioms and three postulates. To consider one such outcome, suppose that both H1 and H2 have been judged true in the "A" sense and H3 false in the "B" sense. The outcome for H3 could be construed as indicating that P2, A2, and/or P3 is false in the "B" sense, but that interpretation is inconsistent with the fact that H1 and H2 have been judged true in the "A" sense. No combination of axioms and postulates could produce such an outcome, and that would be the case even if H3 had been judged false in the "C" rather than the "B" sense. Such outcomes are designated as "anomalous" to signify that they are contrary to the rules of derivation. However, in actual tests such anomalies may occur,[15] and they indicate that some relations are not nearly as close as the judgments of hypotheses would indicate. For example, if H1 and H2 have been judged true in the "A" sense and H3 false in the "B" sense, then it indicates that H1 and/or H2 barely qualified as true in the "A" sense and/or that the average predictive values for H1 and/or H2 is low. That indication should be confirmed by inspection of the correlation coefficients in the descriptive statements. If a type of hypothesis barely qualifies as true in the "A" sense or false in the "C" sense, or if its average predictive value is low, then the judgment of that type should be changed to false in the "B" sense. Thus, returning to the foregoing illustration, the judgment of H1 and/or H2 would be changed from true in the "A" sense to false in the "B" sense.

Rules of revision.   Observe that columns 5-9 of Table 9-1 give only an assessment of the axioms and postulates, not directions for revising the theory. Such directions are given in column 10, which stipulates the minimal changes needed to make the theory consistent with the judgment of the hypotheses.

The entries in column 10 are the products of thirteen revision rules.

1. Do not modify any axiom, postulate, or referential formula if all types of hypotheses have been judged true in the "A" sense.

2. Do not reverse any axiom or postulate (that is, change the direction of the asserted relation) unless some types of hypotheses have been judged false in the "C" sense.

3. Do not revise a referential formula if *any* cognate type of hypotheses (a type derived in part through a theorem from the transformational statement that designates the referential formula) has been judged true in the "A" sense or false in the "C" sense.

4. If only one type of hypothesis has been judged false in the "C" sense, reverse one of its cognate axioms or cognate postulates (one of the axioms or postulates that entered into the derivation of the type of hypothesis through a theorem).

5. If two or more types of hypotheses have been judged false in the "C" sense and they are all cognates (that is, derived in part through a theorem from at least one common axiom or postulate) reverse one of the axioms or postulates that entered into their derivation.

6. Given a choice between reversing an axiom or a postulate, reverse the axiom.

7. Given a choice between reversing more than one axiom or reversing one postulate, reverse the postulate.

8. No axiom or postulate is to be reversed if that change would result in a reversal of the direction predicted for a type of hypothesis that has been judged true in the "A" sense.

9. If two or more types of hypotheses have been judged false in the "C" sense but not all are cognates, group the types into the least possible number of cognate sets and then for each set reverse an axiom or postulate that entered into the derivation of that set.

10. If some but not all types of hypotheses have been judged false in the "B" sense and they have been derived in part from one particular postulate, revise the referential formula corresponding to that postulate.

11. If some types of hypotheses have been judged false in the "B" sense but not all were derived through a theorem from a common postulate, group them into the least possible number of cognate sets by relation to postulates, and for each set revise the referential formula corresponding to the postulate from which the set was derived.

12. If all types of hypotheses judged as false in the "B" sense have been derived in part through a theorem from one axiom but not from one postulate, reject that axiom.

13. If all types of hypotheses have been judged false in the "B" sense, revise all but one referential formula or reject the theory as a whole.

All thirteen rules of revision represent a judgment and hence are debatable. The first rule may appear to be obvious, but observe that the concern is with predictive congruence. Conceivably, even though all types of hypotheses have been judged true in the "A" sense, predictive magnitude could be generally so unimpressive that the theorist elects to reject the theory as whole.

Revision rule 2 is more imperative. The only way an axiom or postulate can be changed is to reverse the direction of the asserted relation, but such modification is pointless unless at least one type of hypotheses has been judged false in the "C" sense.

Rule 3 introduces a complicated consideration. Consider the case where H2 and H3 have been judged false in the "B" sense and

H1 true in the "A" sense. Since H3 is a cognate of T3 (T3 enters into the derivation of theorem 3, which in turn enters into the derivation of H3), the formula designated by the referential in T3 becomes questionable; specifically, it could be that T3 is false in the "B" sense.[16] But H1 is also a cognate of T3, and it has been judged true in the "A" sense. Accordingly, if the referential formula is revised, there is a possibility that after another series of tests H1 will not be judged true in the "A" sense. So it is questionable to modify a referential formula *if any of its cognate hypotheses have been judged true in the "A" sense or false in the "C" sense.*[17] When a hypothesis has been judged false in the "C" sense, suspicion does not fall on a cognate referential formula, for it is unlikely that the actual relation is directly opposite to that asserted in a transformational statement, and it is unlikely that such an "error" occurs frequently in epistemic statements. The analogy is that of estimating the weight of individuals and then comparing the estimates with weights established by scale readings. A close relation between the two sets of values (estimates and scale readings) may not be realized, but it is most improbable that the relation would be inverse.

Application of rules 4 and 5 requires a thorough understanding of the word "cognate." A type of hypothesis and a direct intrinsic statement are cognates if the intrinsic statement enters into the derivation of the theorem from which the hypothesis was derived. Thus we have the following cognates of hypotheses in a type 2-3 theory: T1, P1, A1, A2, P3, and T3 are cognates of H1; T1, P1, A1, P2, and T2 are cognates of H2; T2, P2, A2, P3, and T3 are cognates of H3. Viewed another way, any two types of hypotheses are cognates if linked to one or more direct intrinsic statements in common.[18] Thus, H1 and H2 are cognates, for both are linked to T1, P1, and A1; H1 and H3 are cognates through A2, P3, and T3; while H2 and H3 are cognates only through P2 and T2. It so happens that in a 2-3 theory any two hypotheses are cognates, but for certain other types of theories (for example, three axioms and four postulates) some hypotheses are not cognates.

In applying rule 4, it may happen that more than one axiom could be reversed; and the theorist must rely on intuition to make the choice. Considering a type 2-3 theory as shown in Figure 9-2, if only H1 has been judged false in the "C" sense, then either A1 or A2 is to be reversed, as both are linked to H1.[19] However, if only H2 has been judged false in the "C" sense, there is no choice; because only one axiom, A1, is a cognate of H2. Similarly, if only H3 has been judged false in the "C" sense, then A2 should be reversed.

Where more than one type of hypothesis has been judged false in the "C" sense and all of them are cognates, rule 5 applies. If H2 and H3 have been judged false in the "C" sense, then P2 is to be reversed, since no axiom or other postulate is linked to both hypotheses. Similarly, if H1 and H2 have been judged false in the "C" sense, then A1 should be reversed, as it is the only axiom linked to both H1 and H2.

In the application of rule 4 or 5, the theorist may have a choice between reversing an axiom or a postulate, and that choice should be governed by rules 6 and 7. To illustrate, when only H1 and H2 have been judged false in the "C" sense, the theorist could reverse P1 or A1. In accordance with rule 6, he should reverse A1. The rule reflects a judgment—that a theorist is much more likely to be wrong in stating an axiom than a postulate. However, there are instances where, consistent with the principle or conservation, a postulate is reversed rather than two or more axioms. If H2 and H3 have been judged false in the "C" sense, then the only choice is between reversing P2 or reversing both A1 and A2. Reversing both axioms would not be contrary to rule 8 (H1 would still be true in the "A" sense), but it would be contrary to rule 7. The rule minimizes changes and reflects a judgment—that a theorist is unlikely to be wrong in stating two axioms.

Although the revision rules were designed for all theories, rule 9 is not relevant for a type 2-3 theory. It does not apply because any two hypotheses in a type 2-3 theory are cognates. However, all three types of hypotheses taken together are not cognates, as no postulate or axiom is linked to all three. Accordingly, taken together, the least number of cognate sets of hypotheses is two, that is, H1 and H2 as one set and H1 and H3 as the other.

Of rules 10 and 11, only 10 applies in the case of a type 2-3 theory. Any two hypotheses in such a theory are linked to a common postulate. To illustrate, if H2 and H3 have been judged false in the "B" sense, both of them are linked to P2 and P2 is the only postulate linked to both hypotheses; hence, in accordance with rule 10, the theorist should revise the referential formula corresponding to P2, that is, the formula designated by the referential in transformational statement 2.

The structure of a theory may be such that only two theorems are cognates through a common axiom and a common postulate; if so, rule 12 does not apply. However, when the rule does apply, it is especially important because it identifies the only situation where an axiom should be rejected rather than modified (reversed).

Rule 13 may give the theorist some guidance but little consolation. It covers the case where each type of hypothesis is false in the "B" sense. Of course, contrary to the rule, the theorist may elect to retain and modify the theory but, ordinarily, such a course is most questionable. After all, if rule 13 applies, the theory does not even satisfy the criterion of predictive congruence, and the theorist has no choice except to revise all but one of the referential formulas.

Some uncertainties.   As noted previously, some of the revision rules force the theorist to make a choice in selecting statements for modification. That choice introduces an uncertainty that no rule can eliminate. Indeed, the thirteen rules may suggest a certainty that is totally unrealistic, and an extended commentary on the subject is in order.

While a theory can be modified without reference to columns 5-9 of Table 9-1, those entries should remind the theorist that he can never be certain that his revisions are "correct." To illustrate, given a congruence value of 7 for tests of a type 2-3 theory as shown in Figure 9-2, the theorist may elect to modify A1 rather than A2. Now suppose that in the next set of tests the congruence value is 18. That outcome would not prove that before modification A1 was false in the "C" sense and P1, A2, and P3 were true in the "A" sense. Any of the latter could still be false in the "C" sense, and that will not be known (if at all) until the theory is expanded so as to derive additional theorems. If some of the additional theorems are derived in part from A1 but not from P1, A2, and P3, then the probable truth of A1 can be distinguished from the probable truth of P1, A2, and P3.

Even when the rules of revision stipulate the modification of a particular axiom or postulate, the theorist can never be certain that he has modified the right statement.[20] Consider a congruence value of 13 in Table 9-1. Reversing P2 would be consistent with the principle of conservation, but it *could be* that both A1 and A2 are false in the "C" sense, while P1, P2, and P3 are true in the "A" sense, a condition that would never be revealed by tests.

In columns 5-9 of Table 9-1 there are no entries pertaining to transformational or epistemic statements, but some of the suggested revisions in column 10 call for the modification of a referential formula when evidence indicates that a postulate is false in the "B" sense. There is no paradox, for evidence that a postulate could be false in the "B" sense also indicates the same for the complementary transformational statement and the complementary epistemic statements.[21] All three statements could be false in the "B" sense, but a revision of the theory in light of such evidence should focus on the referential formula designated in the transformational statement. As indicated previously, a postulate can be modified only by reversing the direction of the asserted relation, but a reversal would not be consistent with evidence that the postulate is false in the "B" sense. So the theorist has no alternative but to assume that the complementary transformational statement and/or instances of the complementary type of epistemic statement are false in the "B" sense. Either of the three possibilities could be rectified by modifying the referential formula. Some or all instances of the type of epistemic statement could be false in the "B" sense because the referential formula is not sufficiently empirically applicable (for example, some or all investigators made an error in its application), and for that reason tests should not be undertaken without confidence in the empirical applicability of the referential formulas. In any case, given evidence that a transformational statement and/or the complementary epistemic statements are false in the "B" sense and given doubts about empirical applicability, the theorist should modify his explication of the referential formula.[22] However, even if there are no errors in the application of a referential formula, epistemic statements could be false in the "B" sense because the referential formula was applied to incomparable or unreliable data. With that possibility in mind, the theorist may stipulate a different kind of data, a different source of data, and/or a different procedure for gathering data. Finally, if he doubts the truth of the transformational statement, the theorist can modify the referential formula itself, a change that may or may not alter stipulations pertaining to data.

Given various alternative ways of modifying a referential formula, uncertainty is inevitable, and there is even a greater source of doubt. When the theorist modifies a referential formula, subsequent tests may not substantiate the modification. It is entirely possible that the postulate and not the complementary transformational or epistemic statement is false in the "B" sense. Hence, modification of the referential formula may not enhance predictive accuracy in subsequent tests. True, if the subsequent tests do reveal much greater accuracy, it suggests that the transformational statement and/or the epistemic statements were false in the "B" sense. But if a modification does not alter test results (that is, they continue to be negative), the theorist is in a quandary. Consider a congruence value of 8a. If after repeated modifications of the referential formula, H2 and H3 are still judged false in the "B" sense, the theorist may be tempted to conclude that P2 must be false in the "B" sense, but that may not be the case. Rather, P2 may be true in the "A" sense, but modification of the referential formulas has not made T2 true in the "A" sense or, alternatively, it has not made the epistemic statements true in the "A" sense.

The inevitability of uncertainty is clearly revealed by considering a congruence value of 18. Even though all types of hypotheses have been judged true in the "A" sense, an even number of the axioms and postulates (P1, A1, P2, A2, and P3) could be false in the "C" sense. It is most unlikely that the theorist will be wrong in two or more axioms and postulates, but certainty should not be equated with likelihood.

A variation on the procedure.   The foregoing procedure for revising a theory presumes that types of hypotheses have been judged as true or false by reference to predictive congruence. But the procedure is essentially the same even if the theorist considers average predictive values (for instance, a type of hypothesis is judged as true in the "A" sense if the average of predictive values in the corresponding descriptive statements is greater than + .74). In that connection, observe again that the revisions stipulated in column 10 (Table 9-1) are contingent on the theorist being willing to settle for the predictive magnitude indicated by the descriptive statements. If they indicate an insufficient degree of predictive magnitude, the theorist may elect to reject the theory as a whole; he may do so even though each type of hypotheses has been judged true in the "A" sense.[23]

Even though all hypotheses have been judged true in the "A" sense, the theorist can modify referential formulas with a view to increasing predictive magnitude, but that is a judgment which only the theorist can make. If the average predictive values are unacceptably low and the theorist cannot modify the referential formulas one way or another, then he should consider rejecting the theory as a whole. The same is true when extensive revisions are needed, as in the case where Cv is 7, and especially when the theorist has no basis for reversing one axiom rather than another or cannot think of some way to modify the referential formulas.

Other Types of Theories

The foregoing procedure applies in much the same way to theories other than type 2-3. However, if only to further understanding of the procedure, two additional types are considered.

A type 0-3 theory.   Figure 9-3 shows the structure of a type 0-3 theory, with three postulates and no axioms. The structure is simpler than that of a 2-3 theory, but the procedure for interpreting tests is virtually identical. Moreover, the problems and uncertainties are the same.

Columns 1-3 of Table 9-2 show eleven outcomes in judgments of hypotheses. Those outcomes were determined by considering all combinations of postulates as to their truth or falsity, one instance being: P1 is true in the "A" sense, P2 is false in the "B" sense, and P3 is false in the "C" sense. The outcome of each

Figure 9-3
THE STRUCTURE OF A TYPE 0-3 THEORY AND A SET OF TESTS

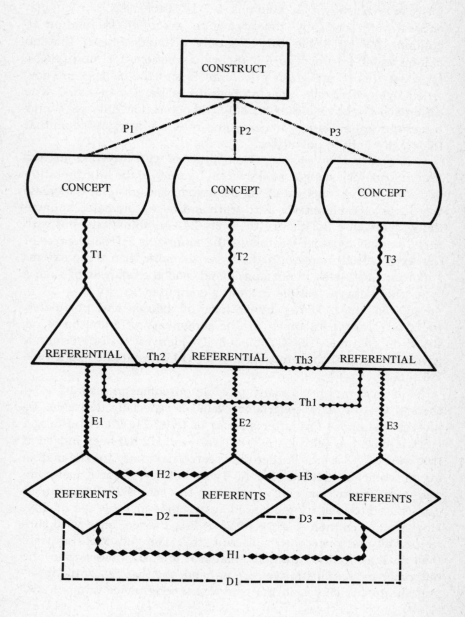

combination is governed by principles I-IV as set forth in analyzing a type 2-3 theory, with references to axioms excluded.

Given tests of a type 0-3 theory, the theorist evaluates the postulates as shown in columns 5-7 of Table 9-2. The column entries were made by the procedure described in relation to columns 5-9 of Table 9-1, the only difference being that no axioms enter into a type 0-3 theory. Directions for modifying a type 0-3 theory are given in column 8, and the changes are governed by the thirteen rules of revision as previously stated. The difference in the case of a type 0-3 theory is that rules pertaining to axioms are excluded, meaning the rules are interpreted so that they apply only to postulates.

A type 3-4 theory. Given a theory with three axioms and four postulates arranged as shown in Figure 9-4, the interpretation of tests is very complicated. As the axioms and postulates increase the possible combinations as to truth and falsity increase exponentially. Thus, in a 2-3 theory there are 243 combinations, but with three axioms and four postulates the number is 2,187. Because of the exponential increase, a theory with more than three axioms and four postulates is not considered, and the analysis of such a type would not be feasible without a computer.

Given the 2,187 combinations of axioms and postulates, there are forty outcomes in the judgments of hypotheses, as shown in columns 1-3 of Table 9-3. The derivation of outcomes is governed by the same four principles (I-IV) as applied to a type 2-3 theory.

Since there are so many possible outcomes in tests of a 3-4 theory, the use of a congruence value is especially desirable. To illustrate, suppose that after a series of tests H1, H3, H4, H5, and H6 are judged as false in the "B" sense and H2 has been judged as true in the "A" sense. Rather than search randomly for the matching outcome in columns 1-3 of Table 9-3, the theorist can compute a congruence value (Cv), which in this case is 25. The number can be used to identify the matching outcome in Table 9-3 readily. However, in column 4 a Cv of 25 is listed twice, once with subscript "a" and once with subscript "b". The subscripts indicate that 25 is a "hybrid" number, that is it identifies more than one outcome. In the illustrative case at hand, the theorist would interpret the test as shown in the row with a congruence value of 25^a.

Table 9-2

ASSESSMENT OF POSTULATES IN A TYPE 0-3 THEORY, AS SHOWN IN FIGURE 9-3, BASED ON JUDGMENTS OF HYPOTHESES FOLLOWING A SERIES OF TESTS

Judgment of Hypotheses Types 1, 2, and 3				Assessment of Postulates*			
False in "B" Sense*	False in "C" Sense*	True in "A" Sense*	Congruence Value (Cv)	P-1	P-2	P-3	Minimal Changes Needed in Modification of Theory
Col. 1	Col. 2	Col. 3	Col. 4	Col. 5 ABC	Col. 6 ABC	Col. 7 ABC	Col. 8
1,2,3			6	ABC	ABC	ABC	Revise two referential formulas or reject the theory as a whole
2,3	1		7	A C	B	A C	Revise T2 referential formula and reverse P1 or P3
2,3		1	8a	A C	B	A C	Revise T2 referential formula
1,3	2		8b	A C	A C	B	Revise T3 referential formula and reverse P1 or P2
1,2	3		9	B	A C	A C	Revise T1 referential formula and reverse P2 or P3
1,3		2	10	A C	A C	B	Revise T3 referential formula
1,2		3	12	B	A C	A C	Revise T1 referential formula
	2,3	1	13	A C	A C	A C	Reverse P2
	1,3	2	14	A C	A C	A C	Reverse P3
	1,2	3	15	A C	A C	A C	Reverse P1
		1,2,3	18	A C	A C	A C	None

*A = actual relation may be in direction asserted and close; B = actual relation may not be close; C = actual relation may be close but not in asserted direction.

Figure 9-4
THE STRUCTURE OF A TYPE 3-4 THEORY AND A SET OF TESTS

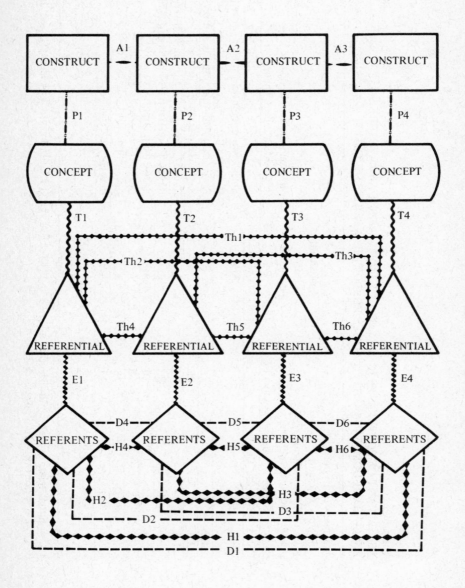

As in previous tables, not all possible actual outcomes are shown in Table 9-3. The outcomes not shown are anomalies, and they should be dealt with as indicated by previous observations on a type 2-3 theory.

The entries in columns 5-11 were made by the same procedure as described in reference to columns 5-9 of Table 9-1. They do not indicate how the theory is to be revised, but they do serve to remind the theorist of uncertainty. As for revisions, directions are given in column 12 of Table 9-3, and those directions are governed by the thirteen revision rules previously considered.

Extension of the Procedure

The procedure for interpreting tests can be extended to theories with structures that differ from those shown in Figures 9-2, 9-3, and 9-4.[24] Regardless of the structure, the theorist should construct a table similar to 9-1, 9-2, or 9-3, with the entries governed by principles I-IV and revision rules 1-13.

Two major complications may be involved in the extension of the procedure. If one of the constructs is linked to more than one concept, then some of the theorems are derived only from postulates and transformational statements. The other complication is introduced when the number of axioms equals or exceeds the number of postulates, as shown in Figure 9-5. The probable truth or falsity of A2 and A3 cannot be distinguished because the axioms are complementary (that is, they enter into the derivation of the same theorems). Accordingly, when interpreting tests, A2 and A3 would be treated as though they were one statement. If the rules call for the reversal of one of the relations asserted in A2 or A3, then the theorist must choose between reversing A2 or A3. However if the rules call for rejection of either A2 or A3, both must be rejected. What has been said of A2 and A3 in a type 4-4 theory would apply to any theory where two or more axioms are complementary.

Diversity in Structure

Figures 9-5 and 9-6 have implications beyond the extension of the procedure for revising theories. Both indicate that formal theories

Table 9-3

ASSESSMENT OF AXIOMS AND POSTULATES IN A TYPE 3-4 THEORY WITH A STRUCTURE AS SHOWN IN FIGURE 9-4, BASED ON JUDGMENTS OF HYPOTHESES FOLLOWING A SERIES OF TESTS

Judgment of Hypotheses Types 1,2,3,4,5, and 6			Congruence Value (Cv)	Assessment of Axioms and Postulates*							Minimal Changes Needed in Modification of the Theory
False in "B" Sense*	False in "C" Sense*	True in "A" Sense*		A1	A2	A3	P1	P2	P3	P4	
1,2,3,4,5,6			21	ABC	ABC	ABC	ABC	ABC	ABC	ABC	Revise three referential formulas or reject the theory as a whole
2,3,4,5,6	1		22	A C	A C	A C	A C	B	B	A C	Revise T2 and T3 referential formulas and reverse A1, A2, or A3
2,3,4,5,6		1	23a	A C	A C	A C	A C	B	B	A C	Revise T2 and T3 referential formulas
1,3,4,5,6	2		23b	A C	A C	ABC	A C	B	A C	ABC	Revise T2 and T4 referential formulas and reverse A1 or A2
1,2,4,5,6	3		24	ABC	A C	A C	ABC	A C	B	A C	Revise T1 and T3 referential formulas and reverse A2 or A3
1,3,4,5,6		2	25a	A C	A C	ABC	A C	B	A C	ABC	Revise T2 and T4 referential formulas
1,2,3,5,6	4		25b	A C	ABC	ABC	A C	A C	ABC	ABC	Revise T3 and T4 referential formulas and reverse A1
1,2,3,4,6	5		26	ABC	A C	ABC	ABC	A C	A C	ABC	Revise T1 and T4 referential formulas and reverse A2
1,2,4,5,6		3	27a	ABC	A C	A C	ABC	A C	B	A C	Revise T1 and T3 referential formulas

No.	Set 1	Set 2	Set 3	Pattern	Action
27b	1,2,3,4,5		6	ABC ABC ABC ABC A C ABC A C A C	Revise T1 and T2 referential formulas and reverse A3
29	1,2,3,5,6		4	A C ABC ABC A C ABC ABC	Revise T3 and T4 referential formulas
30	2,5,6	3,4	1	A C A C A C A C A C B A C	Revise T3 referential formula and reverse P2
31a	3,4,5	2,6	1	A C A C A C A C A C B A C	Revise T3 referential formula and reverse P3
31b	1,2,3,5	4,6		A C B A C A C A C A C A C	Reject A2, reverse A1 and A3
32a	2,5,6	1,4	3	A C A C A C A C B A C	Revise T3 referential formula and reverse A1
32b	3,4,5	1,6	2	A C A C A C A C B A C A C	Revise T2 referential formula and reverse A3
33a	1,2,3,4,5		6	ABC ABC ABC A C A C A C A C	Revise T1 and T2 referential formulas
33b	2,5,6	1,3	4	A C A C A C A C A C B A C	Revise T3 referential formula and reverse A2 or A3
34	1,3,6	4,5	2	A C A C A C ABC ABC A C ABC	Revise T4 referential formula and reverse P2
35	1,2,3,5	6	4	A C B A C A C A C A C A C	Reject A2 and reverse A3
36a	3,4,5	1,2	6	A C A C A C A C A C B A C	Revise T2 referential formula and reverse A1 or A2
36b	1,3,6	2,5	4	A C A C A C A C ABC A C ABC	Revise T4 referential formula and reverse A2
37a	1,5,6	1,3,4		A C A C A C A C A C A C B A C	Revise T3 referential formula
37b	1,2	4	6	A C B A C A C A C A C A C A C	Reject A2 and reverse A1

			No.								Conclusion
1,3,6	2,4	5	37c	A C	A C	ABC	A C	A C	A C	ABC	Revise T4 referential formula and reverse A1
1,2,4	5,6	3	38	ABC	A C	ABC	A C	A C	A C (s)		Revise T1 referential formula and reverse P3
3,4,5		1,2,6	39	A C	A C	A C	B	A C			Revise T2 referential formula
1,2,4	3,6	5	40	ABC	A C	ABC	A C	A C	A C		Revise T1 referential formula and reverse A3
1,2,4	3,5	6	41a	ABC	A C	ABC	A C	A C	A C		Revise T1 referential formula and reverse A2
1,2, 3,5		4,6	41b	A C	B	A C	A C	A C	A C		Reject A2
1,3,6		2,4,5	43	A C	A C	ABC	A C	A C	ABC		Revise T4 referential formula
1,4, 5,6	2,3		47	A C	A C	A C	A C	A C	A C		Reverse A1, A2, and A3
2,3, 4,6	1,5		48	A C	A C	A C	A C	A C	A C		Reverse A1 and A3
1,2,4	3,5,6		49	ABC	A C	ABC	A C	A C	A C		Revise T1 referential formula
2,5,6	1,3,4		50	A C	A C	A C	A C	A C	A C		Reverse P3
3,4,5	1,2,6		51	A C	A C	A C	C	A C	A C		Reverse P2
1,2, 3,5	4,6		52	A C	A C	A C	C	A C	A C		Reverse A2
1,3,6	2,4,5		53	A C	A C	A C	A C	A C	C		Reverse A3
1,2,4	3,5,6		56	A C	A C	C	A C	A C	A C		Reverse A1
1,2,3, 4,5,6			63	A C	A C	A C	A C	A C	A C		None

*A = actual relation may be in direction asserted and close; B = actual relation may not be close; C = actual relation may be close but not in asserted direction.

may be more diverse in structure than is suggested by Figures 9-1, 9-2, 9-3, and 9-4. In those previous figures each construct is linked to one and only one concept and no construct is linked to more than two other constructs, but neither structural feature is necessary.

Figure 9-6 depicts a structure radically different from those previously considered. The type is 3-3, signifying three axioms but only three postulates. When the number of axioms equals or exceeds the number of postulates, then at least one of the constructs is not linked to a concept. That is the case in both Figure 9-5 and 9-6, but note the difference in the structure of the two theories.

In Figure 9-5, some of the constructs are linked to two other constructs, but in Figure 9-6 one of the constructs is linked to three other constructs. Nonetheless, the rules of revision apply to a type 3-3 theory in much the same way as to a type 2-3 theory. To illustrate, suppose that after a series of tests of a type 3-3 theory (as shown in Figure 9-6), the following judgments of the types of hypotheses have been made: H1 (derived in part from P1, A1, A3, and P3) is true in the "A" sense, while both H2 (from P1, A1, A2, and P2) and H3 (from P2, A2, A3, and P3) are false in the "C" sense. Given such judgments in the case of a type 2-3 theory, then according to revision rules 5 and 6 the theorist would reverse P2; but in the case of a type 3-3 theory with a structure as shown in Figure 9-6, the same rules would require reversal of A2, for it is linked to both H2 and H3.

Finally, suppose that two additional axioms are added to the theory shown in Figure 9-6, Ax linking constructs II and III and Ay linking constructs III and IV. The additions would be contrary to the principle of parsimony, according to which the addition of a direct intrinsic statement is superfluous if it does not result in the derivation of more theorems. The addition of the contemplated axioms, Ax and Ay, would not increase the number of theorems, and hence they would be superfluous. True, the substitution of Ax and Ay for A1, A2, and A3 would result in a more parsimonious theory (the ratio of theorems to axioms and postulates would change from .50 to .60); but the theorist may think of constructs II, III, and IV as related only through A1, A2, and A3; if so, he would not add Ax and Ay or substitute them for A1, A2,

Figure 9-5
PARTIAL STRUCTURE OF A TYPE 4-4 THEORY

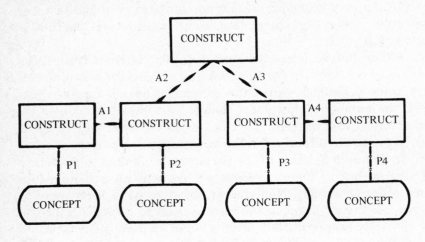

Figure 9-6
PARTIAL STRUCTURE OF A TYPE 3-3 THEORY

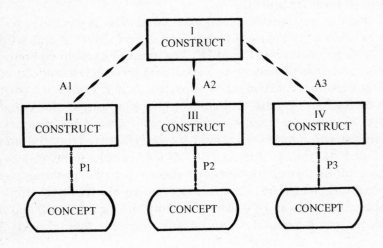

and A3. The substitution would be superfluous, for the theorems would not be altered. Similarly, if the theorist thinks of constructs II, III, and IV as related only through Ax and Ay he would not introduce construct I or axioms A1, A2, and A3, as they would not alter the theorems.

So in the context of theory construction the notion of "superfluous" has to do with the addition or substitution of axioms and not the structure of the theory per se. The structure of a theory is contrary to the principle of parsimony if any intrinsic statement can be derived from others, but that condition does not reveal which intrinsic statement is superfluous. Thus, if the theorist had included Ax (an axiom linking constructs II and III), inspection of the structure would reveal that the principle of parsimony was violated in the construction of the theory, but it would not be possible to designate one of the three axioms, Ax, A1, and A2, as superfluous; any one of them could be considered as superfluous.

Propositions as intrinsic statements.   Very little attention has been devoted to propositions (a direct intrinsic statement that links concepts), but a theory can be constructed without constructs, which is to say that it would comprise propositions, transformational statements, and theorems but no axioms or postulates. Such a theory is depicted in Figure 9-7. Observe that the theory is designated as a type 0-0-3-3, signifying no axioms, no postulates, three propositions, and three transformational statements.

If no axioms or postulates enter into a theory, tests can be interpreted so as to assess the probable truth of transformational statements. As explained previously, if postulates enter into a theory then their probable truth cannot be distinguished from that of complementary transformational statements, and the assessment initially focuses on the postulates, the presumption being that a postulate is more likely to be false in the "C" sense than is a transformational statement. However, in a type 0-0-N theory (no axioms or postulates and one or more propositions), the interpretation of tests may lead to the conclusion that a transformational statement is false in the "C" sense. Further, no special principles, interpretive rules, or revision rules are needed. Given a type 0-0-N

Figure 9-7
PARTIAL STRUCTURE OF A TYPE 0-0-3-3 THEORY

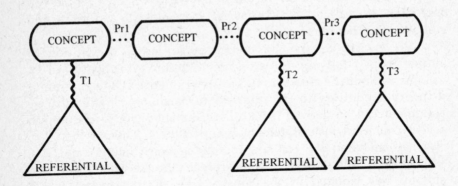

Figure 9-8
PARTIAL STRUCTURE OF A TYPE 2-3-2-3 THEORY

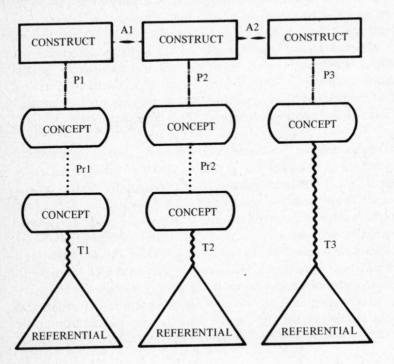

theory, the propostions are treated as axioms and the transformational statements are treated as postulates; so the assessment procedure is like that for any theory that comprises axioms and postulates (for instance, a type 2-3 theory).

The use of axioms or postulates does not exclude propositions; all three types of intrinsic statements may enter into a theory. Figure 9-8 depicts an instance in the form of a type 2-3-2-3 theory, meaning two axioms, three postulates, two propositions, and three transformational statements. Whenever possible, each concept should be linked to a referential; however, in some cases the theorist may not be able to make that link. He may be unable or unwilling to stipulate a formula or, having stipulated one, he may realize that limited resources preclude its application. In either event, he can continue to regard the term as a concept and link it to another concept through a proposition, as shown in Figure 9-8. However, if a proposition is used in that manner, the tests cannot be interpreted so as to assess its probable truth. If one or more of the hypotheses derived in part from the proposition is false in the "B" or "C" sense, the theorist has no choice but to assume that the result is due to an axiom, a postulate, or a referential formula.

Revision or Rejection versus Extension

Throughout this chapter the extension of a second-order theory has not been considered as an alternative to its revision or rejection. They are distinct alternatives, but there can be no rules to guide the theorist's decisions. At most only a few suggestions can be made, and the theorists must rely on his own judgment in each particular case.

If predictive congruence and/or magnitude is more or less uniform from one set of tests to the next, the theorist has no basis other than intuition for contemplating an extension of a second-order theory. In other words, an extension is suggested only when predictive congruence and/or magnitude is substantially greater for one set of tests than others. But such a finding would not reveal contingent properties.

To illustrate, suppose that predictive congruence and magnitude are much greater in a set of tests for India than in a set for the United States. The finding would suggest that at least one intrinsic relation could be contingent on some sociocultural condition and that the condition is markedly different for the two countries. But India and the United States differ in various ways, and the theorist may be unable to identify the relevant difference. Certainly he should not assume that just any obvious contrast between India and the United States can be taken as the contingent property in extending the theory, for tests of the extension would be based on other countries.

The knack of "guessing right" is at a premium in contemplating an extension of a theory, but a theorist is not likely to be successful if he relies on intuition alone. It may be that a comparison of only two sets of tests suggests no particular contingent property; if so, the theorist has no choice (other than revision or rejection) but to continue tests of the second-order theory in the hope that additional tests will suggest some contingent property. For that matter, the extension of a theory is a major step, and it should not be undertaken after considering only two sets of tests. Thus, should test results differ markedly for two countries and should the contrast suggest some contingent property, an extension of the second-order theory would be questionable without a test of it on a third country, one believed to differ from both of the initial two countries as regards the suspected contingent property.

When a theorist selects units for additional tests of a second-order theory with a view to its extension, a thorough knowledge of human culture and history is at a premium, for the goal is to select units that differ markedly with regard to a suspected contingent property. So the theorist should not rely on intuition alone in contemplating an extension of a theory. Indeed, without some knowledge of the social units considered in tests of a second-order theory, not even a marked contrast in sets of tests will suggest a contingent property.

It is conceivable that satisfactory predictive congruence can be realized only by extension of a second-order theory; however, ordinarily, an extension of a theory is undertaken to maximize predictive magnitude, and it is delayed until a sufficient predictive congruence has been realized through revisions of the second-order theory. Two considerations make that strategy all the more desirable. First, if tests of a second-order theory suggest that some of the intrinsic statements are false in the "C" sense, then predictive congruence can be increased by a simple revision. Second, the extension of a second-order theory should not be undertaken without confidence in the truth of the epistemic statements in tests of the theory.

The second consideration poses a difficult problem. Each epistemic statement in tests of a second-order theory could be true in the "A" sense even though all hypotheses are judged false in the "B" sense, but a theorist cannot readily distinguish the truth or falsity of epistemic statements from the truth or falsity of the intrinsic statements. If all hypotheses have been judged as false in the "B" sense, it could be that some of the relations asserted in the intrinsic statements are true in the "A" sense only in a certain type of condition. When that is the case, then an extension of the theory would be in order, but the theorist should be reluctant to undertake an extension without confidence in the truth of the epistemic statements. The dilemma underscores the need for preliminary research on concordance among independent sets of data and referents. Unless such research is undertaken before tests of a second-order theory, the theorist will be haunted by crippling doubts in contemplating an extension of the theory. However, he can take comfort in the possible outcome of an extension. Even if tests of a second-order theory have been uniformly negative, all tests of an extension of the theory could be positive. So an extension of a theory is always an alternative to a rejection, and an audacious theorist can undertake it by intuition alone.

The Question of Discriminatory Power

Although a theory can be revised so as to enhance its discriminatory power, revisions to that end only modify its structure and have no bearing on either predictive magnitude or predictive accuracy. Accordingly, the theorist should not concern himself with discriminatory power until he is satisfied with the predictive accuracy of the theory.

Reconsidering a type 2-3 theory, as depicted in Figure 9-2, suppose that after a series of tests the theorist computes average predictive values for each type of hypothesis with the following results: H1, +.85; H2, +.82; and H3, +.80. Since the types of hypotheses differ very little as to the number of substantive terms that enter directly or indirectly into their derivations, the minute differences in the average predictive values would not be surprising. Since there can be no specific or inflexible criterion of "insufficient" discriminatory power, only the theorist can pass judgment. But suppose that in this particular case he does contemplate a revision of the theory so as to enhance its discriminatory power. Observe that the average predictive value of H1 should be lower, for seven substantive terms enter directly or indirectly into derivations of its instances, as compared to six each in the cases of H2 and H3. The discrepancy could be eliminated by changing the theory from type 2-3 to type 0-3 (see Figure 9-3). In so doing two constructs would be eliminated, one of which enters into P1 and the other enters into P3. Given a type 0-3 theory, five substantive terms enter directly or indirectly into the derivation of each type of hypothesis, so the average predictive values of the three types would be expected to differ very little.

A change from a type 2-3 to a type 0-3 theory results in greater parsimony, and that consideration gives rise to a question. Since any type 2-3 theory can be changed to type 0-3 by eliminating the axioms and two constructs, why would a theorist ever sacrifice parsimony by formulating a type 2-3 theory? The answer is simple—in formulating a theory the theorist is guided by his conception of the nature of the relations (including, possibly, causal notions) and not by a preoccupation with parsimony. However, confronted with insufficient discriminatory power, a theorist may alter his conception of the relations and revise the theory by eliminating what were initially conceived as intervening variables.

As a final illustration, suppose the average predictive values (again for a type 2-3 theory) are: H1, +.85; H2, +.85; and H3, +.50. In such a case the discriminatory power of the theory may be less than the theorist is willing to settle for, but the situation would not be rectified by eliminating the axioms and two of the constructs (that is, by changing from a type 2-3 to a type 0-3 theory). However, the theorist can alter the theory so that what was once postulate 2 and transformational statement 2 become postulate 1 and transformational statement 1. In other words, he can simply change the position of the two pairs of statements in the structure of the theory. The change would not alter predictive accuracy, but it would enhance discriminatory power.

Notes for Chapter 9

[1] Theoretic relations include all relations asserted in the intrinsic and epistemic statements. Even though the *actual* theoretic relations are unknown and unknowable, they should be distinguished from *asserted* theoretic relations (the assertions in the intrinsic and epistemic statements).

[2] Popper has been the major spokesman for the view that theories cannot be proven as true or verified, but his idea that theories can be "corroborated" (1965:chap.X and 387-419) is not accepted. The idea is vague at best, and certainly Popper does not stipulate an explicit and specific criterion for judging a theory as "corroborated." Indeed, accepting Popper's notion of *degree* of corroboration, it is difficult to imagine any criterion that would not be arbitrary. Further, in taking the dilemma of induction and probability theory as central considerations, Popper ignores the dilemma of *deduction*. Specifically, he does not speak to the following question: Presuming that all tests of a theory are positive and presuming that all future tests will be positive, how can one know that the theorems are derivable only from that particular set of direct intrinsic statements?

[3] The point is still lost on some writers. "Some personality theories are true, some functional theories are true, and some ecological theories are true" (Stinchcombe, 1968:4). That statement does more than contradict Popper's thesis; additionally, it represents an uncritical acceptance of the discursive mode of theory construction in sociology and an indifference to dissensus in the field as to the appropriate criteria for assessing theories.

[4] Popper himself recognizes so many questions, problems, issues, and dilemmas (see 1965:108-111 and 278) that it is difficult to imagine anyone having the audacity to stipulate a *definite* criterion for declaring a theory as "falsified." Certainly Popper does not formulate such a criterion. Consider one of his statements: "Thus a few stray basic statements contradicting a theory will hardly induce us to reject it as falsified. We shall take it as falsified only if we discover a *reproducible effect* which refutes the theory" (1965:86). The statement suggests that a theory is falsified only if instances of a particular *type* of finding are consistently negative evidence for the theory. But what constitutes *consistently* negative evidence or, to put it another way, when is an effect to be considered as "reproducible"? The question is especially pertinent for sociology where the idea of *consistently* negative evidence is hardly more far-fetched than the idea of *consistently* positive evidence.

[5] If descriptive statements are taken as akin to "protocol sentences," several issues are introduced. At one time Carnap held that protocol sentences are not in need of confirmation; that is, they are simply accepted as descriptions of immediate experience (see Popper, 1965:96). The notion of immediate experience is scarcely relevant in characterizing descriptive statements; and, in any case, their truth is never beyond question. Comparisons of descriptive statements made by different investigators about the same sets of referents may alleviate doubts but not to the point of certainty. So the notion of a descriptive statement is closely akin to Neurath's conception of a protocol sentence (see Popper, 1965:97), a conception that recognizes uncertainty. However, Popper objects to Neurath's conception on the ground that it provides no rules for accepting or rejecting a particular protocol sentence. His objection suggests that there can be a specific and nonarbitrary criterion for accepting or rejecting a test of a theory, and that suggestion is here denied. Significantly, Popper presents no definite criterion for accepting or rejecting what he designates as "basic" statements, and any criterion would be arbitrary. Hence it is not surprising that Popper is vague in speaking to the question, but his suggestion that certainty can be realized in assessing "test" statements is surprising. The suggestion is contrary to his admission (see 1965:33 and 37) that uncertainty is inescapable in the assessment of scientific theories. The inconsistency may reflect Popper's determination to eliminate "psychologism" from the logic of science. However, it is difficult to see how a descriptive statement is "psychologistic," and the same may be said of of epistemic statements. If theorists or investigators were free to reject a descriptive statement as a matter of conviction, then a psychological element would be introduced; but no descriptive statement is to be rejected. If Popper should argue that the acceptance of all descriptive statements in tests may lead to the falsification of a "true" theory, the argument presupposes that the "truth" of a theory can be known. The presumption is denied. For that matter, while a theory should be testable, tests never falsify a theory, nor should they be undertaken with that end in mind. The purpose of tests is to assess the predictive accuracy of a theory, and such an assessment never ends with the declaration that the theory has been falsified.

[6] Zetterberg's statement on the subject is incredible: "it is usually very hard to confirm a single hypothesis but quite easy to confirm a theory, a system of hypotheses" (1965:103). Since the judgment of a theory depends on the interpretation of tests of hypotheses, it is inconceivable that the former can be easier than the latter. Zetterberg persistently blurs the distinction between theorem and hypothesis (evidently using the terms as interchangeable), but even so his statement is still incredible, if only because it ignores the dilemma of induction and the fallacy of affirming the consequent.

[7] Disagreement is all the more likely since predictive power is not unidimensional; that is, an assessment of predictive power should consider testability, scope, range, intensity, parsimony, and discrimination in addition to predictive accuracy. However, agreement can be furthered if predictive accuracy is given priority over other aspects of predictive power. As such, other aspects of predictive power are considered only when contending theories appear to have approximately the same predictive accuracy.

[8] When a theory comprises divisions, then interpretation of tests may result in the retention of some divisions and the rejection of others, which is, needless to say, preferable to rejecting the theory totally. In that connection, tests of each division should be interpreted separately from tests of other divisions, so that some divisions may be rejected and others retained. If so, the conclusion is obvious—the predictive accuracy of the theory is sufficient only for particular types of units and/or particular types of temporal relations. Note also that the prescribed procedures for tests and interpretation of tests apply regardless of the order of the theory, even though illustrations of the procedures are limited to second-order theories.

[9] A theorem and a direct intrinsic statement are cognates if the latter enters into the derivation of the former. Similarly, two direct intrinsic statements are cognates if both of them enter into the derivation of at least one theorem in common.

[10] The word "probably" is used here in the sense of "could be," and that qualification is especially important in recognition that the dilemma of induction precludes the conclusion that an intrinsic statement is "true."

[11] Consider a type 2-3 theory. Even though D1, D2, and D3 could be taken as known, none of the "actual" theoretic relations can be estimated with any certainty. For example, if the maximum coefficient of correlation in any D2 descriptive statement is $+.90$, it suggests that the relations asserted in E1, T1, P1, A1, P2, T2, and E2 are all very close. But there is no way to prove that each actual relation is direct, let alone close. Further, if the coefficient of correlation in each D2 descriptive statement is $.00$, it does not prove that none of the actual theoretic relations are close. A consideration of the D1 and D3 coefficients of correlation might clarify the situation somewhat but, even if the problem of affirming the consequent could be ignored, there are so many unknown theoretic relations that any conclusion would be conjectural.

[12] Two rules of interpretation have been stipulated in Chapter 8.

[13] The rationale can be explained by considering the first component of the rule: $[Nx - (Ns/P)] > Ny$. By chance alone Nx would approximate Ns/P, so the latter value is subtracted from the former. Hence the question: After eliminating instances of Nx attributable to chance, does Nx exceed Ny? The same rationale is considered in the second component of the rule: $[Ny - (Ns/P)] > Nx$. It may be paradoxical to retain a theory when a "disproportionate" number of tests are negative, but a formally constructed theory can be modified readily if tests are consistently negative. Finally, the third component of the rule, $(Ns/P) > (Nx + Ny)$, anticipates the possibility that tests are disproportionately positive for some hypotheses but disproportionately negative for others.

[14] The one exception is a theory in which there are more transformational statements than concepts, with some or all concepts linked to more than one referential. Such a theory is formulated when the theorist cannot choose among two or more alternative referential formulas without conducting tests.

Given such a theory, the first step after a series of tests is to eliminate some transformational statements so that each concept is linked to only one referential. The choice is made by comparing the predictive values of the types of hypotheses. When a concept is linked to more than one referential, there are alternative transformational statements. Each one enters indirectly (through a cognate theorem) into the derivation of a group of hypotheses, and in that sense there are alternative groups of hypotheses in the initial tests of the theory. The transformational statement that enters into the derivation of the group of hypotheses with the highest predictive value should be selected over the alternatives. What has been said of alternative transformational statements also applies when tests of the theory are based on different versions of the same referential formula (two versions differ as to either the kind of data prescribed or the stipulated procedure for gathering data). For each version there is a distinct group of hypotheses, even though each concept is linked to only one referential formula. After the initial tests, the theorist should make a choice among alternative versions of each referential formula, choosing that version which produced the highest predictive values. All subsequent observations on the interpretation of tests presume that the theorist has made a choice among alternative transformational statements and among alternative versions of referential formulas. The choice may be tentative, but the procedure of interpretation requires that (1) no concept is linked to more than one referential and (2) the tests are based on only one version of each referential formula.

[15] Anomalies can be identified readily by the computation of a Cv value. If the value obtained is not shown in column 4 of Table 9-1, the outcome is an anomaly. However, regardless of the Cv value, an outcome is an anomaly if it does not match one of the outcomes shown in columns 1-3 of Table 9-1.

[16] T3 is not the only direct intrinsic statement that becomes questionable when H3 is judged as false in the "B" sense. T2, P2, A2, and P3 also would be suspect.

[17] Such a modification may be undertaken to enhance the predictive accuracy of the theory but not its predictive congruence.

[18] Each type of hypothesis is linked to direct intrinsic statements through a theorem. Thus, in Figure 9-1, H1 is linked to A1 through Th1. Of course, Th1 and H1 are cognates, since the theorem enters into the derivation of the hypothesis. But in recognition that a theorem enters into the derivation of one and only one type of hypothesis, it is appropriate to speak of a theorem as "corresponding" to a particular type of hypothesis.

[19] Note that if H1 has been judged as false in the "C" sense and the other two types have been judged as true in the "A" sense, a reversal of either A1 or A2 would violate rule 8. However, such an outcome is anomalous, and H1 probably would only barely qualify as false in the "C" sense. So if H1 has been judged as false in the "C" sense, a reversal of A1 or A2 is in order only when H2 and H3 have not been judged as true in the "A" sense. This par-

ticular case should serve as a reminder that some of the illustrations are oversimplifications.

[20] None of the uncertainties and complexities are recognized by Zetterberg in his observations (1965:157-174) on tests of axiomatically formulated theories. The subject is ignored altogether by Dubin (1969), Stinchcombe (1968), and Willer (1967).

[21] Two theoretic statements are "complementary" if they enter into the derivation of the *same theorems or the same types of hypotheses*. To illustrate by reference to Figure 9-2, P2, T2, and E2 are complementary. Two theoretic statements may be cognates but not complementary. A1 and P2 are cognates, for they enter into the derivation of Th2, but only A1 enters into the derivation of Th1 and only P2 enters into the derivation of Th3.

[22] Even when two investigators apply a referential formula to the same body of data, the referents may not agree. Such disagreement would suggest that one or both of the investigators misunderstood the formula or made an error when applying it. So revision of the explication is intended to eliminate ambiguities and, hopefully, reduce errors in applications of the formula. As such, the revision does not alter the kind of data or the prescribed procedure for gathering data.

[23] If the theorist can stipulate an acceptable degree of predictive magnitude, then he judges the truth of hypotheses by an absolute criterion (for example, an average predictive value of at least +.75).

[24] The exception is a theory in which the number of transformational statements exceeds the number of postulates or, viewed another way, the number of concepts. Such a structure means that the theorist has not been able to choose among alternative transformational statements or, saying the same thing, alternative referential formulas. Again the procedure for interpreting tests presumes that after initial tests the theorist has eliminated alternative transformational statements. It also presumes that the theorist has made a choice among alternative versions of the same referential formula, so that all tests considered in the interpretation are based on one version of each referential formula.

PART FOUR: GENERAL OBSERVATIONS ON THE SCHEME

CHAPTER 10 SOME ANTICIPATED OBJECTIONS

Previous writers have not considered the antipathy of sociologists to formal theory construction. As suggested in Chapter 1, the reasons for the antipathy are diverse; but no sociologist is likely to state that he is opposed to formal theory in principle. Rather, opposition will be manifested in reactions to particular modes of construction. It should not be otherwise, as debates over formal theory construction in the abstract are only prelusive. Eventually, debates should focus on contending modes of formal theory construction, and various objections to the proposed scheme are anticipated.

The Burden of the Theorist

No theorist can use the proposed mode of formalization unless he is able and willing to make numerous judgments and decisions, most of which have been considered as "technical" rather than "theoretical" throughout the history of sociology. Specifically, to use the scheme the theorist must (1) judge the empirical applicability of sociological terms, (2) assess the research resources of the field, (3) consider alternative types of space-time relations in formulating empirical assertions, (4) limit assertions to particular types of social units, (5) stipulate formulas for tests of the theory, (6) evaluate the reliability and comparability of various kinds of data, and (7) stipulate procedures for gathering or otherwise obtaining data.

Needless to say, the requisite judgments and decisions impose an enormous burden on the theorist. Indeed, the use of the scheme requires aptitudes that are largely alien to most contemporary theorists, but a theorist is not required to engage in research or even have experience in research. However, he should know what researchers have done, for that indicates what they can do, which is the crucial consideration in attempting to formulate a testable theory. So no theorist is likely to use the scheme effectively unless he masters the research literature of at least one sociological specialty.

The imposition of burdens on theorists can be defended best by a question: If the theorist does not make the requisite judgments and decisions, who will make them? It is pointless to argue that the judgments and decisions will be made by someone else; the history of sociological theory clearly indicates otherwise. If that pronouncement be doubted, then the reader should contemplate another question: How often do you find a report that you consider to be a defensible test of any "grand" sociological theory? Presuming that the readers recognize a distinction between tests and criticisms of a theory, the answer will be "few, if any." But the paucity of tests poses no mystery. Granted that conducting tests of sociological theories is a thankless task, there is an additional reason why so few are reported—sociological theories are not formulated so as to facilitate tests. However, sociological theories do not just happen to be "untestable"; they are untestable precisely because theorists do not make certain kinds of judgments and decisions.

Of course, tests of sociological theories could be realized without imposing responsibilities on theorists. All that investigators need do is to make the judgments and decisions that the theorist refrained from making. But given the great number of untested sociological theories, it is obvious that few investigators respond to the opportunity. Again, there is no mystery. Researchers may not be imaginative, but they are neither dumb nor insensitive to the reward system. They recognize the painfully obvious—if investigators must make numerous decisions and judgments to test a theory, the findings will not be accepted as critical evidence, and rightly so. After all, if investigators must make decisions and judgments to test a theory, what assurance is there that they will agree in their decisions and judgments? If they do not, in what sense are the tests comparable? Indeed, if they do not, in what sense have they tested the same theory? The latter question is all the more important since the judgments and decisions necessary for tests of a sociological theory cannot be separated from the theory itself. It is not even accurate to say that the judgments and decisions "complete" the theory; if anything, they alter it. As such, when test results are negative, the theorist and/or his disciples will argue that the judgments and decisions made by the investigator were "inappropriate." But if "appropriateness" can be judged apart from test results, then why is it necessary for an investigator to make judgments and decisions in the first place? Of course, if the judgments and decisions made by an investigator are appropriate only when test results are positive, then that criterion precludes negative evidence; and, if that criterion is adopted, sociological theories will become even more hoary.

The foregoing comments may be countered by the argument that investigators can agree in making independent judgments and decisions necessary for tests of sociological theories, and so the tests need not be personalized, unique, and idiosyncratic. There is, however, little evidence in the sociological literature to support that argument. There are even instances where the theorist has made the decisions and judgments necessary for tests of his theory, but the critics in effect regard them as inappropriate (see Chambliss and Steele, 1966). Then there is an overriding consideration—if investigators can agree in formulating a procedure to test a theory, then surely the theorist should have had no difficulty in stipulating the procedure at the outset. The truth is that agreement among investigators is unlikely precisely because judgments and decisions pertaining to procedure are difficult to make; and it is idle to suppose that someone other than the theorist can, will, or should make them. So if one longs for the day when an enterprising investigator conducts "critical" tests of the theories of Marx, Durkheim, Weber, or Pareto, then he should be prepared to live indefinitely. It will not happen.

To summarize, the proposed scheme imposes burdens on the theorist for one reason—it is the only way to realize testable theories. But note that the term "testable" is not used vaguely or uncritically, as it often is in critiques of sociological theory. The proposed scheme sets forth a specific procedure for conducting tests.

As a final consideration, it is hoped that critics will refrain from arguing that the burden imposed on the sociological theorist is greater than in the more advanced sciences. That may be so, but it is irrelevant. The burdens of the theoretical physicists may tell us something about physics, but it has no bearing on contemporary sociology.

The imagination of the theorist.   Another possible objection is that adoption of the scheme would have a repressive effect on the imagination of theorists. The issue hinges on the demand that a theorist formulate testable theories, which some sociologists are certain to regard as impractical. Suppose that a theorist has taken his responsibilities seriously but falls short of a testable theory only because the requisite data cannot be gathered or obtained in published form. In such a case, barring promulgation of the theory for that reason alone may appear unrealistic. After all, so the argument will go, if the theory is promulgated, then sociologists of subsequent generations may be able to obtain the stipulated data. But "sociological theories for subsequent generations" has been the tradition in the field for over a century, and the product is untested and seemingly untestable theories.

As for the idea that the scheme would repress the imagination of theorists, just the opposite is true. There could be no better way to stimulate the imagination of a theorist than to insist that he must formulate at least one version of his theory that is testable now. To be sure, the theorist may be forced to stipulate less than ideal data, with the risk being unimpressive predictive accuracy and a premature rejection of the theory. Accordingly, critics may see fit to reject not only the demand for tests but also predictive accuracy as a criterion for assessing theories. Yet the predictive accuracy of a theory is to be judged relatively rather than absolutely, and that alone works against a premature rejection. Moreover, nothing prevents the theorist from presenting two versions of his theory—one purportedly testable and the other only potentially testable. Tests of one version would not yield evidence for or against the other, but requiring a testable version is a carrot to prod the theorist. The testable version is the price paid for the right to be heard.

For some sociologists the demand for testable theories raises the specter of radical operationalism, and they fear that its acceptance would stultify the imagination of theorists. If radical operationalism demands that all terms of a theory be defined "precisely," or by reference to instrumentation, or be such that they are empirically applicable, then the fear of it is defensible. Such a demand would be grossly unrealistic, and it would indeed sterilize theory construction; but the proposed scheme does not even require that all terms be defined, let alone "operationally" (whatever that word may mean). So the theorist is free to use terms that by his own admission designate vague notions or, to use more conventional expressions, unobservable and unmeasurable phenomena. The advantage is especially important in making assertions about values, norms, national character, attitudes, and ideology. All such "predispositional" terms can be used in formal theory construction without any pretense of "operationalization," but the outcome would be testable theorems rather than the usual tautologies that result when such terms are used in a discursively formulated theory. Stated another way, the scheme can be used to develop either an abstractively formulated theory or one that postulates unobservable entities and mechanisms (see Cohen and

Nagel, 1934:397). What more license could a theorist possibly want? True, he must link some of his concepts to referentials, but that step calls for imagination rather than repressing it. However, if by "imagination" critics mean the capacity to beguile an audience, then the scheme hopefully will stifle it.

The Break with Tradition

As argued in Chapter 1, formal theory construction is a break with sociological tradition, and hence the proposed scheme may be viewed by critics as too radical. However, contrary to what previous observations may have suggested, the scheme does not represent a complete break with the past, nor is a complete break considered as desirable or feasible. Rightly or wrongly, sociologists are committed to a terminology, and they are reluctant to be done with the grand theories. However, in both respects the proposed scheme is more conservative than might appear.

Terminology. Unlike the principles attributed to operationalism, adoption of the proposed scheme would not alter the substantive terminology of sociology. The scheme pertains to the form of theories, not their content. So no plea is made for "purification" of the field's vocabulary, and the only terminology proposed is a formal language. The point may appear trivial, but a truly radical alteration of the field's vocabulary would be construed (perhaps rightly so) as a rejection of traditional substantive interests. Certainly the proposed scheme does not imply that sociologists should abandon their substantive interests; they can formulate theories about any phenomenon, and the scheme in no way suggests that some quantitative properties are more important that others.

Although the substantive terminology of sociology is not questioned, the scheme does not preclude innovations. If anything, it facilitates them, for a theorist can devise new terms and use them effectively without offering what purports to be complete or empirically applicable definitions.

Theories in the grand tradition. The terminology of sociology is, of course, only a part of the field's heritage. No less important, one could argue, are those theories formulated by the great men in sociology's history, notably, Marx, Durkheim, Weber, and Pareto. Now it might appear that acceptance of the proposed scheme entails a rejection of the "grand" theories, for they are the antithesis of formal theory construction. Such is not the case.

Although the discursive mode of their exposition will forever detract from the value of the grand theories, the proposed scheme would not entail rejection of their *content*. On the contrary, if those theories are to be used rather than perpetuated as relics, the scheme is a distinct alternative to exegetical sociology. Throughout the works of the grand theorists we find insights as to the interrelations among the properties of social units, and in some instances those insights are expressed as explicit assertions. Those assertions are not likely to be identified as other than axioms, and evidence can be brought to bear on them indirectly only through additional assertions in the form of postulates and transformational statements. Accordingly, additional assertions would have to be made by a contemporary theorist, and they may appear alien to the original grand theory. Nonetheless, contemporary sociology could *use* the grand theories for something other than symbols of ancestor worship.

The grand theories have provided continuity in substantive interests; but the substantive interests of sociology now appear established, so the utility of grand theories in that direction is finished.[1] Hence there are only two alternatives—retain the grand theories as relics or attempt to use *parts* of them in formally constructed theories. Certainly it is difficult to imagine any defensible use of them without formal theory construction. Again, most universal assertions made by grand theorists are not likely to be identified as other than axioms, and axioms have no utility outside of a formally constructed theory.

Two issues are entailed in the prescribed utilization of grand theories. Some critics will argue that the assertions of the grand theorists cannot or should not be lifted out of context, the argument being that each theory must be taken as a whole. However, as previously noted, the theories cannot be tested, and it is not clear what is gained by leaving them intact. Moreover, the argument suggests an incredible principle—that no component assertion of a theory can be valid unless all are valid.

The other issue has to do with the modification of the assertions made by the grand theorists. Even in those few instances where their statements appear to be genuine empirical assertions rather than tautologies, the statements may contain no unit terms or temporal quantifiers, and they are unlikely to contain relational terms that are consistent with formal theory construction. So, typically, such statements will have to be modified before incorporation in a formally constructed theory. That modification is inherently debatable, for someone is certain to allege that the grand theorist has been misinterpreted. Such an objection perpetuates exegetical sociology, and no enterprise is more sterile. We shall never know what Marx, Durkheim, Weber, or Pareto "really meant," and only a pedant should care. The central question is not what they meant in a literal sense but, rather, how their assertions can be used in theory construction. In any case, what better way to resolve debates than to compare the outcome as regards predictive power when different interpretations of a grand theorist's assertions are incorporated in formally constructed theories? Surely anything is preferable to further exegetical debates.

The objection will be, of course, that taking only isolated assertions from the grand theories precludes a sweeping synthesis. But consider what attempts at a synthesis have accomplished. Parsons, for example, offers his work as an integration of Durkheim, Weber, and Pareto; but his discursive mode of formalization perpetuates the major defect of the grand theories. In particular, Parsons comes no closer to a testable theory than did his illustrious predecessors. Indeed, as suggested earlier, the grand theorists did set forth some rather clear-cut empirical assertions; but they do not survive Parsons' synthesis, nor does Durkheim's comparative method, Weber's sense of history, or Pareto's appreciation of social conflict. Sociology cannot survive many more such syntheses.

Positivism

It goes without saying that critics will label the proposed scheme as "positivistic." That label is taken as honorific, but it is controversial. Some critics (see Sjoberg and Nett, 1968) have an abiding passion for categorizing sociological works as positivistic, reductionistic, or mechanistic, as though a label alone is an adequate assessment. Labels are not adequate because they conceal specific issues, and their meaning is far from clear.

The question of meaning is especially critical for the label "positivistic," as there are obviously many divergent conceptions of positivism and the divergence transcends sociology. For example, A. J. Ayer (regarded by some as an arch positivist in philosophy) makes the following observation: "we dissent also from the positivist doctrine with regard to the significance of particular symbols. For it is characteristic of a positivist to hold that all symbols, other than logical constants, must either themselves stand for sense-contents or else be explicitly definable in terms of symbols which stand for sense-contents" (1946:136). By Ayer's criterion the proposed scheme is the very antithesis of positivism. Constructs need not be defined at all, and it is inconceivable that they stand for "sense-contents." Nor is the relation between a construct and a concept or that between a concept and a referential fixed by definition. For that matter, the scheme represents a denial that major sociological terms stand for sense-contents or can be so defined.

The foregoing is not an attempt to dodge the label of positivism. The scheme is positivistic but only in that it represents a demand for testable theories and the assessment of theories by reference to their predictive power. Critics may choose to ignore that conception of positivism, but hopefully they will consider a question: If the goal is consensus in the assessment of theories, what is the alternative to predictive power as the primary criterion? An answer to that question is preferred to the usual wailing about the evils of positivism.

Operationalism

If critics reject the scheme because it is "positivistic," they will have strange company. There is every reason to anticipate that sociologists who are commonly identified as operationalists, especially the followers of Lundberg (1939), will not find the scheme to their liking. They will take exception for several reasons: (1)

the scheme does not even require definitions of all component terms of a theory; (2) the definitions of some terms need not make reference to any particular instrument, operation, or procedure; and (3) the relevance of the notion of "observability" for sociology is questioned. The third consideration is the most crucial. Since it is claimed that neither constructs nor concepts are elements of an empirical language, the orthodox operationalist will dismiss them along with axioms and postulates as fictions.

Debates over realism, idealism, nominalism, and so on, have not been conclusive in any field; and there is no reason to suppose that it would be different in the context of formal theory construction. So only two comments will be made. One is free to think of axioms and postulates, or even transformational statements and epistemic statements, as fictions, that is, products of the theorist's imagination. Alternatively, one may think of them as designating phenomena that somehow transcend immediate experience. As far as the proposed scheme is concerned, it makes no difference whatever. There is only one consideration in assessing a term or a statement, and that is its utility. If in the judgment of a theorist a particular term or statement is necessary to construct a theory, that is the only justification needed.

If that argument is not accepted and all terms are excluded except those which designate "observable" phenomena, it would sterilize theory construction in sociology and strip the field's vocabulary bare. Moreover, however "fictitious" axioms, postulates, and transformational statements may be, it is primarily through them that a synthesis of sociological theories will be achieved. If two theories employ any substantive term (construct, concept, or referential) in common, a synthesis can be realized by one or more *implied* intrinsic statements. For that matter, a synthesis of two theories can be achieved readily even if they do not share any substantive term in common. All that a theorist need do is to make an assertion in the form of an axiom, postulate, or transformational statement that links a substantive term in one theory with a substantive term in the other. So the scheme facilitates a synthesis of theories (provided they are constructed formally), but that would not be so if axioms, postulates, and transformational statements are dismissed as fictions.

Theory without Empirical Assertions

Rather than speak of theory construction, some sociologists (see Buckley, 1967) describe their activities as "system analysis" or "model building." The difference between the meaning of either term and "theory" is by no means clear (see Rudner, 1966:23). If one argues that model building or system analysis encompasses more variables than are ordinarily treated in theory construction, surely the distinction is arbitrary. Another possibility is that in system analysis or model building interrelations among variables are interpreted in a special way;[2] but unless such "interpretations" lead to empirical assertions about the space-time relations among variables, they are not testable. Of course, if model builders and system analysts refrain from making empirical assertions, it suggests that their works are not to be judged by predictive accuracy, but surely they are obliged to stipulate an alternative criterion. In any case, the issue cannot be circumvented by arguing that the proposed scheme does not enable one to describe "systems" or build "models." It does not, for those activities require a substantive terminology, a consideration divorced from formal theory construction. Nonetheless, a question survives the argument: What is the ultimate goal in describing systems or building models? If it is not the articulation of empirical assertions in the form of generalizations about the space-time relations among properties of actual social units, then the proposed mode of formal theory construction has no utility. Stated another way, the scheme has no utility if the goal is prescriptive propositions, the description of the particular and the unique, or an inventory of logical possibilities.

All that has been said of model building and systems analysis applies to another possible objection—that the scheme does not enable one to analyze social "processes" or "dynamics." The issue hinges on the meaning of "analyze." To be sure, the scheme does not provide a substantive terminology for any kind of analysis; so the pivotal question is the goal of analysis. If it is not a set of empirical assertions about an infinite universe, then the scheme has no utility. However, objections to the scheme may extend to a declaration that it cannot be used to articulate empirical assertions about dynamics or processes. That declaration would make the words dynamics and process esoteric, as though they designate phenomena without space-time coordinates. No sociological phenomenon is that exotic. Finally, anticipating the argument that dynamics and process refer to special types of change, empirical assertions about any type of change in quantitative properties can be made by the stipulation of longitudinal comparisons and temporal quantifiers.

Some of the foregoing observations may suggest that any sociological work other than the articulation of empirical assertions is indefensible. On the contrary, a wide range of other activities—conceptual analysis, exploratory research, and the development of research methods—are essential steps toward the articulation of empirical assertions. The issue is not invidious distinctions as regards the division of labor but, rather, the indiscriminate use of the label "theory." For generations sociologists have equated theory with what they consider to be brilliant, profound, imaginative, insightful works. But many of the works so characterized represent a studied evasion of any empirical assertion in the form of a generalization about space-time relations among properties of social units, and yet the sociological audience persists in labeling those works as "theory." The matter is beyond constructive argument, so it can only be emphasized that the notion of theory without explicit empirical assertions is simply alien to the proposed scheme.

Notes for Chapter 10

[1] Of course, some sociologists (for example, Nisbet, 1966) might not agree, as they would have the field continually returning to the "masters" (such as Durkheim, Weber) for inspiration and guidance.

[2] "It is possible to achieve high precision in predicting *when* changes in system states will occur and *what* states will succeed each other, without possessing knowledge of how the system operates" (Dubin, 1969:17). The dictum is pure logomachy. How could such predictions possibly be construed as ignorance about how a system "operates?"

CHAPTER 11 SOME LIMITATIONS

In contemplating reactions to the proposed mode of formal theory construction, one distinction among sociologists is especially relevant: those who are content with the field and those who despair. Of course, it does not follow that sociologists who despair will accept formal theory construction, but it is a virtual certainty that those who are content will not be enthusiastic. Moreover, even if a substantial number do react favorably, they will be sorely disappointed on realizing that formal theory construction is not a panacea, and that realization should not be left to discovery. The proposed scheme has numerous limitations, some of which are probably true of formal theory construction in general.

Terminological Problems

As stressed in Chapter 2, sociology's terminological problems are truly formidable, and the proposed scheme offers very little in the way of solutions. Whatever the inadequacies of the field's substantive terminology, the scheme does not even suggest an alternative; but that limitation should not be surprising, as formal theory construction is divorced from substantive considerations.

The only way the scheme lessens the terminological problem is by enabling theorists to avoid a preoccupation with semantics in certain steps of theory construction. Since the theorist does not consider his definition of constructs as complete or empirically applicable, he is free to use terms that would be precluded by the tenets of radical operationalism. So, viewed that way, the scheme provides a syntactical solution for semantical problems.

Of course, semantical problems are not avoided altogether. They are crucial in the stipulation of referential formulas, and the scheme offers the theorist no directions at that level; he is on his own. Since the stipulation of referential formulas is a substantive consideration, a theorist should not rely on a mode of formal theory construction to guide his judgment. However, methodological and substantive considerations never are divorced completely, and some modes of theory construction may exacerbate terminological problems more than others. The proposed scheme may appear to do so by its emphasis on empirical applicability, but it is inconceivable that any mode of formalization can be used to construct testable theories unless it emphasizes empirical applicability or a similar notion. Moreover, the scheme makes the stipulation of referential formulas less difficult by rejecting the pernicious notion that a referential formula or a procedure for gathering data must somehow be "validated" or "justified" before it is used in theory construction. Even the requirement that a theorist regard his referential formulas as empirically applicable is tempered; he need not demonstrate empirical applicability and, in any case, no particular degree is demanded.

Of all aspects of empirical applicability, the emphasis on feasibility is most likely to be construed as exacerbating the terminological problem. Even if satisfactory in all other respects, the use of a particular referential formula may be precluded by limited research resources; and, so the argument will go, it is unreasonable to demand that a theorist not only solve terminological problems but do so within the limits imposed by the field's resources. Of course, there is an alternative—a mode of theory construction that ignores feasibility—but it would be inconsistent to ignore it and at the same time stress testability. Further, the proposed scheme offers the theorist a special—and crucial—flexibility in making transformational statements. Since the link between a concept and a referential (as expressed in a transformational statement) is not fixed by definition, a theorist can link a concept with several referential formulas. As such, if limited resources preclude the application of some referential formulas, the others can be retained and the outcome is still a testable theory. For that matter, if limited resources preclude application of all contemplated referential formulas, the theorist can link the concept to another concept through a proposition and contemplate other referential formulas. Of course, the insertion of propositions reduces parsimony, but it is nonetheless one strategy for coping with the problem of feasibility.

The stress on universality (another aspect of empirical applicability) is not likely to be as controversial as feasibility. Unfortunately, however, the scheme provides the theorist with no guidance in his attempts to solve what is perhaps the most difficult of all terminological problems—that stemming from cross-cultural and historical relativity. The problem is acute when stipulating referential formulas, but it cannot be avoided entirely in defining concepts and constructs. Even though the relations among constructs, concepts, and referentials are not fixed by definition, a theorist's direct intrinsic statements may reflect his definitions of constructs and concepts. If the theorist's definitions are ethnocentric, then he is likely to stipulate referential formulas that, culturally and historically speaking, have limited applicability.

While admitting that the proposed scheme offers no directions for devising universally applicable referential formulas, some mitigating observations can be made. Since the problem is substantive, it is doubtful that any mode of formal theory construction can solve the problem. In any case, the stress in the proposed scheme will not have a crippling effect on theory construction. A theorist need not demonstrate that a referential formula is universally applicable, or even present evidence on the question. If it were known that the applicability of a referential formula is limited, the scheme does not require that the theorist abandon it, provided he is willing to sacrifice some testability. In that connection, a theory that can be tested only in a particular culture or historical period would be superior to extant sociological theories.

Causation and Explanation

Some critics will regard the exclusion of causation as the major limitation of the scheme; however, what one makes of that limitation is another matter. The issue hinges on the different ways that causal notions can enter into the formulation of a theory. The proposed scheme cannot be used to construct a theory which somehow leads to or constitutes a demonstration of causation, since there is no way to demonstrate causation, certainly not to the satisfaction of all critics.

A more likely criticism of the scheme will allege that it does not permit causal inferences. That criticism is correct in only one respect—the scheme does not stipulate any particular criterion or method for causal inferences. Reasons for that omission have been given in Chapters 2 and 6. Summarizing briefly, any criterion or method for inferring causation would be (1) arbitrary, (2) unacceptable to a vast number of sociologists, and/or (3) alien to the conditions of work in the field. So, far from enhancing the scheme, the inclusion of some criterion or method for inferring causation would diminish its utility.

But the omission is likely to be misconstrued. It does not mean that a theorist should refrain from using the scheme if he thinks in causal terms. On the contrary, a theorist can stipulate a criterion or method for causal inferences and use it within the context of the scheme. If the association between properties is presumed to be causal, then surely the theorist can argue that causation will be reflected in a particular type of space-time relation among referents. The scheme permits the theorist to consider various types of space-time relations, but it leaves the identifica-

tion of "causally relevant" types to his judgment. If the sociological audience does not agree, that is hardly a shortcoming of the
scheme. But if the theorist is of the opinion that causation cannot
be inferred from space-time relations, the scheme does have a
decided limitation—it is alien to his conception of causation.

What has been said of causation applies also to explanation.
Briefly, a theorist may use the scheme to formulate a theory in
such a way that he regards it as an adequate explanation of one or
more of the properties designated by the substantive terms. Note
especially that the scheme requires deductions or derivations,
which are commonly considered as essential in explanation. Further, insofar as the theorist thinks of an explanation as an assertion of a particular type of space-time relation between one
phenomenon (the explanandum) and another phenomenon (the
explicandum), he can make such assertions readily in the context
of the scheme. However, if the theorist considers explanation as
unrelated to assertions of space-time relations, the scheme has no
utility. Moreover, it provides no assurance that the sociological
audience will regard a theory as an adequate explanation, and that
is true regardless of the predictive accuracy indicated by tests.
Indeed, if a theorist expects all sociologists to accept the idea of a
connection between predictive accuracy and explanatory adequacy, he is in for a shock.

All of the foregoing comments come down to one point—the
scheme does not purport to end the debate over causation and
explanation. If anything, it is predicated on the assumption that
the debate will never be resolved; accordingly, sociologists should
cease judging theories by criteria that are not only controversial
but also beyond constructive argument. Their *public* assessments
of theories should consider only predictive power, with the question of causation and explanation left to private opinion. Any
other course will perpetuate anarchy.

The Quest for Certainty

Within the context of the proposed scheme, the goal of theory
construction is the creation or identification of order, and success
should be judged by the predictive power of the theory. As one

aspect of predictive power, accuracy can be assessed only through tests. However, since tests are always finite, a perennial question hovers over the judgment of any theory: How can one be sure that future test results will not be markedly different from those in the past? The answer is that one cannot. Of course, the question introduces the dilemma of induction, and viewed that way the answer is not radical, which is to say that numerous philosophers and scientists regard the dilemma as inescapable. So the scheme represents an unqualified acceptance of the idea that uncertainty is inevitable, but the idea is generalized beyond the recognition that a theory can never be verified. On the whole, sociologists are willing to entertain Popper's dictum (1965) that a theory cannot be conclusively verified, but they are likely to cling to the notion that it can be falsified conclusively. Should they demand that theories be constructed so that they are falsifiable not only in principle but in practice, then the proposed scheme will not be to their liking. The scheme extends the principle of uncertainty to the outer limit, meaning that it does not provide a method or criterion by which a theory can be either verified or falsified conclusively. Hence, if a theorist demands a sense of certainty, he should not even attempt to use the scheme. It does not purport to resolve the dilemma of induction, and that limitation is admitted readily; but at the same time it is categorically denied that any mode of theory construction can resolve the dilemma. As for the idea that theories can be so constructed that they can at least be falsified conclusively, it is conceivable only if one denies the possibility of dishonesty or human error.

The foregoing extension of the uncertainty principle is not limited to interpretation of tests. In using the scheme a theorist will be plagued with uncertainty at each step. There is no way by which the truth of any intrinsic statement, even theorems, can be known. Moreover, even implied assertions, such as the empirical applicability of referential formulas and the unit terms, can never be substantiated conclusively. These additional uncertainties may be more than sociologists will tolerate, but they are reminded again that uncertainties are not created by the scheme. The uncertainties are present in the construction or tests of any theory, and the proposed scheme differs from the discursive mode of theory construction by making recognition of them inescapable.

Qualitative Properties and Macroscopic Sociology

The most conspicuous limitation of the scheme is that it does not permit a theorist to make empirical assertions about relations among qualitative properties. No attempt is made to belittle that limitation by arguing that all sociological terms actually designate quantitative properties. The recognition of unit terms is in itself an admission of the importance of qualitative properties, and the stipulation of referential formulas entails reference to qualitative units.

The only mitigation is that virtually all sociological theories actually make assertions about relations between quantitative properties of social units, even though the distinction may not have been recognized by the theorist. Moreover, there is reason to believe that theorists will become increasingly cognizant of the quantitative character of most major sociological terms. Nonetheless, there is a need for another mode of theory construction, one that applies to qualitative properties.

Although explication of the scheme is slanted in the direction of sociological specialties that deal largely with macroscopic phenomena and especially those that make extensive use of published data, that focus should not be misconstrued. It may be that the scheme's utility is limited to such specialties, but in the final analysis its limits will be revealed only by usage. Certainly there is no compelling reason why the scheme cannot be used in the formulation of theories about microscopic phenomena. Again, the choice of unit terms is a substantive consideration, and referential formulas can be devised so that they apply to data gathered by experimental procedures or field studies. In light of those considerations, the scheme is not even limited to sociology.

Elegance and Crude Predictions

Only those sociologists who take physics as the model of science will truly appreciate some major limitations of the scheme. The scheme is cumbersome, and theories constructed in accordance with it will not be elegant. A theoretical physicist would be appalled by the prescribed form of intrinsic statements, especially the relational terms and temporal quantifiers. He may appreciate the purpose of the extrinsic part of a theory, but it would strike him as primitive. His greatest shock would come in recognizing the second limitation—the relational terms and the system of derivation permit only predictions of ordinal differences.

Now, obviously, the ultimate in an elegant theory is a set of equations, and the proposed scheme is a far cry from that ideal. But some reasons for the divergence are not obvious. Given the terminological problems of sociology, the extrinsic part of a theory is simply a necessary inelegance. The defense of temporal quantifiers is less complicated. Briefly, since there are no related conventions in sociology, an intrinsic statement without temporal quantifiers virtually defies interpretation. Finally, the prescribed relational terms and mode of derivation are questionable for reasons other than inelegance. They permit only ordinal predictions, which is surely the most uninformative theory one can devise, short of settling for the virtually vacuous assertions of "some difference regardless of direction or magnitude." Ordinal predictions are defended on the ground that the present quality of sociological data make any higher aspiration unrealistic. Indeed, higher aspirations might well lead to the premature and therefore counterproductive rejection of sociological theories.

It may be that the quality of sociological data will improve to the point that sociologists can justify the statement of theories in the form of equations, and when that point is reached the utility of some aspects of the scheme will come to an end. The scheme was devised with a view to the conditions of work in some sociological specialties and, once these conditions change, the scheme should be modified. But continuity will not be lost, for the theories constructed in accordance with the present scheme can be restated in a more elegant form.

REFERENCES

1967 Abel, Theodore, "A Reply to Professor Wax," *Sociology and Social Research*, 51 (April):334-336.

1970 Abrahamsson, Bengt, "Homans on Exchange: Hedonism Revived," *American Journal of Sociology*, 76 (September):273-285.

1968 Aiken, Michael, and Jerald Hage,"Organizational Interdependence and Intra-Organizational Structure," *American Sociological Review*, 33 (December):912-930.

1970 Ajzen, Icek, *et al.*,"Looking Backward Revisited: A Reply to Deutscher," *American Sociologist* 5 (August):267-273.

1965 Aron, Raymond, *Main Currents in Sociological Thought*, Vol. I (New York: Basic Books).

1970 ——*Main Currents in Sociological Thought*, Vol. II (New York: Doubleday).

1946 Ayer, Alfred Jules, *Language, Truth and Logic* (New York: Dover Publications).

1959 ——(ed.), *Logical Positivism* (New York: Free Press).

1967 Baar, Carl, "Max Weber and the Process of Social Understanding," *Sociology and Social Research*, 51 (April): 337-346.

1957 Becker, Howard, and Alvin Boskoff (eds.), *Modern Sociological Theory* (New York: Holt, Rinehart and Winston).

1970 Becker, Marshall H., "Sociometric Location and Innovativeness: Reformulation and Extension of the Diffusion Model," *American Sociological Review*, 35 (April): 267-282.

1964 Berelson, Bernard, and Gary A. Steiner, *Human Behavior: An Inventory of Scientific Findings* (New York: Harcourt).

1966 Berger, Joseph, *et al.* (eds.), *Sociological Theories in Progress*, Vol. I (Boston: Houghton Mifflin).

1954 Bergmann, Gustav, *The Metaphysics of Logical Positivism* (New York: McKay).

1958 ——*Philosophy of Science* (Madison: University of Wisconsin Press).

1937 Bierstedt, Robert, "The Logico-Meaningful Method of P. A. Sorokin," *American Sociological Review*, 2 (December): 813-823.

1960 ——"Sociology and Humane Learning," *American Sociological Review*, 25 (February): 3-9.

1961 Black, Max, (ed.), *The Social Theories of Talcott Parsons* (Englewood Cliffs, N. J.: Prentice-Hall).

1964 **Blalock, Hubert M., Jr.,** *Causal Inferences in Nonexperimental Research* (Chapel Hill: University of North Carolina Press).

1967 ——*Toward a Theory of Minority-Group Relations* (New York: Wiley).

1968a ——"The Measurement Problem: A Gap Between the Languages of Theory and Research," in Hubert M. and Ann B. Blalock (eds.), *Methodology in Social Research* (New York: McGraw-Hill), chap. 1.

1968b ——"Theory Building and Causal Inferences," in Hubert M. and Ann B. Blalock (eds.), *Methodology in Social Research* (New York: McGraw-Hill), chap. 5.

1969 ——*Theory Construction* (Englewood Cliffs, N.J.: Prentice-Hall).

1970 ——"The Formalization of Sociological Theory," in John C. McKinney and Edward A. Tiryakian (eds.), *Theoretical Sociology* (New York: Appleton), pp. 272-300.

1962 **Blanche, R.,** *Axiomatics* (New York: Free Press).

1964 **Blau, Peter M.,** *Exchange and Power in Social Life* (New York: Wiley).

1970 ——"A Formal Theory of Differentiation in Organizations," *American Sociological Review*, 35 (April): 201-218.

1954 **Blumer, Herbert,** "What Is Wrong with Social Theory?," *American Sociological Review*, 19 (February): 3-10.

1956 ——"Sociological Analysis and the 'Variable,' " *American Sociological Review*, 21 (December): 683-690.

1961 **Bochenski, I. M.,** *A History of Formal Logic* (Notre Dame, Ind.: University of Notre Dame Press).

1969 **Borgatta, Edgar F.,** (ed.), *Sociological Methodology 1969* (San Francisco: Jossey-Bass).

1956 **Borgatta, Edgar F.,** and **Henry J. Meyer** (eds.), *Sociological Theory* (New York: Knopf).

1953 **Braithwaite, Richard Bevan,** *Scientific Explanation* (London: Cambridge University Press).

1962 **Brodbeck, May,** "Explanation, Prediction, and 'Imperfect' Knowledge," in Herbert Feigl and Grover Maxwell (eds.), *Minnesota Studies in the Philosophy of Science,* Vol. III (Minneapolis: University of Minnesota Press), pp. 231-272.

1968 ——"Meaning and Action," in May Brodbeck (ed.), *Readings in the Philosophy of the Social Sciences* (New York: Crowell-Collier-Macmillan), pp. 58-78.

1963 **Brown, Robert,** *Explanation in Social Science* (Chicago: Aldine).

1968 **Buchler, Ira R.,** and **Henry A. Selby,** *Kinship and Social Organization* (New York: Crowell-Collier-Macmillan).

1967 **Buckley, Walter,** *Sociology and Modern Systems Theory* (Englewood Cliffs, N. J.: Prentice-Hall).

1959 **Bunge, Mario,** *Causality* (Cambridge, Mass.: Harvard University Press).

1942 **Calhoun, Donald W.,** Review of R. M. MacIver's *Social Causation* in *American Sociological Review,* 7 (October): 714-719.

1953 **Cambell, Norman R.,** "The Structure of Theories," in Herbert Feigl and May Brodbeck (eds.), *Readings in the Philosophy of Science* (New York: Appleton), pp. 288-308.

1966 **Catton, William R., Jr.,** *From Animistic to Naturalistic Sociology* (New York: McGraw-Hill).

1967 ——"Flaws in the Structure and Functioning of Functional Analysis," *Pacific Sociological Review,* 10 (Spring): 3-12.

1965 **Caws, Peter,** *The Philosophy of Science* (Princeton, N. J.: Van Nostrand).

1966 **Chambliss, William J.,** and **Marion F. Steele,** "Status Integration and Suicide: An Assessment," *American Sociological Review,* 31 (August): 524-532.

1964 **Cicourel, Aaron V.,** *Method and Measurement in Sociology*, (New York: Free Press).

1968 **Clark, Terry N.,** "Community Structure, Decision-Making, Budget Expenditures, and Urban Renewal in 51 American Communities," *American Sociological Review*, 33 (August): 576-593.

1934 **Cohen, Morris R.** and **Ernest Nagel,** *An Introduction to Logic and Scientific Method* (New York: Harcourt).

1968 **Cohen, Percy S.,** *Modern Social Theory* (London: Heinemann).

1956 **Coser, Lewis A.,** *The Functions of Social Conflict* (New York: Free Press).

1964 **Coser, Lewis A.,** and **Bernard Rosenberg** (eds.), *Sociological Theory: A Book of Readings* (New York: Crowell-Collier-Macmillan).

1969 **Costner, Herbert L.,** "Theory, Deduction, and Rules of Correspondence," *American Journal of Sociology*, 75 (September): 245-263.

1964 **Costner, Herbert L.** and **Robert K. Leik,** "Deductions from 'Axiomatic Theory,'" *American Sociological Review*, 29 (December): 819-835.

1959 **Dahrendorf, Ralf,** *Class and Class Conflict in Industrial Society* (Stanford, Calif.: Stanford University Press).

1963 **Davis, James A.,** "Structural Balance, Mechanical Solidarity, and Interpersonal Relations," *American Journal of Sociology*, 68 (January): 444-462.

1955 **Davis, Kingsley,** "Malthus and the Theory of Population," in Paul F. Lazarsfeld and Morris Rosenberg (eds.), *The Language of Social Research* (New York: Free Press), pp. 540-553.

1959 ——"The Myth of Functional Analysis as a Special Method in Sociology and Antropology," *American Sociological Review*, 24 (December): 757-772.

1945 **Davis, Kingsley,** and **Wilbert E. Moore,** "Some Principles of

Stratification," *American Sociological Review*, 10 (April): 242-249.

1969 Deutscher, Irwin, "Looking Backward: Case Studies on the Progress of Methodology in Sociological Research," *American Sociologist*, 4 (February): 35-41.

1966 DiRenzo, Gordon J. (ed.), *Concepts, Theory, and Explanation in the Behavioral Sciences* (New York: Random House).

1969 Doby, John T., "Logic and Levels of Scientific Explanation," in Edgar F. Borgatta (ed.), *Sociological Methodology, 1969* (San Francisco: Jossey-Bass), pp. 137-154.

1969 Dubin, Robert, *Theory Building* (New York: Free Press).

1967 Duke, James T., "Theoretical Alternatives and Social Research," *Social Forces*, 45 (June): 571-582.

1967 Dumont, Richard G., and William J. Wilson, "Aspects of Concept Formation, Explication, and Theory Construction in Sociology," *American Sociological Review*, 32 (December): 985-995.

1949 Durkheim, Emile, *The Division of Labor in Society* (New York: Free Press).

1950 ——*The Rules of Sociological Method* (New York: Free Press).

1951 ——*Suicide* (New York: Free Press).

1939 Eddington, Arthur, *The Philosophy of Physical Science* (New York: Crowell-Collier-Macmillan).

1954 Einstein, Albert, *Ideas and Opinions* (London: Alvin Redman).

1968 Etzioni, Amitai, "Basic Human Needs, Alienation and Inauthenticity," *American Sociological Review*, 33 (December): 870-885.

1966 Faunce, William A., and M. Joseph Smucker, "Industrialization and Community Status Structure," *American Sociological Review*, 31 (June): 390-399.

1956 **Feigl, Herbert,** "Some Major Issues and Developments in the Philosophy of Science of Logical Empiricism," in Herbert Feigl and Michael Scriven (eds.), *Minnesota Studies in the Philosophy of Science*, Vol. I (Minneapolis: University of Minnesota Press), pp. 3-37.

1953 **Feigl, Herbert,** and **May Brodbeck** (eds.), *Readings in the Philosophy of Science* (New York: Appleton).

1956 **Feigl, Herbert,** and **Michael Scriven** (eds.), *Minnesota Studies in the Philosophy of Science*, Vol. I (Minneapolis: University of Minnesota Press).

1958 **Feigl, Herbert,** *et al.* (eds.), *Minnesota Studies in the Philosophy of Science*, Vol. II (Minneapolis: University of Minnesota Press).

1961 **Feigl, Herbert,** and **Grover Maxwell** (eds.), *Current Issues in the Philosophy of Science* (New York: Holt, Rinehart and Winston).

1962 ——(eds.), *Minnesota Studies in the Philosophy of Science*, Vol. III (Minneapolis: University of Minnesota Press).

1961 **Feyerabend, Paul K.,** "Comment on Hanson's 'Is There a Logic of Scientific Discovery?,'" in Herbert Feigl and Grover Maxwell (eds.), *Current Issues in the Philosophy of Science* (New York: Holt, Rinehart and Winston), pp. 35-39.

1956 **Frank, Philipp G.** (ed.), *The Validation of Scientific Theories* (Boston: Beacon).

1966 **Galle, Omer R.,** and **Karl E. Taeuber,** "Metropolitan Migration and Intervening Opportunities," *American Sociological Review*, 31 (February): 5-13.

1956 **Garfinkel, Harold,** "Conditions of Successful Degradation Ceremonies," *American Journal of Sociology*, 61 (March): 420-424.

1967 ——*Studies in Ethnomethodology* (Englewood Cliffs, N.J.: Prentice-Hall).

1968 **Gibbs, Jack P.,** "Definitions of Law and Empirical Questions," *Law and Society Review,* 2 (May): 429-446.

1969 ———"Marital Status and Suicide in the United States: A Special Test of the Status Integration Theory," *American Journal of Sociology,* 74 (March): 521-533.

1964 **Gibbs, Jack P.,** and **Walter T. Martin,** *Status Integration and Suicide: A Sociological Study* (Eugene: University of Oregon Press).

1967 **Glaser, Barney G.,** and **Anselm L. Strauss,** *The Discovery of Grounded Theory* (Chicago: Aldine).

1961a **Goffman, Erving,** *Asylums* (New York: Doubleday).

1961b ———*Encounters* (Indianapolis: Bobbs-Merrill).

1963 ———*Stigma* (Englewood Cliffs, N. J.: Prentice-Hall).

1960 **Goode, William J.,** "A Theory of Role Strain," *American Sociological Review,* 25 (August): 483-496.

1970 **Gorskij, D. P.,** "On the Types of Definition and Their Importance for Science," in P. V. Tavanec (ed.), *Problems of the Logic of Scientific Knowledge* (Dodrecht, Holland: D. Reidel), pp. 312-375.

1970 **Gouldner, Alvin W.,** *The Coming Crisis of Western Sociology* (New York: Basic Books).

1969 **Greer, Scott,** *The Logic of Social Inquiry* (Chicago: Aldine).

1959a **Gross, Llewellyn** (ed.), *Symposium on Sociological Theory* (New York: Harper & Row).

1959b ———"Theory Construction in Sociology: A Methodological Inquiry," in Llewellyn Gross (ed.), *Symposium on Sociological Theory* (New York: Harper & Row) chap. 17.

1961 ———"Preface to a Metatheoretical Framework for Sociology," *American Journal of Sociology,* 67 (September): 125-136.

1967 ———(ed.), *Sociological Theory: Inquiries and Paradigms* (New York: Harper & Row).

1961 Hanson, Norwood Russell, "Is There a Logic of Scientific Discovery?," in Herbert Feigl and Grover Maxwell (eds.), *Current Issues in the Philosophy of Science* (New York: Holt, Rinehart and Winston), pp. 20-35.

1964 Harris, Marvin, *The Nature of Cultural Things* (New York: Random House).

1955 Hayek, F. A., *The Counter-Revolution of Science* (New York: Free Press).

1959 Heer, David M., "The Sentiment of White Supremacy: An Ecological Study," *American Journal of Sociology*, 64 (May): 592-598.

1952 Hempel, Carl G., *Fundamentals of Concept Formation in Empirical Science* (Chicago: University of Chicago Press).

1958 ——"The Theoretician's Dilemma: A Study in the Logic of Theory Construction," in Herbert Feigl, *et al.* (eds.), *Minnesota Studies in the Philosophy of Science*, Vol. II (Minneapolis: University of Minnesota Press), pp. 37-98.

1959 ——"The Logic of Functional Analysis," in Llewellyn Gross (ed.), *Symposium on Sociological Theory* (New York: Harper & Row), chap. 9.

1965 ——*Aspects of Scientific Explanation* (New York: Free Press).

1970 Hirschi, Travis, and Hanan C. Selvin, "False Criteria of Causality in Delinquency Research," in Norman K. Denzin (ed.), *Sociological Methods* (Chicago: Aldine), pp. 221-242.

1959 Hochberg, Herbert, "Axiomatic Systems, Formalization, and Scientific Theories," in Llewellyn Gross (ed.), *Symposium on Sociological Theory* (New York: Harper & Row), chap. 13.

1954 Hoebel, E. Adamson, *The Law of Primitive Man* (Cambridge, Mass.: Harvard University Press).

1961 Homans, George Caspar, *Social Behavior: Its Elementary Forms* (New York: Harcourt).

1964a ———"Bringing Men Back In," *American Sociological Review*, 29 (December): 809-818.

1964b ———"Contemporary Theory in Sociology," in Robert E. L. Faris (ed.), *Handbook of Modern Sociology* (Chicago: Rand McNally), chap. 25.

1969 ———"Comments on Blau's Paper," in Robert Bierstedt (ed.), *A Design for Sociology: Scope, Objectives, and Methods*, Monograph 9 (Philadelphia: American Academy of Political and Social Science), pp. 80-85.

1964 Hopkins, Terence K., *The Exercise of Influence in Small Groups* (Totowa, N.J.: Bedminster Press).

1963 Huaco, George A., "A Logical Analysis of the Davis-Moore Theory of Stratification," *American Sociological Review*, 28 (October): 801-804.

1959 Inkeles, Alex, "Personality and Social Structure," in Robert K. Merton, *et al.* (eds.), *Sociology Today* (New York: Basic Books), chap. 9.

1968 Isajiw, Wsevolod W., *Causation and Functionalism in Sociology* (New York: Schocken Books).

1966 Israel, Herman, "Some Religious Factors in the Emergence of Industrial Society in England," *American Sociological Review*, 31 (October): 589-599.

1965 Johnson, Barclay D., "Durkheim's One Cause of Suicide," *American Sociological Review*, 30 (December): 875-886.

1964 Kaplan, Abraham, *The Conduct of Inquiry* (San Francisco: Chandler Publishing Co.).

1965 ———"Noncausal Explanation," in Daniel Lerner (ed.), *Cause and Effect* (New York: Free Press), pp. 145-155.

1963 Kinch, John W., "A Formalized Theory of the Self-Concept," *American Journal of Sociology*, 68 (January): 481-486.

1968 Kyburg, Henry E., Jr., *Philosophy of Science* (New York: Crowell-Collier-Macmillan).

1970 Lastrucci, Carlo L., "Looking Forward: The Case for Hard-Nosed Methodology," *American Sociologist*, 5 (August): 273-275.

1962 Lazarsfeld, Paul F., "Philosophy of Science and Empirical Social Research," in Ernest Nagel, *et al.* (eds.), *Logic, Methodology and Philosophy of Science* (Stanford, Calif.: Stanford University Press), pp. 463-473.

1961 Lazarsfeld, Paul F., and Herbert Menzel, "On the Relation Between Individual and Collective Properties," in Amitai Etzioni (ed.), *Complex Organizations* (New York: Holt, Rinehart and Winston), pp. 422-440.

1965 Lerner, Daniel (ed.), *Cause and Effect* (New York: Free Press).

1964 Lipset, S. M., and Hans Zetterberg, "A Theory of Social Mobility," in Lewis A. Coser and Bernard Rosenberg (eds.), *Sociological Theory: A Book of Readings* (New York: Crowell-Collier-Macmillan), pp. 437-461.

1961 Loomis, Charles P., and Zona K. Loomis, *Modern Social Theories* (Princeton, N. J.: Van Nostrand).

1970 Lopreato, Joseph, and Letita Alston, "Ideal Types and the Idealization Strategy," *American Sociological Review*, 35 (February):88-96.

1970 Lopreato, Joseph, and Janet Saltzman Chafetz, "The Political Orientation of Skidders: A Middle-Range Theory," *American Sociological Review*, 35 (June):440-451.

1939 Lundberg, George A., *Foundations of Sociology* (New York: Crowell-Collier-Macmillan).

1942 MacIver, R. M., *Social Causation* (Boston: Ginn).

1960 Madden, Edward H. (ed.), *The Structure of Scientific Thought* (Boston: Houghton Mifflin).

1959 Malinowski, Bronislaw, *Crime and Custom in Savage Society* (Paterson, N.J.: Littlefield, Adams).

1970 Mann, Michael, "The Social Cohesion of Liberal Democracy," *American Sociological Review*, 35 (June):423-439.

1969 **Maris, Ronald W.,** *Social Forces in Urban Suicide* (Homewood, Ill.: Dorsey Press).

1960 **Martindale, Don,** *The Nature and Types of Sociological Theory* (Boston: Houghton Mifflin).

1965 ——(ed.), *Functionalism in the Social Sciences,* Monograph 5 (Philadelphia: American Academy of Political and Social Science).

1915 **Marx, Karl,** *Capital,* 3 vols. (Chicago: Charles H. Kerr).

1968 **McGinnis, Robert,** "A Stochastic Model of Social Mobility," *American Sociological Review,* 33 (October):712-722.

1966 **McKinney, John C.,** *Constructive Typology and Social Theory* (New York: Appleton).

1970 ——"Sociological Theory and the Process of Typification," in John C. McKinney and Edward A. Tiryakian (eds.), *Theoretical Sociology* (New York: Appleton) chap. 9.

1970 **McKinney, John C., and Edward A. Tiryakian** (eds.), *Theoretical Sociology* (New York: Appleton).

1968 **Meehan, Eugene J.,** *Explanation in Social Science* (Homewood, Ill.: Dorsey Press).

1957a **Merton, Robert K.,** *Social Theory and Social Structure* (New York: Free Press).

1957b ——"The Role-Set: Problems in Sociological Theory," *British Journal of Sociology,* 8 (June):107-120.

1967 ——*On Theoretical Sociology* (New York: Free Press).

1959 **Mills, C. Wright,** *The Sociological Imagination* (New York: Oxford).

1951 **Mises, Richard von,** *Positivism* (New York: Dover).

1964 **Mitchell, David,** *An Introduction to Logic* (London: Hutchinson University Library).

1968 **Murch, Arvin W.,** "Political Integration as an Alternative to Independence in the French Antilles," *American Sociological Review,* 33 (August):544-562.

1961 Nagel, Ernest, *The Structure of Science* (New York: Harcourt).

1970 Nettler, Gwynn, *Explanations* (New York: McGraw-Hill).

1966 Nisbet, Robert A., *The Sociological Tradition* (New York: Basic Books).

1947 Northrop, F. S. C., *The Logic of the Sciences and the Humanities* (New York: Crowell-Collier-Macmillan).

1962 Pap, Arthur, *An Introduction to the Philosophy of Science* (New York: Free Press).

1963 Pareto, Vilfredo, *The Mind and Society: A Treatise on General Sociology*, four volumes bound as two (New York: Dover).

1949 Parsons, Talcott, *The Structure of Social Action* (New York: Free Press).

1951 ———*The Social System* (New York: Free Press).

1955 ———"Family Structure and the Socialization of the Child," in Talcott Parsons and Robert F. Bales, *Family, Socialization and Interaction Process* (New York: Free Press), pp. 35-131.

1951 Parsons, Talcott, and Edward A. Shils (eds.), *Toward a General Theory of Action* (New York: Harper & Row).

1961 Parsons, Talcott, *et al.* (eds.), *Theories of Society* (New York: Free Press).

1964 Pelz, Donald C., and Frank M. Andrews, "Detecting Causal Priorities in Panel Study Data," *American Sociological Review*, 29 (December):836-848.

1966 Phillips, Bernard S., *Social Research: Strategy and Tactics* (New York: Crowell-Collier-Macmillan).

1950 Popper, Karl R., *The Open Society and Its Enemies* (Princeton, N.J.: Princeton University Press).

1965 ———*The Logic of Scientific Discovery* (New York: Harper & Row).

1969 **Portes, Alejandro,** "Dilemmas of a Golden Exile: Integration of Cuban Refugee Families in Milwaukee," *American Sociological Review,* 34 (August):505-518.

1962 **Putnam, Hilary,** "The Analytic and the Synthetic," in Herbert Feigl and Grover Maxwell (eds.), *Minnesota Studies in the Philosophy of Science,* Vol. III (Minneapolis: University of Minnesota Press), pp. 358-397.

1960 **Quine, Willard Van Orman,** *Word and Object* (Cambridge Mass.: M.I.T. Press).

1970 **Rescher, Nicholas,** *Scientific Explanation* (New York: Free Press).

1951 **Robinson, W. S.,** "The Logical Structure of Analytic Induction," *American Sociological Review,* 16 (December):812-818.

1957 ———"The Statistical Measurement of Agreement," *American Sociological Review,* 22 (February):17-25.

1968 **Rubin, Zick,** "Do American Women Marry Up?" *American Sociological Review,* 33 (October):750-760.

1966 **Rudner, Richard S.,** *Philosophy of Social Science* (Englewood Cliffs, N.J.: Prentice-Hall).

1970 **Sadovskij, V. N.,** "The Deductive Method as a Problem of the Logic of Science," in P. V. Tavanec (ed.), *Problems of the Logic of Scientific Knowledge* (Dordrecht, Holland: D. Reidel), pp. 160-211.

1967 **Schrag, Clarence,** "Elements of Theoretical Analysis in Sociology," in Llewellyn Gross (ed.), *Sociological Theory: Inquiries and Paradigms* (New York: Harper & Row), pp. 220-253.

1967 **Schutz, Alfred,** *The Phenomenology of the Social World* (Evanston, Ill.: Northwestern University Press).

1962 **Schwirian, Kent P.,** and **John W. Prehn,** "An Axiomatic Theory of Urbanization," *American Sociological Review,* 27 (December):812-825.

1962 **Scriven, Michael,** "Explanations, Prediction, and Laws," in

Herbert Feigl and Grover Maxwell (eds.), *Minnesota Studies in the Philosophy of Science*, Vol. III (Minneapolis: University of Minnesota Press), pp. 170-230.

1959 Selltiz, Claire, *et al.*, *Research Methods in Social Relations* (New York: Holt, Rinehart and Winston).

1957 Simon, Herbert A., *Models of Man* (New York: Wiley).

1968 ———"Causation," in David L. Sills (ed.), *International Encyclopedia of the Social Sciences*, Vol. 2 (New York: Crowell-Collier-Macmillan), pp. 350-356.

1959 Sjoberg, Gideon, "Operationalism and Social Research," in Llewellyn Gross (ed.), *Symposium on Sociological Theory* (New York: Harper & Row), chap. 19.

1968 Sjoberg, Gideon, and Roger Nett, *A Methodology for Social Research* (New York: Harper & Row).

1968 Smelser, Neil J., *Essays in Sociological Explanation* (Englewood Cliffs, N. J.: Prentice-Hall).

1969 ———"The Optimum Scope of Sociology," in Robert Bierstedt (ed.), *A Design for Sociology: Scope, Objectives, and Methods*, Monograph 9 (Philadelphia: American Academy of Political and Social Science), pp. 1-21.

1805 Smith, Adam, *An Inquiry into the Nature and Causes of the Wealth of Nations*, Vol. I (London: T. Cadell and W. Davies).

1969 Smith, Thomas S., "Structural Cystallization, Status Inconsistency and Political Partisanship," *American Sociological Review*, 34 (December):908-921.

1943 Sorokin, Pitirim A., *Sociocultural Causality, Space, Time* (Durham, N.C.: Duke University Press).

1947 ———*Society, Culture, and Personality* (New York: Harper & Row).

1957 ———*Social and Cultural Dynamics* (Boston: Porter Sargent).

1966 ———*Sociological Theories of Today* (New York: Harper & Row).

1968 Stephens, William N., *Hypotheses and Evidence* (New York: Crowell).

1968 Stinchcombe, Arthur L., *Constructing Social Theories* (New York: Harcourt).

1967 Theodorson, George A., "The Use of Causation in Sociology," in Llewellyn Gross (ed.), *Sociological Theory: Inquiries and Paradigms* (New York: Harper & Row), pp. 131-152.

1961 Timasheff, Nicholas S., *Sociological Theory: Its Nature and Growth* (New York: Random House).

1965 Tiryakian, Edward A., "Existential Phenomenology and the Sociological Tradition," *American Sociological Review*, 30 (October):674-688.

1958 Torgerson, Warren S., *Theory and Methods of Scaling* (New York: Wiley).

1961 Udy, Stanley H., Jr., "Technical and Institutional Factors in Production Organization: A Preliminary Model," *American Journal of Sociology*, 67 (November):247-254.

1963 Walker, Marshall, *The Nature of Scientific Thought* (Englewood Cliffs, N.J.: Prentice-Hall).

1969 Wallace, Walter L. (ed.), *Sociological Theory: An Introduction* (Chicago: Aldine).

1967 Wax, Murray L., "On Misunderstanding Verstehen: A Reply to Abel," *Sociology and Social Research*, 51 (April):323-333.

1930 Weber, Max, *The Protestant Ethic and the Spirit of Capitalism* (London: G. Allen).

1947 ——*The Theory of Social and Economic Organization* (New York: Oxford).

1949 ——*The Methodology of the Social Sciences* (New York: Free Press).

1968 ——*Economy and Society*, 3 Vols. (New York: Bedminster Press).

1967 **Willer, David,** *Scientific Sociology: Theory and Method* (Englewood Cliffs, N.J.: Prentice-Hall).

1970 **Willer, David,** and **Murray Webster, Jr.,** "Theoretical Concepts and Observables," *American Sociological Review*, 34 (August):748-757.

1970 **Wilson, Thomas P.,** "Conceptions of Interaction and Forms of Sociological Explanation," *American Sociological Review*, 35 (August):697-710.

1968 **Wilson, William J.,** and **Richard G. Dumont,** "Rules of Correspondence and Sociological Concepts," *Sociology and Social Research*, 52 (January):217-227.

1939 **Woodger, J.H.,** *The Technique of Theory Construction* (Chicago: University of Chicago Press).

1967 **Young, Frank W.,** "Incest Taboos and Social Solidarity," *American Journal of Sociology*, 72 (May):589-600.

1965 **Zetterberg, Hans L.,** *On Theory and Verification in Sociology* (Totowa, N.J.: Bedminster Press).

1968 ——"Pareto's Theory of the Elites," in Vilfredo Pareto, *The Rise and Fall of the Elites* (Totowa, N.J.: Bedminster Press), pp. 1-22.

1934 **Znaniecki, Florian,** *The Method of Sociology* (New York: Holt, Rinehart and Winston).

NAME INDEX

SUBJECT INDEX